# STUDENT'S SOLUTIONS MANUAL

## Slawomir Kwasik

*Tulane University*

# CALCULUS

# AN INTEGRATED APPROACH TO FUNCTIONS AND THEIR RATES OF CHANGE

## PRELIMINARY EDITION

## Robin J. Gottlieb

*Harvard University*

PEARSON

Addison
Wesley

Boston    San Francisco    New York
London    Toronto    Sydney    Tokyo    Singapore    Madrid
Mexico City    Munich    Paris    Cape Town    Hong Kong    Montreal

MW01538445

Reproduced by Pearson Addison-Wesley from electronic files supplied by the author.

Copyright © 2004 Pearson Education, Inc.
Publishing as Pearson Addison-Wesley, 75 Arlington Street, Boston, MA 02116

ISBN    0-321-22467-1

1 2 3 4 5 6 QEP 06 05 04 03

**PEARSON**
Addison
Wesley

# Contents

# CHAPTER 1

# Functions Are Lurking Everywhere

## Section 1.1    Functions Are Everywhere. Exploratory Problems for Chapter 1

**Problem 1.**

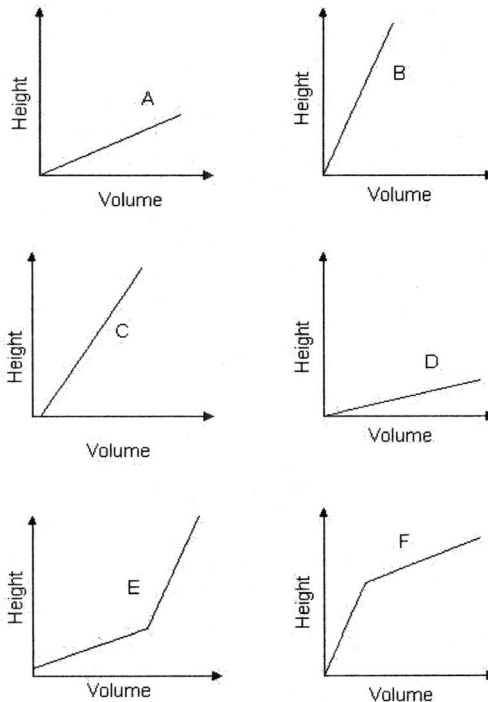

**Problem 2.**

Ink Bottle: (f) ; Conical flask: (d); Evaporating flask: (i); Bucket: (a); Vase (e); Plugged funnel (b). Vertical sides correspond to straight line graphs, which increase at a steady rate. Narrower cross-sections correspond to slower increases in height. Hence, narrower vertical sides correspond to steeper straight lines on the graph. When sides slope outwards, the height will increase slower and slower with respect to volume, and hence the graph will increase at a slower and slower rate. When sides slope inwards, the height will increase faster and faster with respect to volume, and hence the graph will increase at a faster and faster rate.

A common error is to identify the ink bottle with (g) due the notion of "straight" edges on the bottle correspond to straight lines on the graph. A good idea is to begin with the bucket and conical flask and build up from there.

Here are rough sketches of bottles for the remaining graphs.

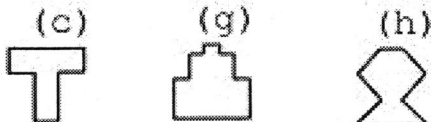

**Problem 3.**

(a)

Where there is a vertical side, the height of the liquid will increase steadily, and hence the graph will be a straight line. Notice that as the "step" bottle looks more like the ink bottle, the graph looks more like a curve.

(b) Answers will vary.

(c) This can be accomplished by constructing two bottles that have the same height and same cross-sectional areas at each height, but have different cross-sectional shapes. For instance, if the base of one bottle has a square base of 2 inches by 2 inches and another has a rectangular base of 1 inch by 4 inches. Alternatively, the cross-sectional shapes could be identical but off-center.

## Section 1.2    What Are Functions ? Basic Vocabulary and Notation

**Problem 2.**

(a) Yes

(b) Yes.

(c) Yes.

(d) Yes.

(e) No. A state can have more than one representative in the House of Representatives.

(f) Yes.

(g) Yes.

(h) Yes.

**Problem 3.**

(a) Functions I, II

(b) domain for I: {a,b,c,d,e}, range for I: {c}
domain for II: {a,b,c,d,e}, range for II: {a,b,d,x,z}

**Problem 5.**

Yes, $P(n)$ is a function because, for any given $n$, there is only one possible value of $P(n)$.

**Problem 7.**

It is a function. The domain is {0,1,2,3,4}, the range is {2,3}. The function is not 1–to–1.

**Problem 8.**

It is a function. The domain is $\{\sqrt{2}, \sqrt{3}, \sqrt{5}, \sqrt{6}\}$, the range is $\{2,3,5,6\}$. The function is 1–to–1.

**Problem 9.**

It is a function. The domain is $\{\sqrt{2}, 2\sqrt{2}, 3\sqrt{2}, 4\sqrt{2}\}$, the range is $\{0\}$. The function is not 1–to–1.

**Problem 10.**

(a) $f(x) = \sqrt{x^2 + 3}$

(b) $f(x) = 2x + 7$

(c) $f(x) = \frac{x-3}{2}$

(d) $f(x) = (x + 10)^3$

**Problem 12.**

(a)   (i) $g(h - 1)$

    (ii) $g(h) - 10$

    (iii) $\frac{h}{2}$

    (iv) $g(h + 6)$

    (v) $f(2)$

    (vi) $f(4) - f(2)$

    (vii) $\frac{f(6) - f(3)}{2}$

    (viii) $\frac{f(5)}{5}$

    (ix) $\frac{f(12) - f(6)}{6}$

(b)   (i) The distance traveled two hours after reaching Gallup.

    (ii) Half of the distance to Gallup.

    (iii) The distance traveled during the first half of the trip from Flagstaff to Gallup.

    (iv) The distance traveled two hours before reaching Gallup.

    (v) 2 miles less than the distance to Gallup.

    (vi) 2 miles more than the distance to Gallup.

    (vii) The car's speed two hours after arriving at Gallup.

    (viii) 2 miles per hour faster than the car's speed upon entering Gallup.

    (ix) 2 miles per hour slower than the car's speed upon entering Gallup.

    (x) Half of the car's speed upon entering Gallup.

    (xi) Half of the car's speed one hour before entering Gallup.

**Problem 14.**

(a) $f(0) = \sqrt{\frac{0}{0+1}} = 0$

(b) $f(3) = \sqrt{\frac{1}{3+1}} = \frac{1}{2}$

(c) $f(-\frac{1}{4}) = \sqrt{\frac{1}{-\frac{1}{4}+1}} = \sqrt{\frac{1}{\frac{3}{4}}} = \frac{2}{\sqrt{3}}$

(d) $f(b) = \sqrt{\frac{1}{b+1}}$

(e) $f(b-1) = \sqrt{\frac{1}{b-1+1}} = \frac{1}{\sqrt{b}}$

(f) $f(b+3) = \sqrt{\frac{1}{b+3+1}} = \frac{1}{\sqrt{b+4}}$

(g) $[f(7)]^2 = (\sqrt{\frac{1}{7+1}})^2 = \frac{1}{8}$

(h) $f(b^2) = \sqrt{\frac{1}{b^2+1}}$

(i) $[f(b)]^2 = (\sqrt{\frac{1}{b^2+1}})^2 = \frac{1}{b^2+1}$

**Problem 16.**

(a) $h(0) = \frac{0^2}{1-2(0)} = 0$

(b) $h(3) = \frac{3^2}{1-2(3)} = -\frac{9}{5}$

(c) $h(p+1) = \frac{(p+1)^2}{1-2(p+1)} = \frac{p^2+2p+1}{-2p-1}$

(d) $h(3p) = \frac{(3p)^2}{1-2(3p)} = \frac{9p^2}{1-6p}$

(e) $2h(3p) = 2(\frac{(3p)^2}{1-2(3p)}) = \frac{18p^2}{1-6p}$

(f) $\frac{1}{h(2p)} = \frac{1}{\frac{(2p)^2}{1-2(2p)}} = \frac{1-4p}{4p^2}$

**Problem 17.**

(a) 1

(b) 2

(c) 6

(d) $3x^2 + 2x + 1$

(e) $3x^2 + 10x + 9$

(f) $9x^2 - 6x + 3$

(g) $27x^2 - 6x + 1$

**Problem 19.**

(a) $f(0) = \frac{3(0)+5}{2} = \frac{5}{2}$;   $f(1) = \frac{3(1)+5}{2} = 4$;   $f(-1) = \frac{3(-1)+5}{2} = 1$.

(b)   (i) $\frac{3x+5}{2} = 0 \Leftrightarrow \frac{3x+5}{2} = 0 \Leftrightarrow x = -\frac{5}{3}$.

   (ii) From above, $f(-1) = 1$, so $x = -1$.

   (iii) $\frac{3x+5}{2} = -1 \Leftrightarrow 3x + 5 = -2 \Leftrightarrow x = -\frac{7}{3}$.

**Problem 21.**

(a) $h(0 = \frac{0^2+2(0)}{3(0)+2} = 0;$   $h(1) = \frac{1^2+2(1)}{3(1)+2} = \frac{3}{5};$   $h(-1) = \frac{(-1)^2+2(1)}{3(-1)+2} = 1.$

(b)  (i) $\frac{x^2+2x}{3x+2} = 0.$ Multiplying both sides by $3x+2$ we obtain $x^2+2x = 0.$ Factoring out $x$ gives $x(x+2) = 0$ and hence $x = 0$ or $x = -2.$

   (ii) $\frac{x^2+2x}{3x+2} = 1.$ Once again $x^2 + 2x = 3x + 2.$ This gives $x^2 - x - 2 = 0.$ This can be factorized out as $(x - 2)(x + 1) = 0$ and hence $x = 2$ or $x = -1.$

   (iii) $\frac{x^2+2x}{3x+2} = -1.$ This leads to $x^2 + 2x = -3x - 2$ or $x^2 + 5x + 2 = 0.$ Recalling the formula for the roots of quadratic gives $x_1 = \frac{-5-\sqrt{25-8}}{2} = \frac{-5-\sqrt{17}}{2}, x_2 = \frac{-5-\sqrt{25-8}}{2} = \frac{-5-\sqrt{17}}{2}$

## Problem 22.

(a) $f(x) = \frac{1}{5x+10}$ is defined for all values of $x$ for which $5x + 10 \neq 0,$ that is $x \neq -2.$ Therefore the domain of $f(x)$ consists of all $x$ with $x \neq 2.$

(b) $g(x) = \sqrt{5x + 10}$ is defined for all values of $x$ for which $5x + 10 \geq 0,$ that is $x \geq -2.$ Therefore the domain of $g(x)$ is $[-2, \infty).$

## Problem 24.

(a) Factor out $x^2 + 3x - 4 = (x + 4)(x - 1).$ Now $x^2 + 3x - 4 = 0$ implies $x = -4$ or $x = 1.$ As a consequence the domain of $f(x) = \frac{3}{x^2+3x-4}$ consists of all $x$ with $x \neq -4,$ $x \neq 1.$

(b) To find the domain we need the restriction $x^2 + 3x - 4 \geq 0.$ This implies $x \leq -4$ or $x \geq 1.$ As a consequence the domain for $g(x)$ consists of values of $x$ for which $x \leq -4$ or $x \geq 1.$

## Problem 26.

(a) $f(x) = \frac{1}{x+2} - \frac{1}{x-1} = \frac{(x-1)-(x+2)}{(x-1)(x+2)} = -\frac{3}{(x-1)(x+2)}$ is defined for all values of $x$ for which $(x - 1)(x + 2) \neq 0.$ Hence the domain of $f(x)$ consists of all $x$ with $x \neq -2$ and $x \neq 1.$

(b) $g(x) = \sqrt{x + 2} - \sqrt{x - 1}$ is defined for all values of $x$ for which $x + 2 \geq 0$ and $x - 1 \geq 0.$ That is, the domain for $g$ is $[1, \infty).$

## Problem 27.

(a) Domain consists of all values of $x$ for which $x \neq 0,$ $x \neq 3,$ $x \neq -1.$

(b) Domain consists of all values of $x$ for which $0 \leq x \leq 3.$

## Problem 29.

(a)  (i) $A(4) = 4^2 = 16$

   (ii) $A(W) = W^2$

   (iii) $A(\sqrt{2} + 3) = (\sqrt{2} + 3)^2 = 11 + 6\sqrt{2}$

   (iv) $A(4 + h) = (4 + h)^2 = h^2 + 8h + 16$

   (v) $A(x - 1) = (x - 1)^2 = x^2 - 2x + 1$

(b) If $S > 1,$ then $A(S - 1) = (S - 1)^2 = S^2 - 2S + 1 = A(S) - 2S + 1.$ Since $S > 1,$ $-2S + 1 < -1.$ Hence $A(S) - 1 > A(S) - 2S + 1 = A(S - 1),$ for all $S > 1.$

(c) The quantity $A(S-1)$ corresponds to the shaded area; it represents the area of the square with side length $S-1$. The quantity $A(S)-1$ corresponds to the sum of the areas of the shaded region and the $1 \times (S-1)$ rectangular regions. In other words, $A(S)-1$ corresponds to the large square minus the tiny square in the upper right-hand corner.

## Section 1.3    Representations of Functions

**Problem 1.**

(a)  (a) Function. Domain: $[-4,5]$. Range: $[-.2,.9]$.

    (b) Not a function.

    (c) Function. Domain: $(-3,7]$. Range: $[0,3]$.

    (d) Function. Domain: $\{-3,-2,-1,0,1,2,3,4\}$. Range: $\{-1,0,1\}$.

    (e) Not a function.

    (f) Function. Domain: $\{0,1,2,3,4\}$. Range: $\{1\}$.

    (g) Function. Domain: $(-\infty,\infty)$. Range: $[0,\infty)$.

    (h) Function. Domain: $[-1,3)$. Range: $\{-1,0,1,2\}$.

    (i) Function. Domain: $[-2,0) \cup (0,2]$. Range: $(-\infty,\infty)$.

(b) None of the functions above is 1–to–1.

**Problem 2.**

(a) Domain: $(-\infty,-2) \cup (-2,\infty)$

(b) Domain: $(-\infty,\infty)$.

**Problem 3.**

(a) Domain: $x \geq 0$,

(b) Domain: $x \geq 3$,

(c) Domain: $x \leq -2$ or $x \geq 2$.

**Problem 4.**

(a) $f(x)$ is defined for $x$ such that $x+1 \neq 0$ and $x+1 > 0$. So, the domain of $f$ is the set $\{x \mid x > -1\} = (-1,\infty)$.

(b) $g(x) = \sqrt{\frac{x}{x+1}}$ is defined for all values of $x$ for which $x+1 \neq 0$ and $\frac{x}{x+1} \geq 0$. Now $\frac{x}{x+1} > 0$ when both the numerator and denominator are either both positive or both negative. Both $x$ and $x+1$ are positive when $x > 0$, and both $x$ and $x+1$ are negative when $x < -1$. Hence the domain of $g$ is the set $\{x \mid x < -1 \text{ or } x \geq 0\} = (-\infty,-1) \cup [0,\infty)$.

**Problem 5.**

(2a) $\{y \mid y \neq 0\} = (-\infty, 0) \cup (0, \infty)$

(2b) $\{y \mid 0 < y < \frac{5}{4}\} = (0, \frac{5}{4})$

(3a) $\{y \mid y \geq 0\} = [0, \infty)$

(3b) $\{y \mid y \geq 0\} = [0, \infty)$

(3c) $\{y \mid y \geq 0\} = [0, \infty)$

(4a) $\{y \mid y > 0\} = (0, \infty)$

(4b) $\{y \mid 0 \leq y < 1, \text{ or } y > 1\} = [0, 1) \cup (1, \infty)$

**Problem 7.**

(a) $i(t)$ and $j(x)$ are equivalent to $f(x)$.

(b) $g(c)$ is equivalent to $f(x)$.

(c) $\phi(m)$ and $\mathcal{T}(x)$ are equivalent to $f(x)$.

**Problem 8.**

(a) $f(-1) = 0$; $f(0) = 2$; $f(1) = 3$.

(b) $x = -1, 2, 4$

(c) $x = 0$, $x \approx 1.6$

(d) $-f(0) + 2f(3) \approx -(2) + 2\left(-\frac{3}{4}\right) = -3.5$

**Problem 11.**

(a) Domain of $k$ : $-1 \leq x < 2$ and $3 \leq x < 4$. Range of $k$ : $0 \leq k \leq 3$.

(b) Since $k(-1) = 2$, $k(1) = \frac{1}{2}$ then $2k(-1) + [k(-1)]^2 + k((-1)^2) = 2 \cdot 2 + 2 \cdot 4 + \frac{1}{2} = 12.5$

**Problem 14.**

(a) yes

(b) yes

(c) no

(d) no

(e) no

(f) yes

(g) no

**Problem 16.**

Domain: $[-4,5)$; Range: $[-3,3)$; 1-to-1.

**Problem 17.**

Domain: $-3 \le x \le 4$, Range: $-1 \le h \le 5$.

**Problem 18.**

Domain: $[-3,4]$; Range: $[-2,6]$; not 1-to-1 because the graph contains horizontal segments.

**Problem 20.**

Domain $-4 < x < 3$, Range $-3 < g \le 3$.

**Problem 22.**

Domain $-3 \le x < 0$ and $0 < x \le 3$, Range $.01 \le l$ and $l \le -.01$

**Problem 24.**

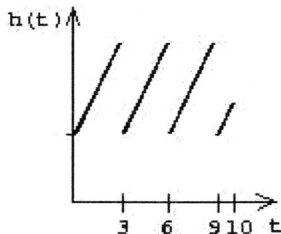

**Problem 26.**

(a) Let $r$ be the radius of the circle. Then the area of the circle is $A(r) = \pi r^2$. Since $d = 2r$, then $r = \frac{d}{2}$ and the area $A(d)$ of the circle is given by $A(d) = \pi(\frac{d}{2})^2 = \frac{\pi}{4}d^2$.

(b) The circumference $c$ of the circle is given by $c = 2\pi r$, therefore $r = \frac{c}{2\pi}$. The area $A(c)$ is then given by $A(c) = \pi(\frac{c}{2\pi})^2 = \frac{\pi c^2}{4\pi^2} = \frac{c^2}{4\pi}$

**Problem 28.**

(a) The volume of a rectangular solid is the product of the area of its base and its height. The area of the square base is $s^2$. Hence the volume is $V = s^2 h$.

(b) The surface of the box consists of six faces, two of which are squares of side length $s$ (the top and bottom) and four of which $s \times h$ rectangles (the four sides). The surface area $A$ is the sum of the areas of these six face: $A = 2s^2 + 4sh$.

(c) If the volume of the box is 120, we have $120 = s^2 h$. Thus $h = \frac{120}{s^2}$. Substituting this expression for $h$ in terms of $s$ into the surface area formula in the previous step gives

$$A(s) = 2s^2 + 4sh = 2s^2 + 4s\left(\frac{120}{s^2}\right) = 2s^2 + \frac{480}{s}.$$

**Problem 31.**

As Lori walks at 150 ft/min, her distance in feet from the meeting point $t$ minutes after departure is $L(t) = 150t$. Similarly, as Nina walks at 320 ft/min, her distance in feet from the meeting point $t$ minutes after departure is $N(t) = 320t$. Let $x$ feet be the distance between the sisters. Consider the right triangle with vertices at the meeting point A, Lori's position $L(t)$ due north of $A$ and Nina's position $N(t)$ due east of $A$. The hypotenuse of this right triangle is $x$ feet and the two legs have lengths $L(t) = 150t$ feet and $N(t) = 320t$ feet, respectively. From the Pythagorean Theorem, we have:

$x^2 = (150t)^2 + (320t)^2$

$x^2 = 22,500t^2 + 102,400t^2$

$x^2 = 124,900t^2$

$x = t \cdot \sqrt{124,900}$

$x = 10t\sqrt{1249}$

Therefore, the two sisters are $x(t) = 10t\sqrt{1249}$ feet away from each other $t$ minutes after they part.

**Problem 36.**

(a) $f(3) = 2(9) + 3 = 21$

(b) $f(2x) = 2(2x)^2 + 2x = 2(4x^2) + 2x = 8x^2 + 2x$

(c) $f(1 + x) = 2(1 + x)^2 + (1 + x) = 2(x^2 + 2x + 1) + (1 + x) = 2x^2 + 5x + 3$

(d) $f\left(\frac{1}{x}\right) = 2\left(\frac{1}{x}\right)^2 + \frac{1}{x} = \frac{2}{x^2} + \frac{1}{x} = \frac{2+x}{x^2}$

(e) $\frac{1}{f(x)} = \frac{1}{2x^2 + x}$

**Problem 38.**

(a) The volume of a circular cone of height $h$ and base radius $r$ is given by $V = \frac{\pi r^2 h}{3}$. If $d$ is the diameter of the base of the cone, $d = 2r$, and hence $h = \frac{1}{3}d = \frac{2r}{3}$ and $r = \frac{3h}{2}$. Therefore, $V(h) = \frac{\pi\left(\frac{3h}{2}\right)^2 h}{3} = \frac{3\pi h^3}{4}$.

(b) Using $h = \frac{3r}{2}$ obtained in part (a), we have $V(r) = \frac{\pi r^2\left(\frac{2r}{3}\right)}{3} = \frac{2\pi r^3}{9}$.

**Problem 41.**

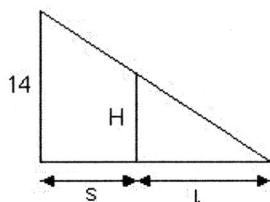

To find $l$ we use similar triangles, i.e. $\frac{l}{H} = \frac{l+s}{14}$. From this $14l = Hl + Hs$ and $l = \frac{Hs}{14-H}$

**Problem 42.**

(a) False. For $f(x) = 3$, we have $3 \cdot 3 = 9 \neq 3$.

(b) True.

(c) False. For $f(x) = 3$, we have $3 + 3 = 6 \neq 3$.

(d) True.

(e) False. For $f(x) = 3$, we have $3 \cdot 3 = 9 \neq 3$.

(f) True.

(g) False. For $f(x) = 3$, we have $3 \cdot 3 = 9 \neq 3$.

## Problem 44.

Let $L$ = the fixed amount of fence the gardener has, let $s$ = the length of one side, and let $a$ = the length of an adjacent side. By calculating the perimeter of the rectangular garden, we have: $L = 2s + 2a$. Now $L = 2s + 2a \Leftrightarrow 2a = L - 2s \Leftrightarrow a = \frac{L-2s}{2}$. As the garden is rectangular, the area of the garden $A(s)$ as a function of the side length $s$ is $A(s) = sa = s\left(\frac{L-2s}{2}\right))$.

## Problem 47.

The volume of the box is given by the product of the base area and the height of the box. The area of the base is given by $(10 - 2x)(6 - 2x)$. The height is given by $x$. Therefore the volume $V$ is given by $V = x(10 - 2x)(6 - 2x) = 4x^3 - 32x^2 - 60x$.

## Problem 52.

Rachel's fixed hourly wage is $F$ dollars per hour. If she works $h$ hours, she earns $Fh$ dollars plus a percentage of her sales. We now calculate her hourly sales. Coffee costs $a$ dollars per cup and she sells $c$ cups per hour; thus, her coffee sales are $ac$ dollars per hour. As desserts are $b$ dollars per item and she sells $d$ items per hour, her dessert sales are $bd$ dollars per hour. Hence, her total sales are $ac + bd$ dollars per hour. As Rachel makes $k$ cents per dollar of sales, or $\frac{k}{100}$ dollars per dollar of sales, she makes $\frac{k(ac+bd)}{100}$ dollars per hour in tips. Therefore, her total earning $E(h)$ in dollars expressed a function of $h$ is $E(h) = fFh + \frac{kh(ac+bd)}{100}$.

## Problem 55.

(a) The volume of a cylinder of height $h$ inches and base radius $r$ inches is $V = \pi r^2 h$ cubic inches. Solving for $h$, we obtain $h = \frac{88}{\pi r^2}$.

(b) Assume that the cylinder has height $h$ inches and base radius $r$ inches. The surface area of the cylinder is the sum of the areas the top and bottom circles and the side. Unwrapping the side yields a rectangle whose dimensions are $h$ by $2\pi r$, which is the circumference of the circular base. Thus the surface area $S$ is $S = 2\pi r^2 + 2\pi rh = 2\pi r^2 + (2\pi r)\frac{88}{\pi r^2} = 2\pi r^2 + \frac{176}{r}$ square inches.

(c) Each circle requires $(2r)^2 = 4r^2$ square inches of cardboard, and the side requires $\frac{176}{r}$ square inches of cardboard. Hence the total amount of cardboard used is: $2(2r)^2 + \frac{176}{r} = 8r^2 + \frac{176}{r}$ square inches. As cardboard costs $k$ cents per square inch, the cost of the material required for the container is $k\left(8r^2 + \frac{176}{r}\right)$ cents.

(d) The custom-made plastic lids and bottoms will each require $2\pi r^2$ square inches of plastic and will cost $7k$ cents per square inches. Hence the total cost of making the cylindrical container is $7k(2\pi r^2) + \frac{176k}{r} = 14\pi r^2 + \frac{176k}{r}$.

# CHAPTER 2

# Characterizing Functions and Introducing Rates of Change

## Section 2.1    Features of a Function

**Problem 1.**

(a) $h$ and $j$

(b) $g$ and $i$

(c) $h$ and $i$

(d) $g$ and $j$

(e) $f$, $h$, and $i$

**Problem 3.**

(a) $A(x)$ is a function.

(b) $A(x)$ is increasing; as $x$ increases from $x = -2$ to $x = 8$, the total area under the curve from $t = -2$ to $t = x$ increases.

**Problem 4.**

(a) $(-3, 7)$

(b) $[-6, 0]$

(c) $[-6, -2]$, $[2, 5]$, $[9, 10]$

(d) $[-2, 0]$

(e) $[2, 5]$, $[9, 10]$

**Problem 7.**

(a) $x = -1$, $x = 2$

(b) $(-1, 2)$

(c) $[-2, 1]$, $[3, 4]$,

(d) $[-2, -1)$, $[3, 4)$

**Problem 9.**

(a) $x = -2$.

(b) $(-2, 7]$

(c) $[-3, 1)$

(d) $[-3, -2)$

**Problem 10.**

   (a) Yes.

   (b) No, because the graph of $f$ has a vertical asymptote at $x = 2$.

   (c) No, because the graph of $f$ has a hole at $x = 2$.

## Section 2.2   A Pocketful of Functions

**Problem 1.**

   (a) Given that $\frac{PV}{T} = k$, for some constant $k$, and that $T$ is constant, we have that $P = \frac{kT}{V}$, where $P$ and $V$ are the only variables. Thus pressure and volume are inversely proportional.

   (b) If $P$ is held constant, we have that $V = \frac{k}{P} \cdot T$, where $V$ and $T$ are the only variables, and $k$ is constant. Thus volume and temperature are directly proportional.

**Problem 3.**

   (a)  (i) 2

        (ii) $x - 3$, if $x \geq 3$; $-x + 3$, if $x < -3$.

   (b)  (i) $|-5 - \pi|$

        (ii) $|\sqrt{3} - \pi|$

   (b)  (ii) $x$ is more than 2 units from 3

        (iii) $|c| < |b|$

        (iv) $x$ is at most 4 units from 3.

        (v) $|w - d| = 6$

        (vi) $|q - 5| \leq 18$

**Problem 5.**

   (a) Solve the corresponding equation $\left|2x + \frac{1}{2}\right| = 6$.

| Case (1): $2x + \frac{1}{2} \geq 0$ | Case (2): $2x + \frac{1}{2} < 0$ |
|---|---|
| $2x + \frac{1}{2} = 6$ | $-\left(2x + \frac{1}{2}\right) = 6$ |
| $2x = \frac{11}{2}$ | $-2x - \frac{1}{2} = 6$ |
| $x = \frac{11}{4}$ | $-2x = \frac{13}{2}$ |
|  | $x = -\frac{13}{4}$ |

These two solutions partition the number line into three intervals: $\left(-\infty, -\frac{13}{4}\right)$, $\left(-\frac{13}{4}, \frac{11}{4}\right)$, and $\left(\frac{11}{4}, \infty\right)$. Substituting the numbers $-4$, $0$, $3$, which respectively lie in these three intervals, into the original inequality, we see that the solution is $\left[-\frac{13}{4}, \frac{11}{4}\right]$. On a number line, this is represented by:

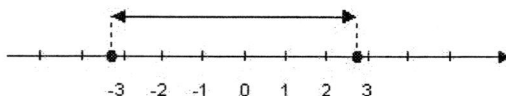

(b) Solve the corresponding equation $|3x - 4| = 8$.

| Case (1): $3x - 4 \geq 0$ | Case (2): $3x - 4 < 0$ |
|---|---|
| $3x - 4 = 8$ | $-(3x - 4) = 8$ |
| $3x = 12$ | $-3x + 4 = 8$ |
| $x = 12$ | $-3x = 4$ |
| | $x = -\frac{4}{3}$ |

These two solutions partition the number line (other than these two solutions) into three intervals: $\left(-\infty, -\frac{4}{3}\right)$, $\left(-\frac{4}{3}, 4\right)$, and $(4, \infty)$. Substituting the number $-2$, $0$, and $5$, which respectively lie in these three intervals, into the original inequality, we see that the solution is $\left(-\infty, -\frac{4}{3}\right) \cup (4, \infty)$. On a number line, this is represented by:

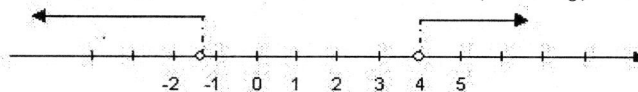

## Problem 6.

(a) $-2x - 7 < -8$ $\Leftrightarrow$ $-2x < -1$ $\Leftrightarrow$ $x > \frac{1}{2}$. In interval notation, the solution is $\left(\frac{1}{2}, \infty\right)$. On a number line, this is represented by:

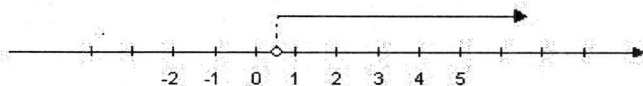

(b) Solve the corresponding equation $|-2x - 8| = 2$.

| Case (1): $-2x - 8 \geq 0$ | Case (2): $-2x - 8 < 0$ |
|---|---|
| $-2x - 8 = 2$ | $-(-2x - 8) = 2$ |
| $-2x = 10$ | $2x + 8 = 2$ |
| $x = -5$ | $2x = -6$ |
| | $x = -3$ |

These two solutions partition the number line (other than these two solutions) into three intervals: $(-\infty, -5)$, $(-5, -3)$, and $(-3, \infty)$. Substituting the numbers $-6$, $-4$, and $0$, which respectively lie in these three intervals, into the original inequality, we see that the solution is $(\infty, -5] \cup [-3, \infty)$. On a number line, this is represented by:

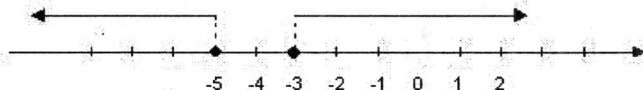

(c) Using the analysis from part (b) and substituting the numbers $-6$, $-4$, and $0$, which respectively lie in these three intervals, into the original inequality, we see that the solution is $(-5, -3)$. On a number line, this is represented by:

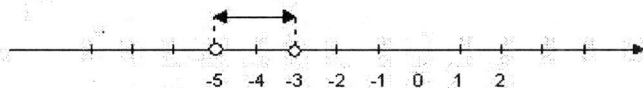

## Problem 8.

(a) True.

(b) True.

(c) False. If $x = 0$ and $y = 1$, $|x - y| = |-1| = 1$, but $|x| - |y| = |0| - |1| = 0 - 1 = -1$.

(d) False. If $x = 1$ and $y = -1$, $|x + y| = |1 - 1| = |0| = 0$, but $|x| + |y| = |1| + |-1| = 1 + 1 = 2$.

(e) True.

(f) True.

**Problem 10.**

(a) $x = -4, -2, 2$

(b) $x = -3, -1, 3, 6$

(c) $x \in (-\infty, -4) \cup [-3, -1) \cup (2, 3) \cup (3, 6)$

**Problem 12.**

(a) $2|x| > 4$ implies $|x| > 2$ and hence $x > 2$ or $x < -2$

(b) $|2x - 1| \leq 3$ implies $-3 \leq 2x - 1 \leq 3$, this gives $-1 \leq x \leq 2$

(c) $|x^2 - 1| \geq 0$ gives $x$ to be arbitrary number

**Problem 13.**

(a) Even: $f(-x) = (-x)^2 + 3(-x)^4 = x^2 + 3x^4 = f(x)$.

(b) Even: $g(-x) = \frac{1}{f(-x)} = \frac{1}{f(x)} = g(x)$.

**Problem 15.**

(a) Odd: $f(-x) = \frac{(-x)^2 - 1}{(-x)^3} = \frac{x^2 - 1}{-x^3} = -\frac{x^2 - 1}{x^3} = -f(x)$.

(b) Even: $g(-x) = \frac{(-x)^2 - 1}{(-x)^4 + 1} = \frac{x^2 - 1}{x^4 + 1} = g(x)$.

**Problem 18.**

(a) Odd: $f(-x) = -x + \frac{1}{-x} = -x - \frac{1}{x} = -\left(x + \frac{1}{x}\right) = -f(x)$.

(b) Neither: $g(-x) = 1 + \frac{1}{-x} = 1 - \frac{1}{x}$ and $-g(x) = -\left(1 + \frac{1}{x}\right) = -1 - \frac{1}{x}$.

**Problem 19.**

(a) $f(x^2) = 1 \Leftrightarrow \frac{1}{x^2} = 1 \Leftrightarrow x^2 = 1 \Leftrightarrow x = -1, 1$.

(b) $-f(x) = f(x - 1) \Leftrightarrow -\frac{1}{x} = \frac{1}{(x-1)} \Leftrightarrow 1 - x = x \Leftrightarrow 2x = 1 \Leftrightarrow x = \frac{1}{2}$.

(c) $2f(x - 2) = f(x + 3) \Leftrightarrow \frac{2}{x-2} = \frac{1}{x-3} \Leftrightarrow 2(x - 3) = x - 2 \Leftrightarrow 2x - 6 = x - 2 \Leftrightarrow x = 8$.

## Section 2.3    Average Rates of Change

**Problem 1.**

In 1970 price was 35 cents. In 2000 price was 89 cents.

(a) Difference: 89-35=54 cents, so the price increase is 54 cents.

(b) $x\% \cdot 35 = 54$, $\frac{x}{100} = \frac{54}{35} \Rightarrow x = \frac{5400}{35} = 154.3$. So the percent increase in price is 154.3%.

(c) Set $t = 0$ for 1970, $t = 30$ for 2000. Average rate of change in price is $\frac{89-35}{30-0} = 1.8 \frac{cents}{year}$

**Problem 3.**

(a) $f(c)$, $f(d)$, $f(b)$, $f(a)$

(b) $f(b) - f(a)$ negative, $f(b) - f(c)$ positive, $f(d) - f(c)$ positive. $f(b) - f(a) < f(d) - f(c) < f(b) - f(c)$.

(c) $\frac{f(b)-f(a)}{b-a}$ negative, $\frac{f(a)-f(b)}{a-b}$ negative, $\frac{f(c)-f(b)}{c-b}$ negative, $\frac{f(d)-f(c)}{d-c}$ positive, $\frac{f(d)-f(b)}{d-b}$ negative, $\frac{f(d)-f(b)}{d-b} < \frac{f(c)-f(b)}{c-b} < \frac{f(b)-f(a)}{b-a} = \frac{f(a)-f(b)}{a-b} < \frac{f(d)-f(c)}{d-c}$

**Problem 5.**

(a) $D(t)$ is positive and decreasing.

(b) The average rate of change from $t = 0$ to $t = 1$ is $\frac{36,792-149,351}{1-0} = -112,559$ deaths per year. The percent change is $100\left(\frac{-112,559\,\text{deaths}}{149,351\,\text{deaths}}\right)\% \approx 75.4\%$.

(c) The average rate of change from $t = 1$ to $t = 2$ is $\frac{36,792-21,222}{2-1} = -15,570$ deaths/year. The percent change is $100\left(\frac{-15,570\,\text{deaths}}{36,792\,\text{deaths}}\right)\% \approx -42.3\%$.

(d) The average rate of change from $t = 2$ to $t = 3$ is $[2,3]$ is $\frac{21,222-17,047}{3-2} = -4175$ deaths/year. The percent change is $100\left(\frac{-4175\,\text{deaths}}{21,222\,\text{deaths}}\right)\% \approx -19.7\%$.

**Problem 7.**

(a) $\frac{h(2.01)-h(2)}{2.01-2}$

(b) $\frac{h(a+0.001)-h(a)}{0.001}$

(c) $\frac{h(a+k)-h(a)}{k}$

**Problem 8.**

(a) $\frac{f(3)-f(1)}{3-1} = \frac{\frac{1}{3^2+1}-\frac{1}{1^2+1}}{3-1} = \frac{\frac{1}{10}-\frac{1}{2}}{2} = -\frac{1}{5} = -0.2$

(b) $\frac{f(1.5)-f(1)}{1.5-1} = \frac{\frac{1}{1.5^2+1}-\frac{1}{1^2+1}}{1.5-1} = \frac{\frac{4}{13}-\frac{1}{2}}{0.5} = -\frac{5}{13}$

(c) $\frac{f(1.01)-f(1)}{1.01-1} = \frac{\frac{1}{1.01^2+1}-\frac{1}{1^2+1}}{1.01-1} = \frac{\frac{10,000}{20,201}-\frac{1}{2}}{0.01} = -\frac{20,100}{40,402} = -\frac{10,050}{20,201}$

(d) $\frac{f(1+h)-f(1)}{h}$; for part (a), $h = 2$; for part (b), $h = 0.5$; for part (c), $h = 0.01$.

(e)

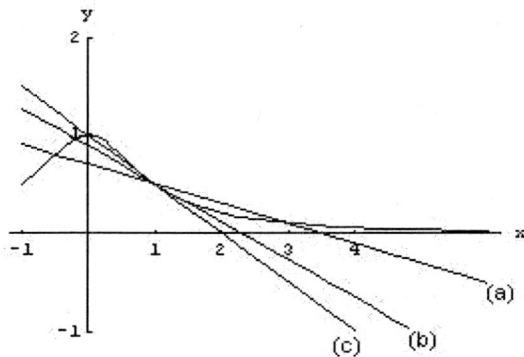

**Problem 10.**

(a) 3

(b) 5

(c) 7

(d) $2a + 3$

(e) $2a + h$

**Problem 12.**

(a)

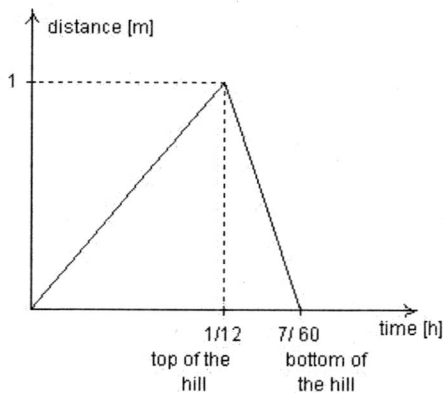

(b) Average speed $= \dfrac{\text{total distance}}{\text{total time}} = \dfrac{2}{7}$ miles per minute $= \left(\dfrac{2\,\text{miles}}{7\,\text{minutes}}\right)\left(\dfrac{60\,\text{minutes}}{1\,\text{hour}}\right) = \dfrac{120\,\text{miles}}{7\,\text{hours}} \approx 17.1$ mph. The average of 12 mph and 30 mph is $\frac{12+30}{2} = 21$ mph. The actual average speed of about 17.1 mph is not simply the average of the two constant speeds because the cyclist spends more time climbing (one twelve of an hour, or 5 minutes) than he spends on the decent ($\frac{1}{30}$ of an hour, or 2 minutes).

**Problem 13.**

(a) The volume of a cylinder of height $h$ feet and base radius $r$ feet is $V = \pi r^2 h$ cubic feet. Here, $r = 10$ feet is constant, and $h$ varies; hence $V(h) = 100\pi h$.

(b)

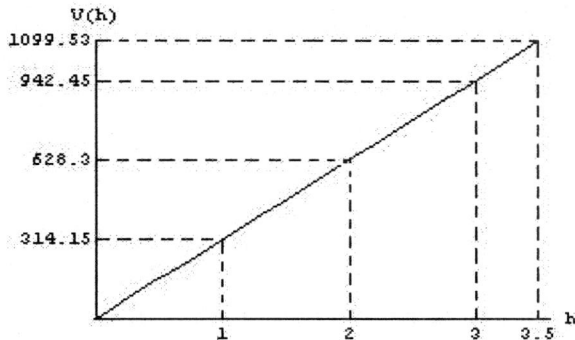

(c) $[0, 350\pi]$

(d) The average rate of change of volume with respect to height, $\frac{\Delta V}{\Delta h} = 100\pi$, as the volume function is linear. Increasing the depth of water by 1 foot requires $100\pi$ additional cubic feet of water; a 1/2-foot increase requires $50\pi$ additional cubic feet of water.

(e) $\frac{\Delta V}{\Delta h} = 100\pi \text{ ft}^3\text{ft} \approx (100\pi \text{ ft}^3\text{ft})(1 \text{ gallon}/0.16043 \text{ ft}^3) \approx 1957 \text{ gallons/ft}$.

(f) Yes, the volume of water in the pool is directly proportional to the height of the water, and the proportionality constant is $100\pi$.

**Problem 15.**

$f(x) = \frac{1}{x^2}$

A: $(3, f(3)) = (3, \frac{1}{9})$

B: $(3 + h, f(3 + h)) = (3 + h, \frac{1}{(3+h)^2})$

So, the answer is (b)

## Section 2.4    Reading a Graph to Get Information About a Function

**Problem 1.**

(a) $t = 2$

(b) There are 3 solutions: $t = -3, -1, 4$.

(c) There are 3 solutions: $t \approx -4, 1, 3$.

(d) $f(0) = 2$

(e) Approximately $(-3.4, -2) \cup (-2, -0.7) \cup (3.8, 4.3)$.

(f) Approximately $(-1, 2] \cup (4.3, 5]$.

**Problem 2.**

(a)  (a) When is the mule farthest north of the temple between noon and 5:00?

   (b) How many times is the mule at the temple? At about what times?

   (c) How many times is the mule exactly 3 miles north of the temple? At about what times?

(d) Where is the mule at noon?

(e) When is the mule within 1 mile of the temple?

(b) (a) When is the mule traveling north the fastest between noon and 5:00?

(b) How many times does the mule stop? At about what times?

(c) How many times is the mule traveling north at exactly 3 mph? At about what times?

(d) What is the mule's velocity at noon?

(e) When is the mule traveling a speed of less than 1 mph?

**Problem 5.**

The sign of velocity is determined by whether the position graph is increasing (positive velocity) or decreasing (negative velocity). Velocity is positive when the graph of position versus time is increasing, and velocity is negative when the graph of position versus time is decreasing.

**Problem 6.**

(a) picture d

(b) picture f

(c) picture e

**Problem 8.**

(a) The object changes direction when $t \approx -1.44, 0, 0.23$, and $1.21$. Zooming near these values of $t$ makes the graph look like a horizontal line.

(b) The object's velocity is negative when $t$ is in $(-1.44, 0) \cup (0.23, 1, 21)$, and is positive when $t$ is in $[-2, -1.44) \cup (0, 0.23) \cup (1.21, 2]$

(c) The object's velocity at $t = 0$ is 0 miles per hour.

**Problem 10.**

(a)

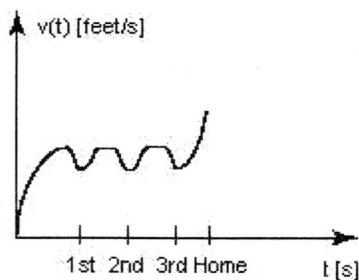

(b)

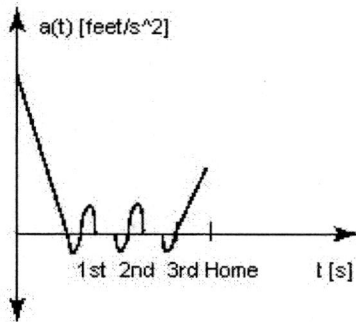

**Problem 12.**

(a) I: We begin 60 miles east of Sturbridge and stay for an hour. Then we travel at 60 mph west to Sturbridge, where we remain for an hour. Then we return to our original position by travelling 60 mph east for an hour, and stay for an hour.

   II: We begin in Sturbridge, where we stay for an hour. Then we travel at 60 mph east for one hour, stay for an hour, and return to Sturbridge by travelling west at 60 mph for one hour.

   III: We begin 20 miles east of Sturbridge travelling east at 60 mph for one hour. We stay for an hour, and then travel west at 40 mph for one hour, and then at 60 mph for one hour, ending 20 miles west of Sturbridge.

(b)

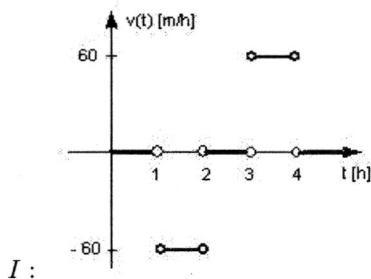

*I :*

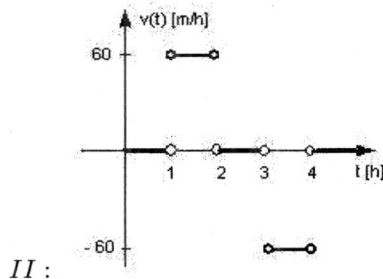

*II :*

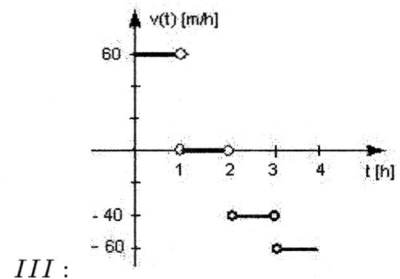

*III :*

(c)

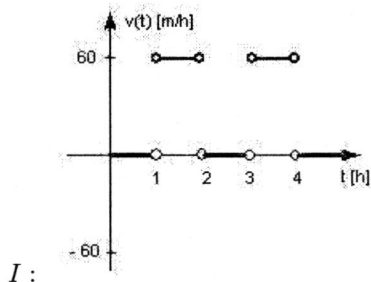

*I :*

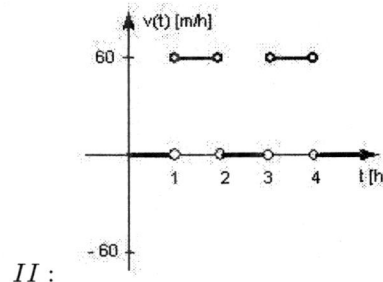

*II :*

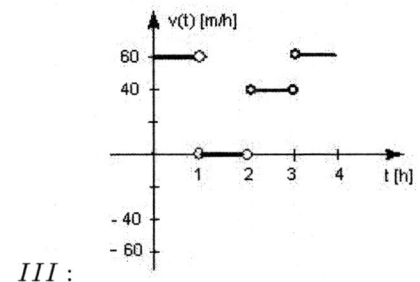

*III :*

**Problem 14.**

(a) $25\frac{km}{h}$

(b) 12:30 P.M.

(c) go home

(d) the graph is a line

(e) $50\frac{m}{h}$

(f)

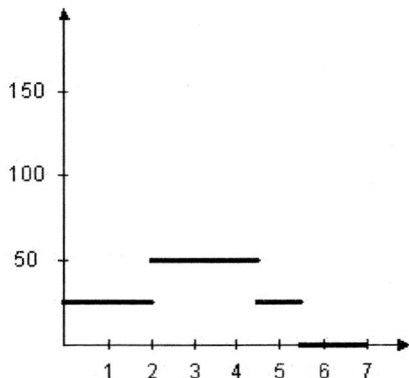

## Section 2.5    The Real Number System: An Excursion

**Problem 1.**

Area of the circle of radius $r$ is given by $A(r) = \pi r^2$. We know that $A(r) = 2\,inches^2$, so $\pi r^2 = 2, \Rightarrow$ $r^2 = \frac{2}{\pi}, \Rightarrow r = \sqrt{\frac{2}{\pi}}$

**Problem 3.**

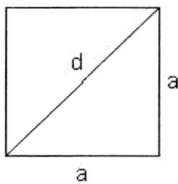

$a^2 + a^2 = d^2$, so $d^2 = 50$ and therefore $d = \sqrt{50} = 5\sqrt{2}$

**Problem 5.**

$x^2 + 1 = 6$
$x^2 - 5 = 0$
$(x - \sqrt{5})(x + \sqrt{5}) = 0$
So, $x - \sqrt{5} = 0$ or $x + \sqrt{5} = 0$, which gives $x = \sqrt{5}$ or $x = -\sqrt{5}$.

**Problem 8.**

(a) There are infinitely many rational numbers in $[2, 2.001]$.

(b) There are infinitely many irrational numbers in $[2, 2.001]$.

**Problem 10.**

The domain of this function is $(-\infty, \infty)$, which contains an infinite number of irrational numbers. Hence, there must be an infinite number of points on the graph with irrational $x-$coordinates.

# CHAPTER 3

# Functions Working Together

## Section 3.1    Combining Outputs

**Problem 1.**
   The function $h(x) = f(x)g(x) = x^2 \left(\frac{1}{x}\right) = x$ is undefined at $x = 0$ because $g(x)$ is undefined at $x = 0$.

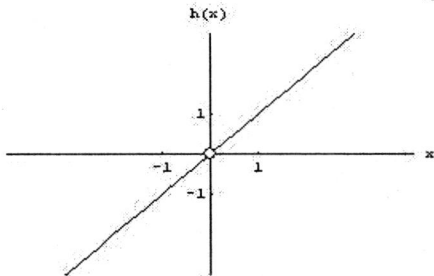

**Problem 3.**

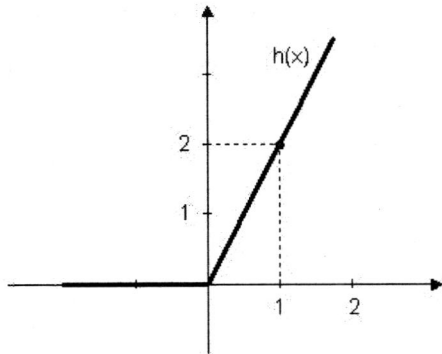

**Problem 5.**

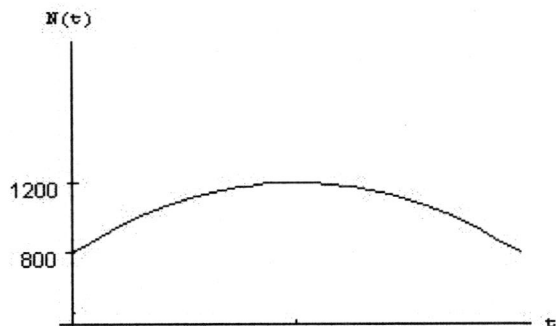

**Problem 8.**

(a)  (i) $f(x) + g(x) = x(x+1) + x^3 + 2x^2 + x = x^2 + x + x^3 + 2x^2 + x = x^3 + 3x^2 + 2x$

(ii) $\frac{f(x)}{g(x)} = \frac{x(x+1)}{x^3+2x^2+x} = \frac{x(x+1)}{x(x^2+2x+1)} = \frac{x(x+1)}{x(x+1)^2} = \frac{1}{x+1}$

(iii) $\frac{g(x)}{f(x)} = \frac{x^3+2x^2+x}{x(x+1)} = \frac{x(x+1)^2}{x(x+1)} = x + 1$

(iv) $\frac{[f(x)]^2}{g(x)} = \frac{[x(x+1)]^2}{x(x+1)^2} = \frac{x^2(x+1)^2}{x(x+1)^2} = x$

(b)  $xf(x) = g(x) \Leftrightarrow x(x(x+1)) = x^3 + 2x^2 + x \Leftrightarrow x^2(x+1) = x(x+1)(x+1) \Leftrightarrow x^2(x+1) - x(x+1)^2 = 0$
$\Leftrightarrow x(x+1)(x-(x+1)) = 0 \Leftrightarrow x(x+1)(-1) = 0 \Leftrightarrow x(x+1) = 0 \Leftrightarrow x = -1, 0.$

**Problem 9.**

(a) $C(x) = 20,000 + 10x.$

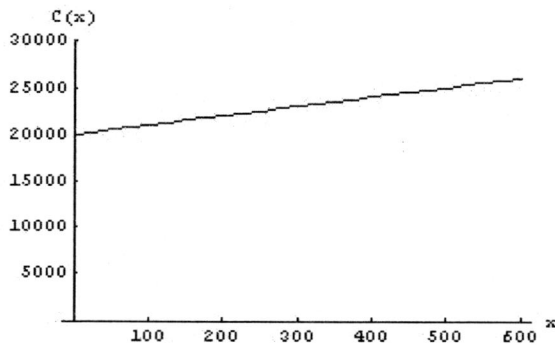

(b) The total cost is increasing by $\frac{\Delta C}{\Delta x} = \frac{C(x+\Delta x)-C(x)}{x+\Delta x-x} = \frac{(20,000+10(x+\Delta x))-(20,000+10x)}{\Delta x} = 10$ dollars per widget.

(c) $R(x) = 50x.$

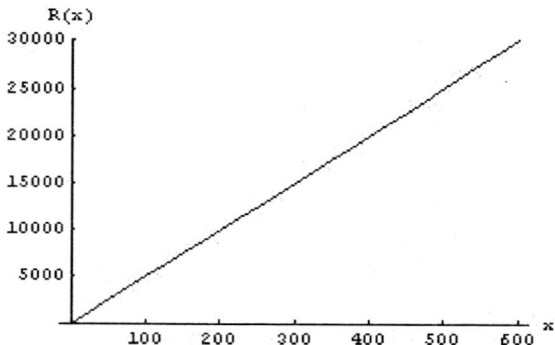

(d) $P(x) = R(x) - C(x) = (50x) - (20,000 + 10x) = 40x - 20,000.$

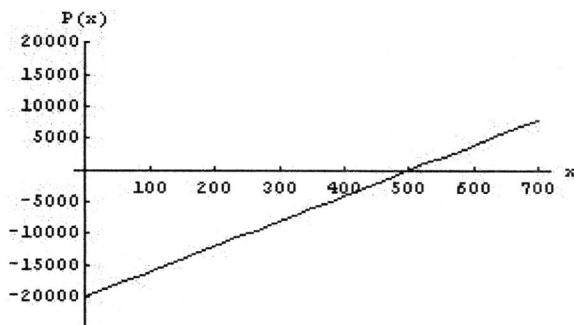

(e) $P(400) = 40(400) - 20,000 = -4000$, which means that if 400 widgets are produced and sold then the company will lose \$4000. $P(700) = 40(700) - 20,000 = 8000$, which means that if 700 widgets are produced and sold then the company will have a profit of \$8000. $P(401) = 40(401) - 20,000 = -3960$ and $P(402) = 40(402) - 20,000 = -3920$. Profit increases by \$40 for each additional widget produced and sold. Thus, $\frac{\Delta P}{\Delta x} == \frac{P(x+\Delta x)-P(x)}{x+\Delta x-x} = \frac{((40(x+\Delta x)-20,000)-((40x-20,000)}{\Delta x} = 40$, which is constant for all values of $x$.

(f) $0 = P(x) = 40x - 20,000 \Leftrightarrow 40x = 20000 \Leftrightarrow x = 500$. The company must sell 500 widgets in order to break even.

(g) If the domain of $P(x)$ is the set of all integers where $0 \leq x \leq 1200$, the range of $P(x)$ is the set of all integers in the interval $= [40(0) - 20,000, 40(1200) - 20,000] = [-20000, 28000]$. Since the profit function is linear and has a positive slope, profit is maximized when production $x$ is maximized; that is the company should sell all 1200 widgets to maximize profits.

## Section 3.2    Composition of Functions

**Problem 1.**

(a) $f(g(x)) = \left(\frac{1}{x}\right)^2 = \frac{1}{x^2}$

(b) $g(f(x)) = \frac{1}{x^2}$

(c) $xh(f(x)) = x\left(3(x^2) + 1\right) = 3x^3 + x$

(d) $f(h(g(x))) = \left(3\left(\frac{1}{x}\right) + 1\right)^2 = \left(\frac{3}{x} + 1\right)^2 = \frac{9}{x^2} + \frac{6}{x} + 1$

(e) $g(g(w)) = \frac{1}{\frac{1}{w}} = w$

(f) $h(h(t)) = 3(3t + 1) + 1 = 9t + 4$

(g) $g(f(\frac{1}{x})) = \frac{1}{\left(\frac{1}{x}\right)^2} = \frac{1}{\frac{1}{x^2}} = x^2$

(h) $g(2h(x - 1)) = \frac{1}{2(3(x-1)+1)} = \frac{1}{2(3x-2)} = \frac{1}{6x-4}$

(i) As calculated in part (e), $g(g(x)) = x$. We also see that $[g(x)]^2 = \left(\frac{1}{x}\right)^2 = \frac{1}{x^2}$. Now $g(g(-1)) = -1 \neq 1 = \frac{1}{(-1)^2} = [g(-1)]^2$. Therefore, $g(g(x)) \neq [g(x)]^2$.

(j) Note that $[h(x)]^2 = (3x+1)^2 = 9x^2 + 6x + 1$ and $h(x^2) = 3x^2 + 1$. Now $[h(1)]^2 = 16 \neq 4 = h(1) = h(1^2)$. Therefore, $[h(x)]^2 \neq h(x^2)$.

**Problem 2.**

(a) $f(g(1)) = f(-2) \approx 3$

(b) $f(g(0)) = f(-3) = 0$

**Problem 4.**

(a) $f(f(2)) = f(-2) = 3$

(b) $f(f(1)) = f(0) = 1$

**Problem 6.**
$2f(x) = 0$, $f(x) = 0 \Rightarrow x = -3$, $x = 1$, $x = 4$.

**Problem 7.**
$f(g(x)) = 0 \Leftrightarrow g(x) = a$ and $f(a) = 0$. From the previous problem, we have that $f(a) = 0$ for $a = -3, 1, 4$; thus, we must solve for the values of $x$ for which $g(x) = -3, 1$, or $4$. We observe from the graph of $g$ that $x = 0$ is the only solution of $g(x) = -3$, $x \approx \pm 2.5$ are the two solutions of $g(x) = 1$, and $x = 4$ is the only solution of $g(x) = 4$. Therefore the solutions to the original equation are $x \approx -2.5$, $x = 0$, $x \approx 2.5$, and $x = 4$.

**Problem 10.**
$f(0)x + f(1) = g(0)x + g(1) \Leftrightarrow (1)x + 0 = (-3)x + (-2) \Leftrightarrow x = -3x - 2 \Leftrightarrow 4x = -2 \Leftrightarrow x = -\frac{1}{2}$.

**Problem 12.**
$f(g(h(x))) = f\left(g\left(\frac{2}{3x}\right)\right) = f\left(\frac{2}{3x} + 1\right) = 3\left(\frac{2}{3x} + 1\right)^2 + \left(\frac{2}{3x} + 1\right)$;
$g(h(f(x))) = g(h(3x^2 + x)) = g\left(\frac{2}{3(3x^2 + x)}\right) = \left(\frac{2}{3(3x^2 + x)}\right) + 1 = \frac{9x^2 + 3x + 2}{9x^2 + 3x}$.

**Problem 14.**

(a) $f(x) + g(x) = (x - 3) + (x^2 - 6x) = x^2 - 5x - 3$

(b) $f(x) - g(x) = (x - 3) - (x^2 - 6x) = -x^2 + 7x - 3$

(c) $f(x)g(x) = (x - 3)(x^2 - 6x) = x^3 - 6x^2 - 3x^2 + 18x = x^3 - 9x^2 + 18x$

(d) $f(g(x)) = f(x^2 - 6x) = x^2 - 6x - 3$

(e) $g(f(x)) = g(x - 3) = (x - 3)^2 - 6(x - 3) = x^2 - 6x + 9 - 6x + 18 = x^2 - 12x + 27$

(f) $\frac{f(x)}{g(x)} = \frac{x-3}{x^2 - 6x}$

**Problem 15.**

(a) $x$-intercepts of $f(x)$ can be found by solving $f(x) = 0$. So, $x - 3 = 0 \Rightarrow x = 3$ is $x$-intercept.

For $y$-intercepts we set $x = 0$, so $y = 0 - 3 = -3$ is $y$-intercept.

(b) $x$-intercepts of $g(x)$: $x^2 - 6 = 0$ $x(x - 6) = 0$, $\Rightarrow$ $x = 0$ or $x = 6$.

For $y$-intercepts we set $x = 0$, so $g(0) = 0^2 - 0 \cdot 6 = 0$

(c) $x$-intercepts of $f(x) \cdot g(x)$ : $x^3 - 9x^2 + 18x = 0$, $x(x^2 - 9x + 18) = 0$, $x(x - 3)(x - 6) = 0 \Rightarrow x = 0$ or $x = 3$ or $x = 6$.

For $y$-intercepts we set $x = 0$, so $f(0)g(0) = 0^3 - 9 \cdot 0^2 + 18 \cdot 0 = 0$ is $y$-intercept.

(d) $\frac{f(x)}{g(x)} = \frac{x-3}{x^2-6x}$, $D : x \neq 0$ and $x \neq 6$. $x$-intercepts: $\frac{x-3}{x^2-6x} = 0 \Leftrightarrow x - 3 = 0$, so $x = 3$ is $x$-intercept.

For $y$-intercepts we set $x = 0$, but 0 is excluded from the Domain, so there are no $y$-intercepts.

## Problem 19.

(a) $f(-1)g(-1) = 2 \cdot 3 = 6$

(b) $f(g(-1)) = f(3) = -2$

(c) $g(f(-1)) = g(2) = 3$

(d) $h(g(f(2))) = h(g(4)) = h(0) = -3$

(e) $\frac{f(0)+2}{g(0)} = \frac{1+2}{4} = \frac{3}{4}$

(f) $5h(3) + f(f(1)) = 5 \cdot 1 + f(3) = 5 + (-2) = 3$

(g) $f(f(f(0))) = f(f(1)) = f(3) = -2$

## Problem 22.

(a) $2f(x+1) = 2\left(\frac{1}{2-(x+1)}\right) = \frac{2}{1-x}$.

(b) $f(2x-2) = \frac{1}{2-(2x-2)} = \frac{1}{4-2x}$.

(c) $g(\sqrt{x}+1) = (\sqrt{x}+1)^2 + 1 = (x + 2\sqrt{x} + 1) + 1 = x + 2\sqrt{x} + 2$.

(d) $f(g(x)) = \frac{1}{2-(x^2+1)} = \frac{1}{1-x^2}$.

(e) $g(f(x)) = \left(\frac{1}{2-x}\right)^2 + 1 = \frac{1}{x^2-4x+4} + 1 = \frac{1}{x^2-4x+4} + \frac{x^2-4x+4}{x^2-4x+4} = \frac{x^2-4x+5}{x^2-4x+4}$.

(f) $f(f(x)) = \frac{1}{2-\left(\frac{1}{2-x}\right)} = \frac{1}{\frac{2(2-x)}{2-x} - \frac{1}{2-x}} = \frac{1}{\frac{4-2x}{2-x} - \frac{1}{2-x}} = \frac{1}{\frac{3-2x}{2-x}} = \frac{2-x}{3-2x}$

(g) $g\left(\frac{1}{f(x)}\right) = g\left(\frac{1}{\frac{1}{2-x}}\right) = g(2-x) = (2-x)^2 + 1 = (x^2 - 4x + 4) + 1 = x^2 - 4x + 5$.

(h) $\frac{g(x)}{f(x)} = \frac{x^2+1}{\frac{1}{2-x}} = (x^2+1)(2-x) = -x^3 + 2x^2 - x + 2$.

## Problem 24.

(a) $f(d) = \frac{k}{d}$, where $k$ is a constant.

(b) Let $P$ be Eli's location at $t = 0$. After $t$ minutes, Max is $100 + 300t$ feet north of $P$, and Eli is $250t$ feet east of $P$. If $d$ is the distance between Max and Eli, the Pythagorean Theorem gives us $d^2 = (100 + 300t)^2 + (250t)^2$. Hence $g(t) = d = 50\sqrt{61t^2 + 24t + 4}$.

(c) $f(g(t)) = \frac{k}{50\sqrt{61t^2+24t+4}} = \frac{k\sqrt{61t^2+24t+4}}{50(61t^2+24t+4)}$. This composite function takes the number of minutes since Max and Eli beginning walking away from each other and gives the quality of the transmission of the walkie-talkies.

## Problem 26.

(a) $f(g(2)) = f(\frac{1}{2+1}) = f(\frac{1}{3}) = \frac{\frac{2}{3}}{\frac{1}{3}+3} = \frac{\frac{2}{3}}{\frac{10}{3}} = \frac{2}{10} = \frac{1}{5}$

(b) $f(g(x)) = f(\frac{1}{x+1}) = \frac{2 \cdot \frac{1}{x+1}}{\frac{1}{x+1}+3} = \frac{\frac{2}{x+1}}{\frac{3x+3+1}{x+1}} = \frac{2}{3x+3+1}$

**Problem 28.**

(a) $f(g(2)) = f(1) \approx -1.5$

(b) $g(f(2)) = g(0) = 3$

(c) $f(g(x)) = 0$ whenever $g(x) = -2$ or $g(x) = 2$. Now $g(x) = -2$ when $x \approx -3.5$ or $x = 4$; and $g(x) = 2$ when $x = -1$ or $x = 1$. Therefore, the zeros to $f(g(x)) = 0$ are $x \approx -3.5$, $x = -1$, $x = 1$, and $x = 4$.

**Problem 30.**

(a) $R(K(L(x))) = \frac{1}{|-5x|^2} = \frac{1}{25x^2}$

(b) $R(L(R(x))) = \frac{1}{\left(\frac{-5}{x^2}\right)^2} = \frac{1}{\frac{25}{x^4}} = \frac{x^4}{25}$

(c) $R(K(x)) = \frac{1}{|x|^2} = \frac{1}{x^2}$

(d) $R(D(R(x))) = \frac{1}{\left(\frac{1}{x^2}+3\right)^2} = \frac{1}{\left(\frac{3x^2+1}{x^2}\right)^2} = \frac{1}{\frac{9x^4+6x^2+1}{x^4}} = \frac{x^4}{9x^4+6x^2+1}$

**Problem 32.**

(a) The domain of $l(x) = g(f(x)) = \sqrt{|x|}$ is $(-\infty, \infty)$.

(b) The domain of $m(x) = g(h(f(x))) = \sqrt{|x|-2}$ is the set of all $x$ such that $|x|-2 \geq 0$, which is $(-\infty, -2] \cup [2, \infty)$.

**Problem 34.**

(a) $\frac{h(4)-h(-4)}{4-(-4)} = \frac{5-5}{8} = 0$

(b) $\frac{h(4)-h(0)}{4-0} = \frac{5-3}{4} = \frac{2}{4} = \frac{1}{2}$

(c) Assume $k \neq 0$. $\frac{h(4+k)-h(4)}{(4+k)-4} = \frac{\sqrt{(4+k)^2+9}-5}{k} = \frac{\sqrt{k^2+8k+25}-5}{k}$

**Problem 36.**

$h(x) = f(g(x)) = \frac{2}{g(x)+2} = \frac{2}{x-2+2} = \frac{2}{x}$. Domain $(-\infty, 0) \cup (0, \infty)$.

$j(x) = g(f(x)) = g\left(\frac{2}{x+2}\right) = \frac{2}{x+2} - 2 = \frac{2}{x+2} - \frac{2x+4}{x+2} = \frac{-2x-2}{x+2}$. Domain: $(-\infty, -2) \cup (2, \infty)$.

**Problem 38.**

$h(x) = f(g(x)) = \frac{g(x)}{g(x)-3} = \frac{\frac{2}{x}}{\frac{2}{x}-3} = \frac{\frac{2}{x}}{\frac{2-3x}{x}} = \frac{2}{2-3x}$. Domain $(-\infty, 0) \cup \left(0, \frac{2}{3}\right) \cup \left(\frac{2}{3}, \infty\right)$.

$j(x) = g(f(x)) = \frac{2}{f(x)} = \frac{2}{\frac{x}{x-3}} = \frac{2x-6}{x}$. Domain $(-\infty, 0) \cup (0, 3) \cup (3, \infty)$.

**Problem 40.**

$(f+g)(x) = (3x+2) + (5x-1) = 8x+1$. Domain: $(-\infty, \infty)$.

$(fg)(x) = (3x+2)(5x-1) = 15x^2 + 7x - 2$. Domain: $(-\infty, \infty)$.

$\left(\frac{f}{g}\right)(x) = \frac{3x+2}{5x-1}$. Domain: $\left(-\infty, \frac{1}{5}\right) \cup \left(\frac{1}{5}, \infty\right)$.

**Problem 42.**

$(f+g)(x) = \frac{3}{x+1} + \frac{2x}{x-5} = \frac{3(x-5)}{(x+1)(x-5)} + \frac{2x(x+1)}{(x+1)(x-5)} = \frac{2x^2+5x-15}{x^2-4x-5}$. Domain: $(-\infty, -1) \cup (-1, 5) \cup (5, \infty)$.

$(fg)(x) = \left(\frac{3}{x+1}\right)\left(\frac{2x}{x-5}\right) = \frac{6x}{x^2-4x-5}$. Domain: $(-\infty, -1) \cup (-1, 5) \cup (5, \infty)$.

$\left(\frac{f}{g}\right)(x) = \left(\frac{3}{x+1}\right)\left(\frac{x-5}{2x}\right) = \frac{3(x-5)}{(x+1)(2x)} = \frac{3x-15}{2x^2+2x}$. Domain: $(-\infty, -1) \cup (-1, 0) \cup (0, 5) \cup (5, \infty)$.

**Problem 44.**

(a) $g(f(2)) = g\left(\frac{1}{2} + 2\right) = g(\frac{5}{2}) = \frac{2\left(\frac{5}{2}\right)}{\left(\frac{5}{2}\right)^2+1} = \frac{5}{\frac{25}{4}+1} = \frac{5}{\frac{29}{4}} = \frac{20}{29}$

(b) $f(g(2)) = f\left(\frac{2(2)}{2^2+1}\right) = f\left(\frac{4}{5}\right) = \frac{1}{\frac{4}{5}} + \frac{4}{5} = \frac{5}{4} + \frac{4}{5} = \frac{25}{20} + \frac{16}{20} = \frac{41}{20}$

**Problem 46.**

(a) $g(f(1)) = g(2) = \frac{4}{5}$

(b) $f(g(1)) = f(1) = 2$

**Problem 48.**

(a) $f(g(x)) = f(\frac{2x}{x^2+1}) = \frac{x^2+1}{2x} + \frac{2x}{x^2+1}$

(b) $g(f(x)) = g(\frac{1}{x} + x) = g(\frac{x+1}{x}) = \frac{2 \cdot \frac{x+1}{x}}{(\frac{x+1}{x})^2+1}$

**Problem 51.**

$h(f(x)) + h(g(x)) = \frac{1}{2x^2} + \frac{1}{x+1} = \frac{x+1}{2x^2(x+1)} + \frac{2x^2}{2x^2(x+1)} = \frac{2x^2+x+1}{2x^2(x+1)}$

**Problem 53.**

$g(x)h(f(x)) = 1$. This gives $(x+1)h(2x^2) = (x+1)\frac{1}{2x^2} = 1$, so $x+1 = 2x^2$ and hence $2x^2 - x - 1 = 0$. We write $2x^2 - x - 1 = (x-1)(2x+1)$. Therefore $x = 1$, or $x = -\frac{1}{2}$

**Problem 55.**

$h(f(x) + 3g(x)) = h(2) \Leftrightarrow \frac{1}{2x^2+3x+3} = \frac{1}{2} \Leftrightarrow 2x^2 + 3x + 3 = 2 \Leftrightarrow 2x^2 + 3x + 1 = 0 \Leftrightarrow (2x+1)(x+1) = 0$
$\Leftrightarrow x = -1$ and $x = -\frac{1}{2}$

**Problem 57.**

$f(g(f(x))) = 2(2x^2+1)^2 = 8 \Leftrightarrow (2x^2+1)^2 = 4 \Leftrightarrow 4x^4 + 4x^2 + 1 = 4 \Leftrightarrow 4x^4 + 4x^2 - 3 = 0 \Leftrightarrow$
$(2x^2+3)(2x^2-1) = 0 \Leftrightarrow x^2 = \frac{-3}{2}, \frac{1}{2} \Rightarrow x = \pm\frac{1}{\sqrt{2}}$

## Section 3.3 Decomposition of Functions

**Problem 1.**

$f(x) = \frac{1}{\sqrt{x}}$ and $g(x) = x^2 + 6$.

**Problem 3.**

$j(x) = h(g(f(x)))$. Put $f(x) = 4x^2 + 3x$, $g(x) = \frac{1}{\sqrt{x}}$, $h(x) = \frac{2}{3}x$

**Problem 5.**
$$g(x) = \frac{1}{x} + 1$$

**Problem 7.**
Put $f(x) = \frac{1}{x}$ and $g(x) = x^2 + 4$.

**Problem 9.**
Put $f(x) = x^3 - 2x + 3$ and $g(x) = \sqrt{x}$.

**Problem 11.**
$h(x) = f(g(x))$. Put $f(x) = 3x^2 + 2x + 3$ and $g(x) = x^2$.

**Problem 13.**
$h(x) = f(g(x))$. Put $f(x) = 2|x|$ and $g(x) = 3x - 4$.

**Problem 16.**
$h(x) = f(g(x))$. Put $f(x) = x^2 + x + 1$ and $g(x) = 3^x$.

**Problem 18.**
$k(x) = f(g(h(x)))$. Put $f(x) = \frac{3}{x}$, $g(x) = \sqrt{x}$, and $h(x) = x^2 + 1$.

**Problem 20.**
$k(x) = f(g(h(x)))$. $f(x) = \sqrt{x}$, $g(x) = x^3 + 5$, and $h(x) = x^2 + 1$.

## Section 3.4    Altered Functions, Altered Graphs

**Problem 2.**

(a) $g(x) = 3|f(x)| \Rightarrow$ the zeros of $g$ and $f$ are the same. Hence, the zeros of $g$ are $x = -5, -2, 0$, and 5.

(b) The only zero of $w(x) = -2x^2$ is $x = 0$. Hence the only zeros of $h(x) = w(f(x))$ are the zeros of $f$, which are $x = -5, -2, 0, 5$.

(c) There is insufficient information because we need to know the values of $x$ for which $f(x) = \frac{1}{3}$ to determine where $p(x) = 3f(x) + 1 = 0$.

(d) $q(x) = 4f(x + 1) \Rightarrow x = a$ is a zero of $q$ if and only if $x + 1 = a + 1$ is a zero of $f$. Hence the zeros of $q$ are $x = -6, -3, -1$, and 4

(e) $m(x) = 4f(-x) \Rightarrow x = a$ is a zero of $m$ if and only if $x = -a$ is a zero of $f$. Hence the zeros of $f$ are $x = -5, 0, 2$, and 5.

(f) $n(x) = -f(x) \Rightarrow$ the zeros of $n$ and $f$ are the same. Hence, the zeros of $n$ are $x = -5, -2, 0$, and 5.

**Problem 5.**

(a) (vi)

(b) (ii)

(c) (vii)

(d) (viii)

(e) (iii)

(f) (i)

**Problem 7.**

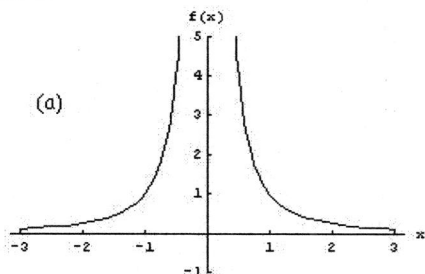

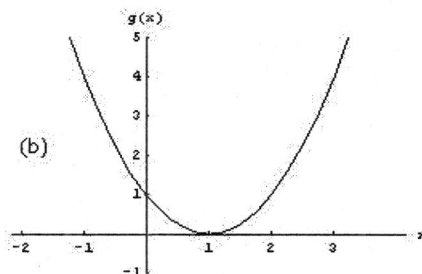

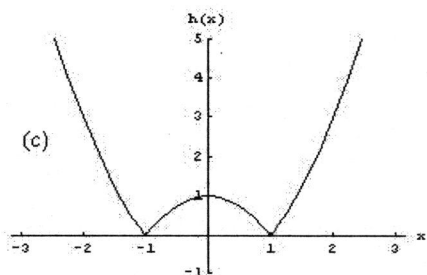

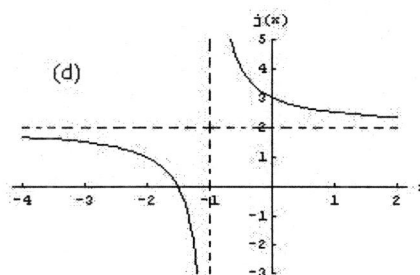

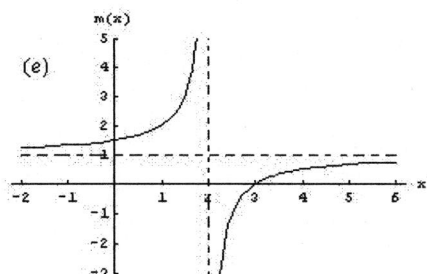

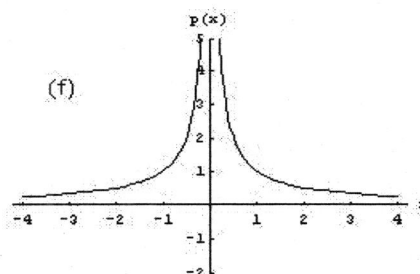

**Problem 10.**

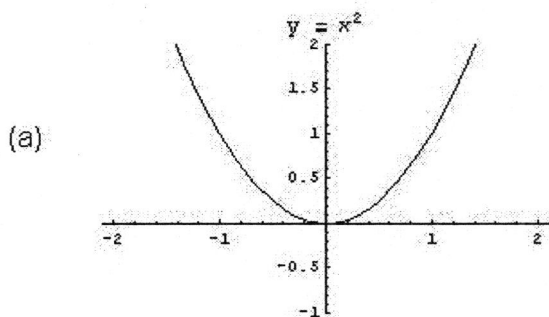

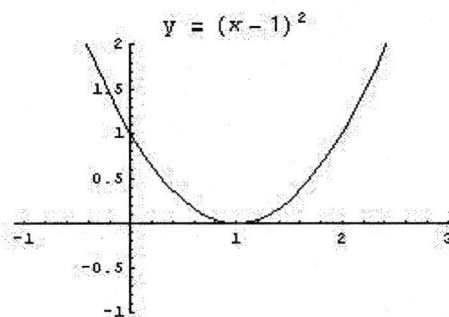

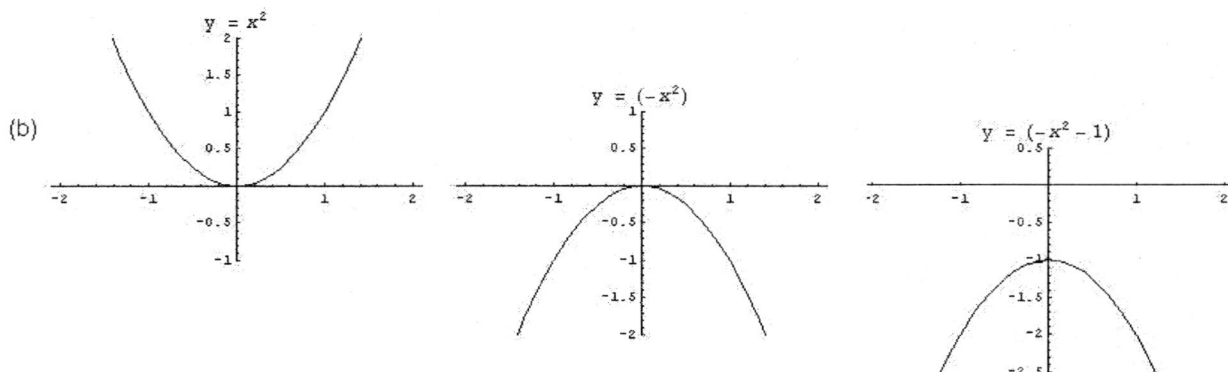

**Problem 12.**

(a) Shift the graph of $y = x^2$ left 3 units, then flip over the $x-$axis, then shift down 1 unit.

(b) Shift the graph of $y = x^2$ right 3 units, then up 1 unit.

**Problem 14.**

(a) Shift the graph of $y = \frac{1}{x}$ left 4 units, then stretch by a factor of 2, then shift up 1 unit.

(b) Shift the graph of $y = \frac{1}{x}$ right $\pi$ units, then flip over the $x-$axis.

**Problem 16.**

(a) Shift the graph of $y = x^2$ right $2\pi$ units, then shift up $\pi$ units.

(b) Shift the graph of $y = x^2$ right $2\pi$ units, then flip over the $x-$axis, then shift up $\pi$ units.

**Problem 17.**

(a) $y = \frac{x+3}{x+2} = \frac{(x+2)+1}{x+2} = 1 + \frac{1}{x+2}$. The graph is obtained by shifting the graph of $y = \frac{1}{x}$ to the left by 2 units and then up by 1 unit.

(b) $y = \frac{x+1}{x-1} = \frac{(x-1)+2}{x-1} = 1 + \frac{2}{x-1}$. The graph is obtained by shifting the graph of $y = \frac{1}{x}$ to the right by 1 unit, vertically stretching by a factor of 2, and then shifting upward by 1 unit.

**Problem 19.**

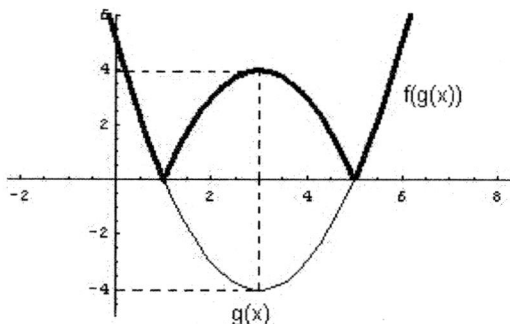

**Problem 22.**

(a)

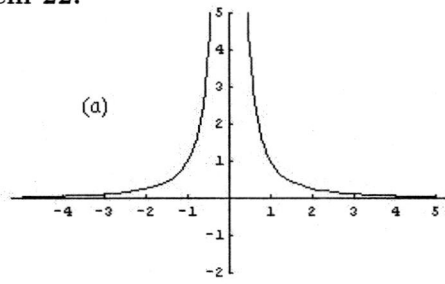

(b)

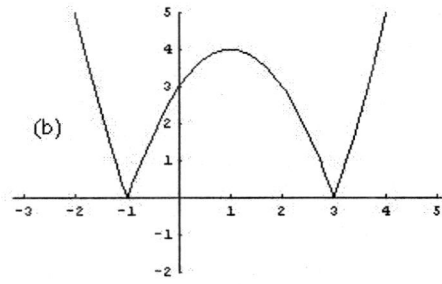

(c)

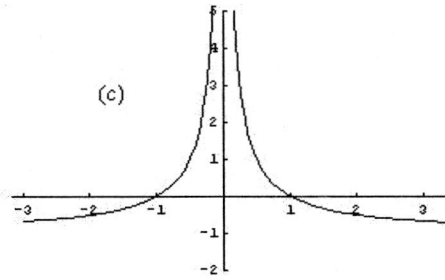

# CHAPTER 4
# Linearity and Local Linearity

## Section 4.1     Making Predictions

**Problem 1.**

Let $L(t)$ be the length of her workout, in pool lengths, after $t$ days: $L(t) = 20 + 2(\frac{t}{4}) = 20 + \frac{1}{2}t$. This is not entirely accurate because she adds 2 lengths after every 4 days; thus her workout is still 20 lengths at $t = 2$, for instance, but $L(2) = 21$.

**Problem 2.**

About $54. She will be spending money at the same rate as during the first 4 days.

**Problem 3.**

We assume that the rate of customer arrival is about the same in this small span of time. If 6 people arrived in the last half hour, 15 minutes $\left( \dfrac{6 \, \text{people}}{30 \, \text{minutes}} \right) = 3$ people will likely arrive in the next 15 minutes.

Over longer spans of time this assumption is not as accurate. For example, he will likely have more than 24 customers between 11:30 and 1:30 because it is lunchtime during which the rate of customer arrival is likely to be higher.

## Section 4.2     Linear Functions

**Problem 1.**

(a) A $(0, f(0))$; B $(b, f(b))$; C $(c, f(c))$; D $(d, f(d))$; E $(e, f(e))$

(b) 0

(c) $y = f(b)$ (or $y = f(d)$)

(d) $f(b)$

(e) $d - b$

(f) $\frac{f(c) - f(0)}{c}$

(g) $y = \left( \frac{f(c) - f(0)}{c} \right) x + f(0)$

(h) $x = d$

(i) Undefined, as the line $\mathcal{L}_3$ is vertical.

**Problem 2.**

(a) $\frac{2}{5}$

(b) 3

(c) No slope. Vertical line $x = \sqrt{2}$

(d)  0 (Horizontal line)

**Problem 4.**

The point slope equation of a line is $y - y_1 = m(x - x_1)$. That is $y - (-3) = -\frac{1}{2}(x - (-2)) \Leftrightarrow y = -\frac{1}{2}x - 4$.

**Problem 6.**

The slope of the line through $(0, a)$ and $(b, 0)$ is $m = \frac{0-a}{b-0} = -\frac{a}{b}$. Because $(0, a)$ lies on the the line, $a$ is the $y$−intercept; hence, the equation of the line is $y = -\frac{a}{b}x + a$.

**Problem 8.**

In slope-intercept form, we can write $3x - 4y = 7$ as $y = \frac{3}{4}x - \frac{7}{4}$. The slope of this line is $m = \frac{3}{4}$. Any line parallel to $3x - 4y = 7$ also has a slope of $m = \frac{3}{4}$. Using the point-slope form, the equation of the required line is $y - \sqrt{2} = \frac{3}{4}(x - \sqrt{3}) \Leftrightarrow y = \frac{3}{4}x + (\sqrt{2} - \frac{3}{4}\sqrt{3})$.

**Problem 9.**

In slope-intercept form, we can write $\pi x - \sqrt{3}y = 12$ as $y = \frac{\pi}{\sqrt{3}} - 4\sqrt{3}$. This line has slope $M = \frac{\pi}{\sqrt{3}}$, and the slope of any line perpendicular to it is $m = -\frac{1}{M} = -\frac{\sqrt{3}}{\pi}$. As the required line passes through the origin, its equation is $y = -\frac{\sqrt{3}}{\pi}x$.

**Problem 11.**

The slope of the line $y - \pi = \pi(x - 1)$ is $M = \pi$, and any line perpendicular to it has slope $m = -\frac{1}{M} = -\frac{1}{\pi}$. As the $y$−intercept of the required line is 3, the equation of the line is $y = -\frac{1}{\pi}x + 3$.

**Problem 12.**

The equation horizontal line passing through the point $(a, b)$ has the form $y = b$. Hence the equation of the horizontal line passing through $(-\sqrt{\pi}, \pi^2)$ is $y = \pi^2$

**Problem 13.**

The equation vertical line passing through the point $(a, b)$ has the form $x = a$. Hence the equation of the vertical line passing through $(-\sqrt{\pi}, \pi^2)$ is $x = -\sqrt{\pi}$.

**Problem 15.**

The $y$-coordinate for the point $P$ is $b^2 + 3b + 1$. The $y$-coordinate for the point $Q$ is $(b+h)^2 + 3(b+h) + 1 = b^2 + 2bh + h^2 + 3b + 3h + 1$. The slope of the secant line through $P, Q$ is $\frac{b^2 + 2bh + h^2 + 3b + 3h + 1 - b^2 - 3b - 1}{h} = \frac{h^2 + 2bh + 3h}{h} = h + 2b + 3$.

**Problem 18.**

The total cost of construction of the mosque in dollars is $c(x) = C + Tx$.

## Section 4.3    Modeling and Interpreting the Slope

**Problem 1.**

(a)  $R(x) = 0.07x$

(b)  $C(x) = 6000 + 0.01x$

(c)  $P(x) = R(x) - C(x) = 0.06x - 6000$

(d)  $P(x) = 0.06x - 6000 = 0 \Rightarrow x = 100,000$ copies must be made to break even.

(e)

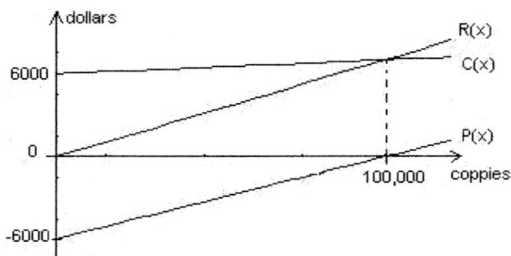

(f) $A(x) = \frac{C(x)}{x} = \frac{6000}{x} + 0.01$

(g)

| $x$ | 0 | 1 | 10 | 100 | 1000 | 10000 |
|---|---|---|---|---|---|---|
| $A(x)$ | $undefined$ | 6000.01 | 600.01 | 60.01 | 6.01 | 0.61 |

(h)

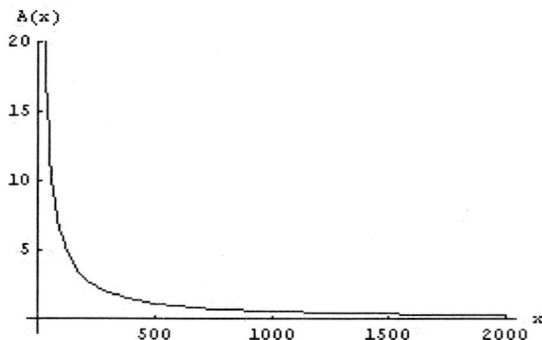

## Problem 3.

(a) Let $w$ be the additional weight added, in pounds and $T(w)$ be the time to complete course, in seconds. Given that $T(0) = K$, and $T(6) = K+22$, and assuming linearity, we can use the slope-intercept equation of a line to find $T(w) = \frac{22}{6}w + K = \frac{11}{3}w + K$.

(b) The rate of change of the linear function $T(w)$ is its slope, which is $\frac{11}{3}$ seconds per pound. Each additional pound of weight added to the bicycle, the rider's time increases by $\frac{11}{3} = 3\frac{2}{3}$ seconds.

## Problem 5.

(a) We want to find the formula for the linear function $F(C)$, where $C$ is degrees Celsius and $F$ is the corresponding degrees Fahrenheit. We have two points on the graph of $F$: $(0, 32)$ and $(100, 212)$. Now the slope of $F(C)$ is $m = \frac{212-32}{100} = \frac{9}{5}$, using the point-slope form of a line, we have $F - 32 = \frac{9}{5}(C - 0)$. Therefore, $F(C) = \frac{9}{5}C + 32$

(b) $F(C) = \frac{9}{5}C + 32 \Rightarrow F - 32 = \frac{9}{5}C \Rightarrow C(F) = \frac{5}{9}(F - 32) = \frac{5}{9}F - \frac{160}{9}$.

(c) Since 1 degree Kelvin = 1 degree Celsius, the slope of the linear function $K(C)$, where $C$ is degrees Celsius and $K$ is degrees Kelvin, is 1. As $(0, 273.15)$ is on graph of $K(C)$, we have $K(C) = c + 273.15$

(d) $K(C(F)) = C(F) + 273.15 = \frac{5}{9}F - \frac{160}{9} + 273.15 = \frac{5}{9}F + \frac{45,967}{180}$.

**Problem 7.**

(a) The points $(-10, 350)$ and $(150, 400)$ are on the graph of $B(T)$, and the slope of this graph is $m = \frac{400-350}{150-(-10)} = \frac{5}{16}$. The point-slope form gives us $B - 400 = \frac{5}{16}(T - 150) \Leftrightarrow. B(T) = \frac{5}{16}T + 353.125$.

(b) The $B$-intercept is $B(0) = 353.125$. A ball whose temperature is 0°F would be hit 353.125 feet by this swing.

(c) The slope is $\frac{5}{16}$. Every increase of 16°F in the ball's temperature increases the ball's travel distance by 5 feet.

## Section 4.4    Applications of Linear Models

**Problem 1.**

Let $W(x)$ be the social worker's weekly wages as a function of the number of hours worked, $x$.
$$W(x) = \begin{cases} Dx & \text{for } 0 \le x \le 40 \\ 40D + 1.5D(x - 40) & \text{for } x > 40, \end{cases}$$

**Problem 3.**

(a) Equation of the supply curve:
the slope $m$ is given by $m = \frac{12}{9} = \frac{4}{3}$. Equation of the line is $y = \frac{4}{3}(x - 9)$ or $y = \frac{4}{3}x - 12$
Equation of the demand curve:
The slope $m$ is given by $m = \frac{0-16}{12} = -\frac{4}{3}$

(b) 6.75
Equation for the line becomes: $y = -\frac{4}{3}(x - 12)$ or $y = -\frac{4}{3}x + 16$.

**Problem 5.**

(a)
| item | 0 | 5 | 10 | 15 | 20 |
|---|---|---|---|---|---|
| A | 100 | 140 | 180 | 220 | 280 |
| B | 80 | 130 | 180 | 230 | 280 |

(b) By graphing the functions representing both salary schemes on the same axes, we see that Company $B$ is better for sales between 10 and 20 items per week.

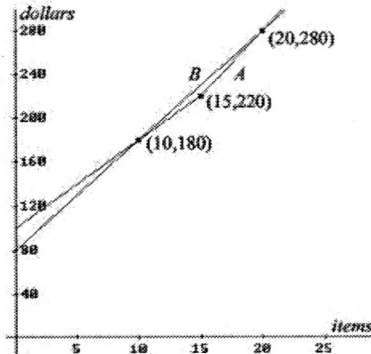

(c) $S_A(x) = \begin{cases} 100 + 8x & \text{for } 0 \le x \le 15 \\ 220 + 12(x - 15) = 40 + 12x & \text{for } x > 15 \end{cases}$
$S_B(x) = 80 + 10x$

To determine when $S_B(x) > S_A(x)$, first solve $80 + 10x > 100 + 8x$ for $0 \le x \le 15$: we get $x > 10$; hence, B is better for $10 < x \le 15$. Now solve $80 + 10x > 40 + 12x$ for $x > 15$; we get $x < 20$, so B is also better for $15 < x < 20$. Combining, $S_B(x) > S_A(x)$ is better for $10 < x < 20$.

**Problem 7.**

(a) $A(x) = \begin{cases} 8000 & \text{for } 0 \le x \le 10,000 \\ 20(.12)(x - 10,000) + 8000 = 2.4x - 16,000 & \text{for } x > 10,000 \end{cases}$

(b) $B(x) = 0.10x$

(c)  (i) For $0 \le x \le 10,000$, $A(x) = B(x) \Rightarrow 8000 = 2x \Rightarrow x = 4000$. For $x > 10,000$, $A(x) = B(x) \Rightarrow 2.4x - 16,000 = 2x \Rightarrow x = 40,000$. The two plans are equivalent for $x = 4000$ and $x = 40,000$.

   (ii) For $0 \le x \le 10,000$, $B(x) > A(x) \Rightarrow 2x > 8000 \Rightarrow x > 4000$. For $x > 10,000$, $B(x) > A(x) \Rightarrow 2x > 2.4x - 16,000 \Rightarrow x < 40,000$. Therefore, plan B is better for all values of $x$ in $(4000, 40,000)$.

   (iii) For $0 \le x < 10,000$, $A(x) > B(x) \Rightarrow 8000 > 2x \Rightarrow x < 4000$. For $x > 10,000$, $A(x) > B(x) \Rightarrow 2.4x - 16.000 > 2x \Rightarrow x > 40,000$. Therefore, plan A is better for all values of $x$ in $[0, 4000) \cup (40,000, \infty)$.

**Problem 10.**

Let $M(t)$ be the total daily mileage after $t$ hours. The two points $(0,3)$ and $\left(\frac{1}{3}, \frac{13}{3}\right)$ are the graph of $M(t)$. As $M(t)$ is linear, we calculate its slope to be $m = \frac{\frac{13}{3} - 3}{\frac{1}{3} - 0} = 4$ and use the slope-intercept form to obtain the equation $M(t) = 3 + 4t$. They stop when $M(t) = 13$, which occurs at $t = 2.5$ hours. The domain of $M(t)$ is $[0, 2.5]$.

**Problem 13.**

(a)

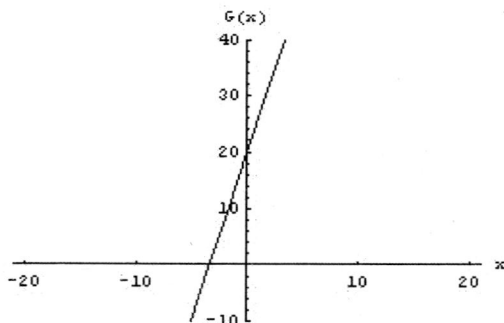
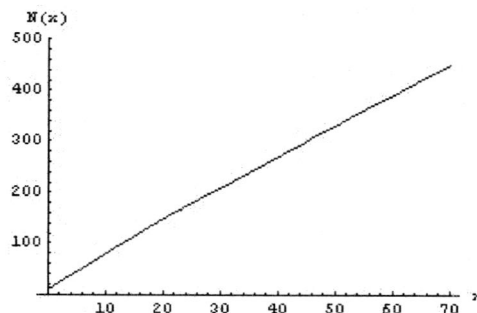

(b) $G(x) = 20 + 6x$

$N(x) = \begin{cases} 10 + 7x & \text{for } 0 < x \le 20 \\ 10 + 20(7) + 6(x - 20) = 30 + 6x & \text{for } 20 < x \le 70 \\ 30 + 6(70) + 5(x - 70) = 100 + 5x & \text{for } x > 70 \end{cases}$

(c) *Great Mugs* is cheaper for orders of between 10 and 80 mugs.

(d) For $x > 80$, the difference in price between the two is $G(x) - N(x) = x - 80$, which becomes arbitrarily large as the size of the order $x$ increases without bound.

# CHAPTER 5

# The Derivative Function

## Section 5.1    Calculating the Slope of a Curve and Instantaneous Rate of Change

**Problem 1.**

(a) By completing the square, we see that $h(t) = -16t^2 + 8t + 48 = -16\left(t^2 - \frac{1}{2}t\right) + 48 = -16\left(t - \frac{1}{4}\right)^2 + 49$. The graph of $h$ is a parabola that opens downward with its vertex at $\left(\frac{1}{4}, 49\right)$. Therefore, since $1 > \frac{1}{4}$, the ball is heading down at $t = 1$.

(b) The average rate of change of height with respect to time over $[0.9, 1] = \frac{\Delta h}{\Delta t} = \frac{h(1)-h(0.9)}{1-0.9} = \frac{-16(1)^2+8(1)+48-(-16(0.9)^2+8(0.9)+48)}{0.1} = -22.4$ ft/sec.
The average rate of change of height with respect to time over $[1, 1.1] = \frac{\Delta h}{\Delta t} = \frac{h(1.1)-h(1)}{1.1-1} = \frac{-16(1.1)^2+8(1.1)+48-(-16(1)^2+8(1)+48)}{0.1} = -25.6$ ft/sec.

The ball's velocity is between -25.6 ft/sec and -22.4 ft/sec.

(c) The average rate of change of height with respect to time over $[0.99, 1] = \frac{\Delta h}{\Delta t} = \frac{h(1)-h(0.99)}{1-0.99} = \frac{-16(1)^2+8(1)+48-(-16(0.99)^2+8(0.99)+48)}{0.01} = -23.84$ ft/sec.
The average rate of change of height with respect to time over $[1, 1.01] = \frac{\Delta h}{\Delta t} = \frac{h(1.01)-h(1)}{1.01-1} = \frac{-16(1.01)^2+8(1.01)+48-(-16(1)^2+8(1)+48)}{0.01} = -24.16$ ft/sec.

The ball's velocity is between -24.16 ft/sec and -23.84 ft/sec.

(d) $h'(1) = \lim_{\Delta t \to 0} \frac{h(1+\Delta t)-h(1)}{\Delta t} = \lim_{\Delta t \to 0} \frac{-16(1+\Delta t)^2+8(1+\Delta t)+48-(-16(1)^2+8(1)+48)}{\Delta t} = \lim_{\Delta t \to 0} \frac{-32\Delta t - 16(\Delta t)^2 + 8\Delta t}{\Delta t} = \lim_{\Delta t \to 0} -16\Delta t - 24 = -24$. The instantaneous velocity of the ball at $t = 1$ is $-24$ ft/sec.

**Problem 3.**

(a) The slope of the secant line through $P$ and $Q = m(h) = \frac{f(1+h)-f(1)}{h} = \frac{(1+h)^3-1^3}{h} = \frac{h^3+3h^2+3h}{h} = h^2 + 3h + 3$. Now $m(-0.1) = 2.71$, $m(-0.01) = 2.9701$, $m(-0.001) = 2.99700$, $m(0.0001) = 3.00030$, $m(0.001) = 3.00300$, $m(0.01) = 3.0301$, and $m(0.1) = 3.31$.

(b) $f'(1) = \lim_{h \to 0} m(h) = \lim_{h \to 0} (h^2 + 3h + 3) = 3$.

(c) The function $f(x) = x^3$ is increasing at an increasing rate; that is, the slopes of tangent lines are nonnegative and increasing as $x$ increases. Thus any secant line with $h > 0$ will have a greater slope that the slope of the tangent line at $x = 1$, and any secant line with $h < 0$ will have a slope less that the slope of the tangent line at $x = 1$. Therefore, the difference quotients for $h > 0$ are greater than $f'(1)$, and the difference quotients for $h < 0$ are less than $f'(1)$.

**Problem 5.**
$f'(9) = \frac{1}{6} \approx 0.16667$

| $2+h$ | 8.9 | 8.99 | 8.999 | 9.0001 | 9.001 | 9.01 | 9.1 |
|---|---|---|---|---|---|---|---|
| $\frac{f(2+h)-f(2)}{h}$ | 0.16713 | 0.16671 | 0.16667 | 0.16667 | 0.16666 | 0.16662 | 0.16621 |

**Problem 7.**

$f'(1) = 4$

| $1 + h$ | 0.9 | 0.99 | 0.999 | 1.0001 | 1.001 | 1.01 | 1.1 |
|---|---|---|---|---|---|---|---|
| $\frac{f(1+h)-f(1)}{h}$ | 3.439 | 3.94040 | 3.99400 | 4.00060 | 4.00600 | 4.06040 | 4.641 |

**Problem 10.**

(a) The average rate of change over $[2, 2.5] = \frac{\Delta w}{\Delta t} = \frac{w(2.5)-w(2)}{2.5-2}\frac{\text{gallons}}{\text{hours}} = \frac{-16(2.5)^2+256-(-16(2)^2+256)}{0.5}\frac{\text{gallons}}{\text{hours}} = -72$ gallons/hour. The quantity $\frac{\Delta w}{\Delta t}$ is negative because the pool is being drained. That is, amount of water $w(t)$ in the pool decreases as time progresses.

(b) The function $w(t) = -16t^2 + 256$ is equal to the function $s(t) = -16t^2 + 256$ from the dropped-rock problem in this section of the text. The results of the that problem tell us that the rate of change of water in the pool at time $t = 2$ is $-64$ gallons per hour.

**Problem 12.**

| $x$−coordinate of $Q$ | 1.99 | 1.999 | 2.0001 | 2.001 | 2.01 |
|---|---|---|---|---|---|
| Slope of $PQ$ | 79.20399 | 79.92004 | 80.00800 | 80.08004 | 80.80401 |

The slope of the tangent line at $P$ is 80.

**Problem 15.**

(a)

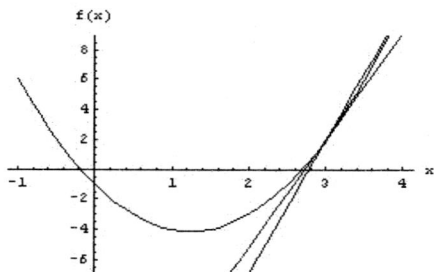

(b) The slope of the secant line through $P = (3, f(3))$ and $Q = (3 + h, f(3 + h))$ is $m(h) = \frac{f(3+h)-f(3)}{h} = \frac{2(3+h)^2-5(3+h)-1-(2(3)^2-5(3)-1)}{h} = \frac{12h+2h^2-5h}{h} = \frac{2h^2+7h}{h} = 2h + 7$. Thus, $m(1) = 2(1) + 7 = 9$, $m(0.1) = 2(0.1) + 7 = 7.2$, and $m(0.01) = 2(0.01) + 7 = 7.02$.

(c) The slope of the tangent line to the graph of $f$ at $P = (3, f(3)) = (3, 2)$ is $m = \lim_{h \to 0} m(h) = \lim_{h \to 0} 2h+7 = 7$.

(d) The point-slope form provides the equation of the tangent line: $y - 2 = 7(x - 3) \Leftrightarrow y = 7x - 19$.

**Problem 16.**

The slope of the tangent line at $x = 1$ is given by $m = -\frac{1}{2}$. The equation of the tangent line is then $y - 1 = -\frac{1}{2}(x - 1)$ or $y = -\frac{1}{2}x + \frac{3}{2}$.

**Problem 19.**

(a)

$$f'(2) = \lim_{h \to 0} \frac{f(2+h) - f(2)}{h} = \lim_{h \to 0} \frac{\frac{3}{2+h-5} - \frac{3}{2-5}}{h} = \lim_{h \to 0} \frac{\frac{3}{h-3} + 1}{h} = \lim_{h \to 0} \frac{3 + h - 3}{(h-3)h} = -\frac{1}{3}$$

**Problem 21.**

$$f'(1) = \lim_{h \to 0} \frac{f(1+h) - f(1)}{h} = \lim_{h \to 0} \frac{\frac{3}{2-(1+h)} - \frac{3}{2-1}}{h} = \lim_{h \to 0} \frac{\frac{3}{1-h} - \frac{3(1-h)}{1-h}}{h} = \lim_{h \to 0} \frac{3h}{h(1-h)} = \lim_{h \to 0} \frac{3}{1-h} = 3$$

**Problem 23.**

$$f'(1) = \lim_{h \to 0} \frac{f(1+h) - f(1)}{h} = \lim_{h \to 0} \frac{\frac{1+h}{2} + \frac{2}{1+h} - \left(\frac{1}{2} + 2\right)}{h} = \lim_{h \to 0} \frac{\frac{h(h-3)}{2(h+1)}}{h} = \lim_{h \to 0} \frac{h-3}{2(h+1)} = -\frac{3}{2}$$

## Section 5.2    The Derivative Function

**Problem 2.**

$$f'(x) = \lim_{h \to 0} \frac{f(x+h) - f(x)}{h} = \lim_{h \to 0} \frac{k(x+h)^2 - kx^2}{h} = \lim_{h \to 0} \frac{k[x^2 + 2xh + h^2 - x^2]}{h}$$

$$= \lim_{h \to 0} \frac{k(2xh + h^2)}{h} = \lim_{h \to 0} \frac{kh(2x+h)}{h} = 2kx$$

**Problem 4.**

$$f'(x) = \lim_{h \to 0} \frac{f(x+h) - f(x)}{h} = \lim_{h \to 0} \frac{(x+h)^3 - x^3}{h} = \lim_{h \to 0} \frac{3x^2 h + 3xh^2 + h^3}{h} = \lim_{h \to 0} 3x^2 + 3xh + h^2 = 3x^2$$

**Problem 6.**

$$g'(x) = \lim_{h \to 0} \frac{g(x+h) - g(x)}{h} = \lim_{h \to 0} \frac{\frac{x+h}{2(x+h)+5} - \frac{x}{2x+5}}{h} = \lim_{h \to 0} \frac{1}{h} \frac{(x+h)(2x+5) - (2(x+h)+5)x}{(2(x+h)+5)(2x+5)}$$

$$= \lim_{h \to 0} \frac{1}{h} \frac{5h}{(2(x+h)+5)(2x+5)} = \lim_{h \to 0} \frac{5}{(2(x+h)+5)(2x+5)} = \frac{5}{(2x+5)^2}$$

**Problem 7.**
From problem 1 we have $f'(x) = 3$. $f'(0) = 3$; $f'(2) = 3$; $f'(-1) = 3$.

**Problem 8.**

$$f'(x) = \lim_{h \to 0} \frac{f(x+h) - f(x)}{h} = \lim_{h \to 0} \frac{\pi(x+h) - \sqrt{3} - (\pi x - \sqrt{3})}{h} = \lim_{h \to 0} \frac{\pi h}{h} = \pi$$

$$f'(0) = \pi; \ f'(2) = \pi; \ f'(-1) = \pi$$

**Problem 11.**
From problem 2, we have $f'(x) = 2x$. $f'(0) = 2(0) = 0$; $f'(2) = 2(2) = 4$; $f'(-1) = 2(-1) = 2$.

**Problem 15.**
From problem 12, we have $f'(x) = -\frac{1}{x^2}$.

(a) The average rate of change on $\left[\frac{1}{2}, 2\right]$ is $\frac{f(2) - f(1/2)}{2 - 1/2} = \frac{1/2 - 2}{2 - 1/2} = -1$. Now $f'(x) = -\frac{1}{x^2} = -1 \Rightarrow x = \pm 1$.

(b) The average rate of change on $[1,4]$ is $\frac{f(4)-f(1)}{4-1} = \frac{1/4-1}{3} = -\frac{1}{4}$. Now $f'(x) = -\frac{1}{x^2} = -\frac{1}{4} \Rightarrow x = \pm 2$.

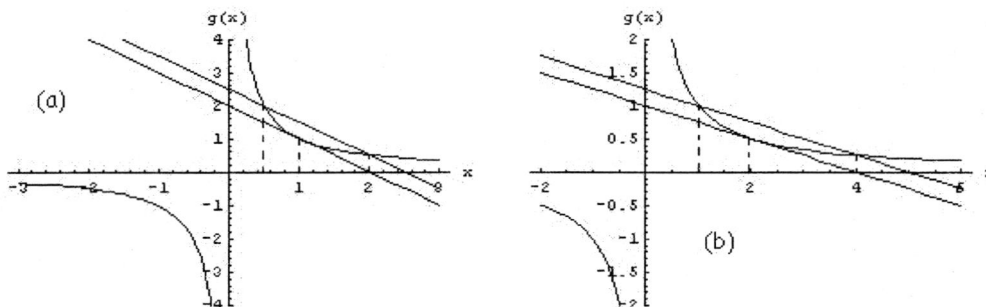

(c) The average rate of change on $[c,d]$ is $\frac{f(d)-f(c)}{d-c} = \frac{\frac{1}{d}-\frac{1}{c}}{d-c} = \frac{\frac{c-d}{cd}}{d-c} = -\frac{1}{cd}$. Now $f'(x) = -\frac{1}{x^2} = -\frac{1}{cd} \Rightarrow$
$x = \pm\sqrt{cd}$.

**Problem 17.**

(a) The graph of $\sqrt{x-1}$ is an one-unit horizontal shift to the right of the graph of $\sqrt{x}$.

(b) The graph of the derivative of $\sqrt{x-1}$ is an one-unit horizontal shift to the right of the graph of the derivative of $\sqrt{x}$.

(c) The equation holds because the graph of the derivative of $\sqrt{x-1}$ is an one-unit horizontal shift to the right of the graph of the derivative of $\sqrt{x}$.

(d) $f'(5) = \lim_{x\to 5} \frac{f(x)-f(5)}{x-5} = \lim_{x\to 5} \frac{\sqrt{x-1}-\sqrt{5-1}}{x-5} = \lim_{x\to 5} \frac{(x-1)-4}{(x-5)(\sqrt{x-1}+2)} = \lim_{x\to 5} \frac{1}{\sqrt{x-1}+2} = \frac{1}{4}$.

**Problem 19.**

$$f'(x) = \lim_{h\to 0} \frac{f(x+h)-f(x)}{h} = \lim_{h\to 0} \frac{(x+h)^{-\frac{1}{2}} - x^{-\frac{1}{2}}}{h} = \lim_{h\to 0} \frac{\frac{1}{\sqrt{x+h}} - \frac{1}{\sqrt{x}}}{h} = \lim_{h\to 0} \frac{\frac{\sqrt{x}-\sqrt{x+h}}{(\sqrt{x+h})\sqrt{x}}}{h}$$

$$= \lim_{h\to 0} \frac{(\sqrt{x}-\sqrt{x+h})(\sqrt{x}+\sqrt{x+h})}{h(\sqrt{x+h})\sqrt{x}(\sqrt{x}+\sqrt{x+h})} = \lim_{h\to 0} \frac{x-x-h}{h(\sqrt{x+h})\sqrt{x}(\sqrt{x}+\sqrt{x+h})} = -\frac{1}{2x\sqrt{x}} = -\frac{1}{2}x^{-\frac{3}{2}}$$

## Section 5.3    Qualitative Interpretation of the Derivative

**Problem 1.**

(a) $(2,5]$

(b) $[0,2)$

(c) $x = 2$

**Problem 3.**

(a) E; (b) B; (c) A; (d) G; (e) H; (f) F; (g) C; (h) D

**Problem 5.**

**Problem 7.**

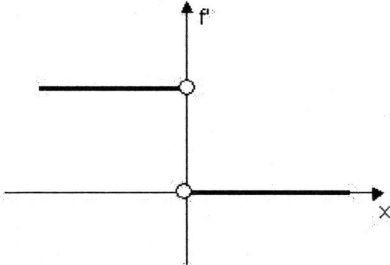

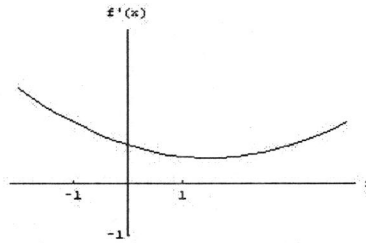

**Problem 11.**

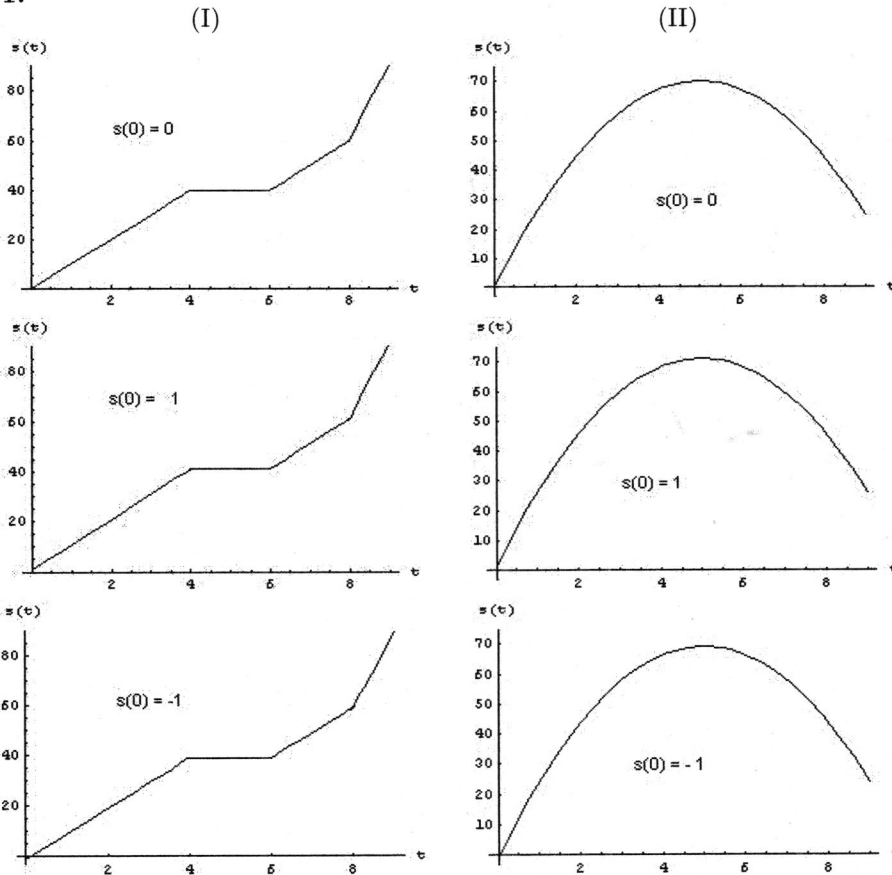

## Section 5.4    Interpreting the Derivative: Meaning and Notation

**Problem 1.**

  (a), (c), (d), (f), (g).

**Problem 3.**

(a)

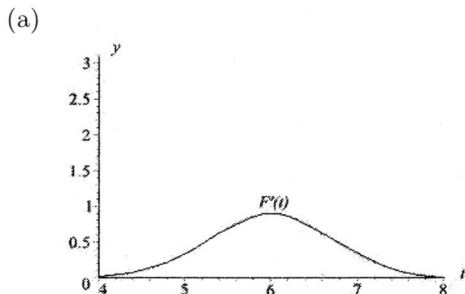

(b) The slope of $F$ is the rate of change of the number of fish in the pond over time; the slope decreases over time as $F(t)$ approaches $C$.

**Problem 5.**

(a) $\frac{dB}{da}$ is the instantaneous rate of change of the length of the drive with respect to change in altitude; $\frac{dB}{da} = \frac{\Delta B}{\Delta a}$ as $\Delta a \to 0$. Its units are feet/feet (change in distance / change in altitude).

(b) $\frac{dB}{da} = \frac{2}{275}$ ft/ft. The slope is constant because $B(a)$ is linear.

(c) $B(a) = mx + b$. We know $m = \frac{2}{275}$ from above. We know that $B(0) = 400$ from the problem, so $m(0) + b = 400$, and $b = 400$. $B(a) = \frac{2}{275}x + 400$.

(d) $B(1050) = \frac{2}{275}(1050) + 400 \approx 407.8$ ft.

(e) $B(5280) = \frac{2}{275}(5280) + 400 \approx 438.4$ ft.

**Problem 7.**

(a) Its units are acres/year.

(b) $A(43) > A(12)$ ($A(t)$ is an increasing function). However, from the problem statement, we can guess that farm size increased faster in 1952 than 1983, so $A'(12) > A'(43)$.

**Problem 9.**

(a) After four hours into the trip, the balloon is rising at a rate of 70 feet per hour.

(b) (i) The input of $h$ is $x$, the balloon's horizontal distance from the mouth of the river. The output of $h$ is the balloon's height above the ground.

(ii) When the balloon is 700 feet from the mouth of the river, it is 100 feet above the ground

(iii) When the balloon is 700 feet from the mouth of the river, it is rising at a rate of 60 feet per additional foot traveled up the river bank.

# CHAPTER 6
# The Quadratics

## Section 6.1    A Profile of Quadratics from a Calculus Perspective

**Problem 1.**

$f'$: (i); $g'$: (iv); $h'$: (ii); $j'$: (iii)

**Problem 3.**

$y' = -2\sqrt{3}x + \sqrt{27} = \sqrt{3}(-2x+3)$; $0 = y' = \sqrt{3}(-2x+3) \Rightarrow x = \frac{3}{2}$ and $y\left(\frac{3}{2}\right) = -\sqrt{3}\left(\frac{3}{2}\right)^2 + \sqrt{27}\left(\frac{3}{2}\right) + 15 = \frac{9}{4}\sqrt{3} + 15$. As the coefficient of $x^2$ is negative, the vertex $\left(\frac{3}{2}, \frac{9}{4}\sqrt{3} + 15\right)$ is the highest point on the curve.

**Problem 5.**

(a) The function is $f(x) = 3x + C$, where $C$ is a constant. If $f(x)$ passes through the origin then we have $0 = C$, and hence $f(x) = 3x$. If $f(x)$ passes through $(0,2)$ then we have $2 = C$, and $f(x) = 3x + 2$.

(b) The function is $f(x) = \pi x + C$. If it passes through the origin, then $C = 0$ and $f(x) = \pi x$. If it passes through $(0,2)$ then $C = 2$ and $f(x) = \pi x + 2$.

**Problem 7.**

(a) $f(x) = 3x^2 - 2x + C$. If it passes through the origin then $C = 0$ and $f(x) = 3x^2 - 2x$. If $f(x)$ passes through $(0,2)$ then $C = 2$ and $f(x) = 3x^2 - 2x + 2$.

(b) $f(x) = \frac{m}{2}x^2 + bx + c$. If $f(x)$ passes through the origin then $C = 0$ and $f(x) = \frac{m}{2}x^2 + bx$. If $f(x)$ passes through $(0,2)$ then $C = 2$ and $f(x) = \frac{m}{2}x^2 + bx + 2$.

## Section 6.2    Quadratics From a Noncalculus Perspective

**Problem 1.**

(a) $h(x) = f(x)g(x) = (x+3)(x-5) = 0 \Rightarrow x = -3$ or $x = 5$.

(b) $h(x) = (x+3)(x-5) = -7 \Leftrightarrow 0 = x^2 - 2x - 8 = (x+2)(x-4) \Rightarrow x = -2$, or $x = 4$.

(c) $h(x) = (x+3)(x-5) = -15 \Leftrightarrow 0 = x^2 - 2x = x(x-2) \Rightarrow x = 0$, or $x = 2$.

(d) $h(x) = (x+3)(x-5) = c \Leftrightarrow 0 = x^2 - 2x - 15 - c$ Using the quadratic formula to solve this last equation for $x$, we have $x = \frac{-(-2) \pm \sqrt{(-2)^2 - 4(-15-c)}}{2} = 1 \pm \sqrt{16 + c}$.

(e) $j(x^2) - 2 = \frac{f(x^2)}{g(x^2)} - 2 = \frac{x^2+3}{x^2-5} - 2 = 0$. $\Leftrightarrow 0 = \frac{x^2+3-2(x^2-5)}{x^2-5} = \frac{-x^2+13}{x^2-5} \Leftrightarrow -x^2 + 13 = 0$. Now $-x^2 + 13 = 0 \Rightarrow x = \pm\sqrt{13}$. Since $(\pm\sqrt{13})^2 - 5 = 13 - 5 = 8 \neq 0$, the solutions are $x = \pm\sqrt{13}$.

(f) $[j(x)]^2 - 1 = \left[\frac{f(x)}{g(x)}\right]^2 - 1 = \frac{(x+3)^2}{(x-5)^2} - 1 = \frac{(x+3)^2 - (x-5)^2}{(x-5)^2} = \frac{16x-16}{(x-5)^2} = 0 \Leftrightarrow 16x - 16 = 0$. Now $16x - 16 = 0$ when $x = 1$, and $(1-5)^2 = 16 \neq 0$. Hence the solution to the equation is $x = 1$.

(g) $h(x) = j(x) \Leftrightarrow f(x)g(x) = \frac{f(x)}{g(x)} \Leftrightarrow (x+3)(x-5) = \frac{x+3}{x-5} \Leftrightarrow \frac{(x+3)((x-5)^2-1)}{x-5} = 0.$ Now $0 = (x+3)((x-5)^2 - 1) = (x+3)(x-4)(x-6) \Rightarrow x = -3$, $x = 4$, or $x = 6$. Notice that none of these three values of $x$ make the denominator from the original equation, $(x-5)$, equal to zero. Hence $x = -3$, $x = 4$, and $x = 6$ are the solutions to the equation.

## Problem 3.

(a) $x^2 - 7 = 0$ gives $x^2 = 7, x = \pm\sqrt{7}$

(b) $x = 5$ or $x = -5$

(c) $(x+1)^2 = 25, x + 1 = \pm 5, x = -6$ or $x = 4$

(d) $x = 4$ or $x = -6$

(e) $x = 0$ or $x = -3$

(f) $x = \frac{\sqrt{7}-1}{3}$ or $x = -\frac{\sqrt{7}-1}{3}$

(g) $x = -3, x = 1$

## Problem 5.

(a) None: $2(x-3)^2 - 5 = -6 \Leftrightarrow 2(x-3)^2 = -1$. This last equation has no solutions because $2(x-3)^2 \geq 0 > -1$.

(b) Two: $-4(x+1)^2 + 3 = -6 \Leftrightarrow (x+1)^2 = \frac{9}{4} \Leftrightarrow x = -1 \pm \frac{3}{2} \Leftrightarrow x = \frac{1}{2}$ or $x = -\frac{5}{2}$.

(c) Two: $|x+1| - 3 = -2 \Leftrightarrow |x+1| = 1 \Leftrightarrow x = 0$ or $x = -2$.

(d) $x^2 - 3 \geq 1 \Leftrightarrow x^2 \geq 4 \Leftrightarrow x \leq -2$ or $x \geq 2$.

(e) Four solutions: $|x^2 - 3| = 1 \Leftrightarrow x^2 - 3 = 1$ or $x^2 - 3 = -1 \Leftrightarrow x^2 = 4$ or $x^2 = 2 \Leftrightarrow x = \pm 2$ or $x = \pm\sqrt{2}$.

## Problem 7.

(a) Maximum at $x = 0$; minimum at $x = 3$

(b) Maximum at $x = 0$; minimum at $x = 4$

(c) Maximum at $x = 4$; minimum at $x = 0$

## Problem 10.

(a) An equation for the curve is $y = k(x+2)(x-2)$, for some constant $k \neq 0$. As $-4 = k(0+2)(0-2)$, $k = 1$. Thus, $y = (x+2)(x-2)$.

## Problem 12.

(a) $x^4 + x^2 = 6$. Put $x^2 = z$. Then $z^2 + x = 6$ or $z^2 + z - 6 = 0$. This gives $(z+3)(z-2) = 0$ and $z = -3, or z = 2$. $x^2 = z$ hence $x = \pm\sqrt{2}$

(b) Put $x^2 = z$. Then $z^2 - 5z + 6 = 0$. This is $(z-3)(z-2) = 0$. So $z = 3$ or $z = 2$. As a consequence $x = \pm\sqrt{(3)}$ or $x = \pm\sqrt{2}$

**Problem 14.**

$(x-2)^4 - 2(x-2)^2 = -1 \Leftrightarrow (x-2)^4 - 2(x-2)^2 + 1 = 0 \Leftrightarrow ((x-2)^2 - 1)^2 = 0 \Leftrightarrow (x-2)^2 = 1 \Leftrightarrow x - 2 = 1$ or $x - 2 = -1 \Leftrightarrow x = 3$ or $x = 1$.

## Section 6.3   Quadratics and Their Graphs

**Problem 1.**

**Problem 3.**

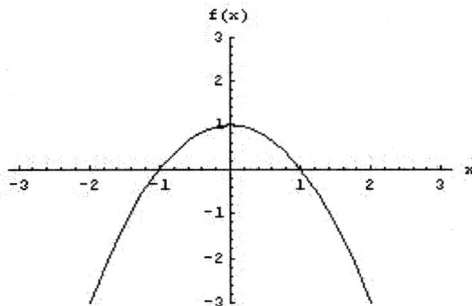

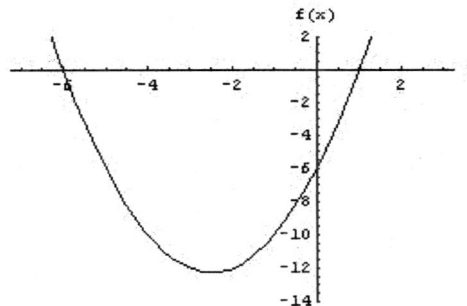

**Problem 5.**

**Problem 7.**

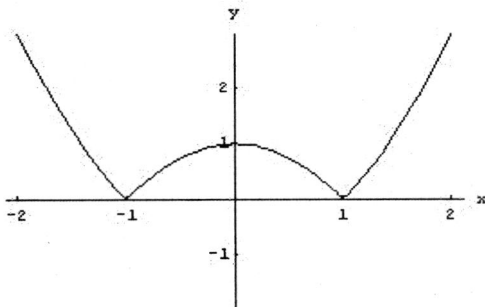

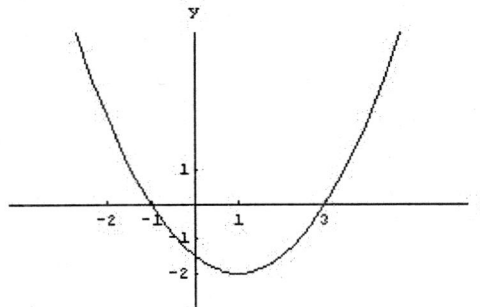

**Problem 9.**

(a) Let $y = k(x+4)(x-2)$; as $(-1,2)$ is on the graph, $2 = k(-1+4)(-1-2)$, from which we obtain $k = -\frac{2}{9}$. Therefore, $y = -\frac{2}{9}(x+4)(x-2)$.

(d) Let $y = k(x+2)(x-1)(x-3)$; as $(0,-2)$ is on the graph, $-2 = k(0+2)(0-1)(0-3)$, from which we obtain $k = -\frac{1}{3}$. Therefore, $y = -\frac{1}{3}(x+2)(x-1)(x-3)$.

**Problem 11.**
Lines with slope 2 have the form $y = 2x + b$. Thus, $f(x) = 2x + b$, for any constant $b$.

**Problem 13.**

The equation of such a parabola has the form $y = k(x+5)(x-1)$, where $k \neq 0$. Thus, $f(x) = k(x+5)(x-1)$ for any constant $k$.

**Problem 15.**

The equation a parabola with vertex $(-1, \pi)$ has the form $y = a(x+1)^2 + \pi$, where $a \neq 0$. Thus, $f(x) = a(x+1)^2 + \pi$, where $a \neq 0$.

**Problem 17.**

$y-$coordinate of the vertex is the maximum value of 1, and the $x-$coordinate of the vertex is the midpoint of the $x-$intercepts 3 and $-2$, which is $\frac{3+(-2)}{2} = \frac{1}{2}$. Thus, the vertex is $\left(\frac{1}{2}, 1\right)$, and the equation this parabola is of the form $y = a\left(x - \frac{1}{2}\right)^2 + 1$, where $a < 0$. As the point $(3, 0)$ lies on this parabola, we have that $0 = a\left(3 - \frac{1}{2}\right)^2 + 1$, from which we obtain $a = -\frac{4}{25}$. Therefore, the equation of the parabola is $y = -\frac{4}{25}\left(x - \frac{1}{2}\right)^2 + 1$

**Problem 19.**

The equation of a parabola with $x-$intercepts $\pi$ and $3\pi$ has the form $y = a(x-\pi)(x-3\pi)$, where $a \neq 0$. Given that the $y-$intercept is $-2$, we have that $-2 = a(0-\pi)(0-3\pi)$, from which we obtain $a = -\frac{2}{3\pi^2}$. Therefore, the equation of the parabola is $y = -\frac{2}{3\pi^2}(x-\pi)(x-3\pi)$.

**Problem 21.**

$f(x) = ax^2 + bx + c$. $x$-coordinate for the vertex is $x = -\frac{b}{2a}$, this gives $-2 = -\frac{b}{2a}$ or $4a = b$. Our equation is then $f(x) = ax^2 + 4ax + c$. Now: $-1 = a + 4a + c$ and $3 = 4a + 8a + c$. From this $4 = -9a$, hence $a = -\frac{4}{9}$, $b = -\frac{16}{9}$ and $c = \frac{11}{9}$. So $f(x) = -\frac{4}{9}x^2 - \frac{16}{9}x + \frac{11}{9}$

**Problem 22.**

**(b)**

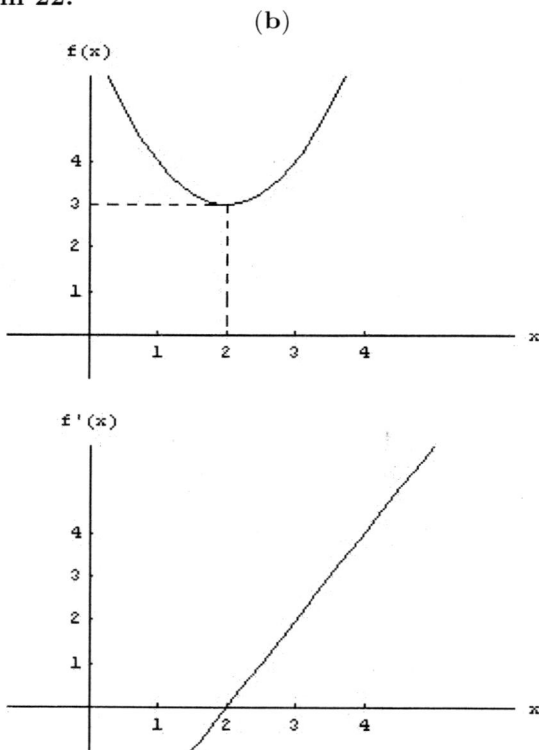

**(d)**

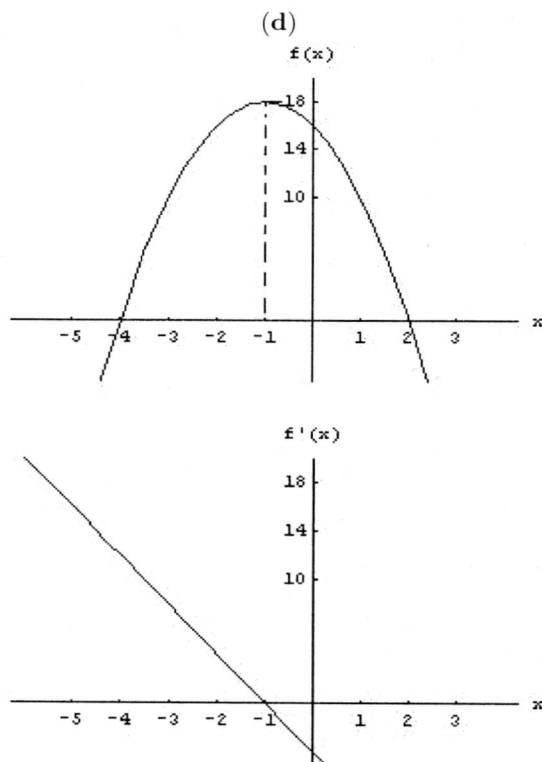

**(f)**

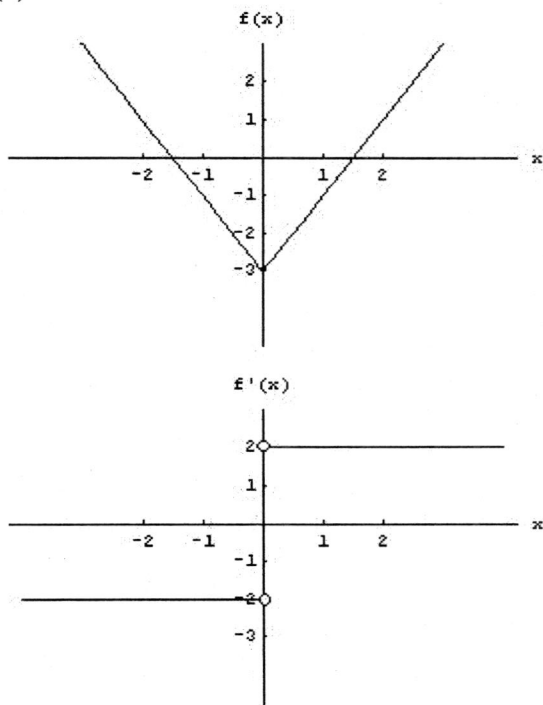

**Problem 24.**

A parabola can be represented by an equation of the form $y = ax^2 + bx + c$, where $a \neq 0$. We substitute the three known points on parabola, $(2, 0)$, $(-1, 9)$, and $(1, -5)$, into this equation to obtain the following system of three equations:

$$\begin{aligned} 0 &= 4a + 2b + c \\ 9 &= a - b + c \\ -5 &= a + b + c \end{aligned}$$

By subtracting the last two equations, we obtain $-14 = 2b$, and hence $b = -7$. By substituting $b = -7$ into the first two equations, we obtain:

$$\begin{aligned} 0 &= 4a - 14 + c \\ 9 &= a + 7 + c \end{aligned}$$

By subtracting these two equations, we obtain $-9 = 3a - 21$, and hence $a = 4$. By substituting $a = 4$ and $b = -7$ into the third equation of the original system, we obtain $c = -2$. The vertex of this parabola is $\left(-\frac{b}{2a}, -\frac{b^2}{4a} + c\right) = \left(\frac{7}{8}, -\frac{81}{16}\right)$.

## Section 6.4    The Free Fall of an Apple

**Problem 1.**

(a) $R(x) = (720 + 10x)(220 - x) = -10x^2 + 1480x + 158,400$

(b) The domain is $[0, 220]$.

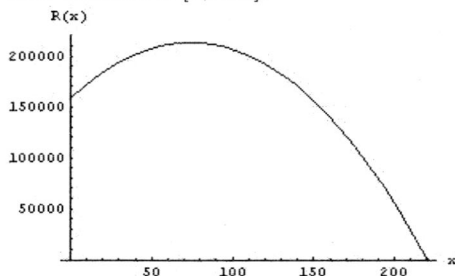

(c) As $R$ is a quadratic function with negative lead coefficient, its graph is a parabola that opens downward. The maximum value of the revenue function $R$ is the $y$-coordinate of the vertex of the parabola. The $x$-coordinate of the vertex is the number of unsold seats that will result in the maximum revenue. We find the $x$-coordinate of the vertex by solving the equation $0 = R'(x) = -20x + 1480$. This solution to this equation is $x = 74$, and $R(74) = -10(74)^2 + 1480(74) + 158,400 = 213,160$. Therefore, 74 unsold seats result in the maximum profit of $213,160$.

**Problem 3.**

(a)

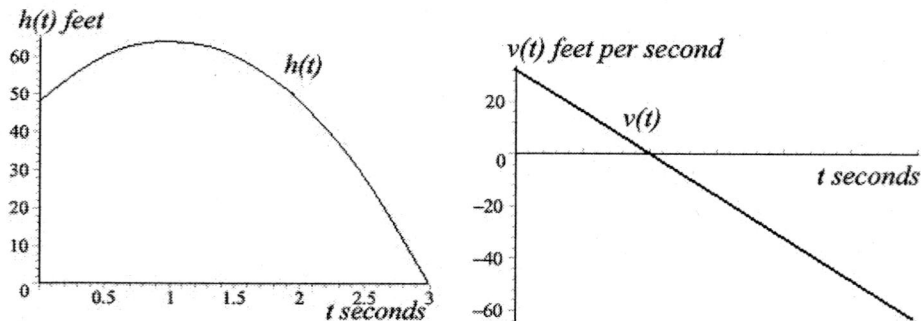

(b) The ball was thrown from a height of $h(0) = 48$ feet.

(c) $v(t) = h'(t) = -32t + 32$; $v(0) = 32$. The initial velocity was 32 ft/sec. The ball was thrown up because $v(0) = 32$ is positive.

(d) The ball's height was decreasing at $t = 2$ because its velocity at $t = 2$, $v(2) = -32(2) + 32 = -32$, is negative.

(e) The ball achieves its maximum height when $v(t) = -32(t) + 32 = 0$, which is when $t = 1$. That is, the ball achieves its maximum height 1 second after being thrown. At $t = 1$, the height of the ball is $h(1) = -16(1)^2 + 32(1) + 48 = 64$ feet, and the velocity is $v(1) = 0$ ft/sec.

(f) The ball hits ground at the first instance after $t > 0$ when $h(t) = -16(t + 1)(t - 3) = 0$, which is when $t = 3$. Therefore the ball was in the air for 3 seconds.

(g) The ball's acceleration was $a(t) = v'(t) = -32$ ft/sec$^2$. Yes; this makes physical senses because acceleration of a falling body due to gravity is -32 ft/sec$^2$.

**Problem 5.**

(a) As as the price, $p$, increases, the demand, $q$, decreases. Hence, $m$ should be negative.

(b) $R(p) = (\text{price})(\text{demand}) = pq = p(mp + b) = mp^2 + bp$

(c)

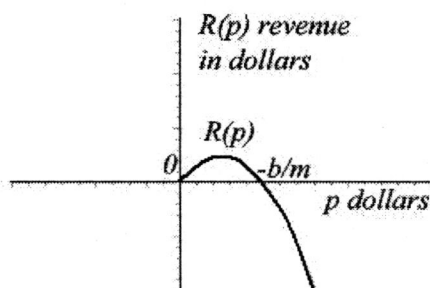

(d) The maximum revenue occurs at the vertex of the graph, which occurs halfway between intercepts at $p = 0$ and $p = -\frac{b}{m}$. Thus the price level of $p = -\frac{b}{2m}$ dollars maximizes the revenue.

(e) $R'(p) = 2mp + b$

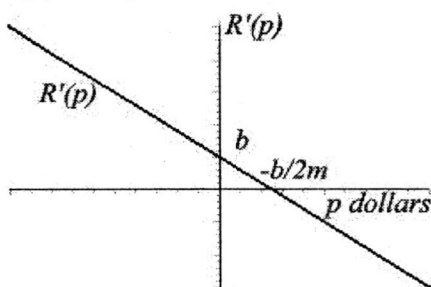

**Problem 7.**

(a) $R(p) = 35p(75 - p)$. The graph of $R(p)$ is a parabola, so its maximum is at its vertex, therefore $p = 37.5$

(b) $R(37.5) = 35 \cdot 37.5(37.5) = 49,218.75$

**Problem 9.**

(a) Amelia sells $q(x) = 120 - 5x$ bowls per week at a price of $x$ dollar per bowl. Now $q(x + 1) - q(x) = (120 - 5(x + 1)) - (120 - 5x) = -5$. By raising the price of a bowl by 1 dollar, she sells 5 fewer bowls.

(b) $R(x) = (\text{price})(\text{quantity}) = xq(x) = x(120 - 5x)$.

(c) As her revenue function $R$ is quadratic, her revenue will be maximzed at the sales level of $x$ bowls for which $R'(x) = 0$. Now $R'(x) = 120 - 10x$, and hence $R'(x) = 0$ when $x = 12$. Therefore her revenue is maximized when she charges \$12 per bowl.

(d) The maximum weekly revenue is $R(12) = \$720$.

# CHAPTER 7

# The Theoretical Backbone: Limits and Continuity

## Section 7.1     Investigating Limits-Methods of Inquiry and Definition

**Problem 1.**

(a) $\displaystyle\lim_{x \to \infty} (1.1)^x = \infty$

(b) $\displaystyle\lim_{x \to \infty} (0.9)^x = 0$

(c) $\displaystyle\lim_{x \to 0} (1.1)^x = 1$

(d) $\displaystyle\lim_{x \to -\infty} (1.1)^x = 0$

(e) $\displaystyle\lim_{x \to -\infty} (0.9)^x = \infty$

$\qquad$ if $0 < b < 1$ then $\displaystyle\lim_{x \to \infty} b^x = 0$

$\qquad$ if $b > 1$ then $\displaystyle\lim_{x \to \infty} b^x = \infty$

**Problem 3.**

(a) $\displaystyle\lim_{h \to -2} \frac{(h-3)(h+2)}{(h+2)} = \lim_{h \to -2}(h-3) = -5$

$\quad f(x) = x - 3,\ D : x \neq -2$

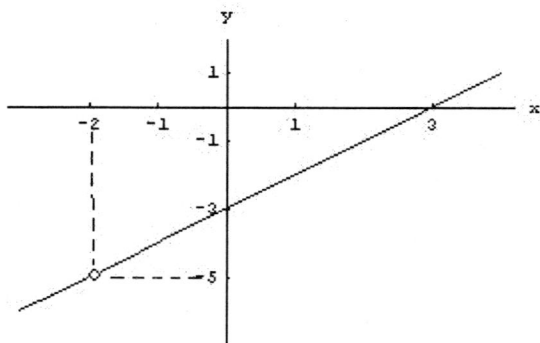

(b) $\displaystyle\lim_{x \to 5} \frac{x^2 - 25}{x - 5} = \lim_{x \to 5} \frac{(x-5)(x+5)}{x-5} = \lim_{x \to 5}(x+5) = 10$

$\quad f(x) = \frac{x^2-25}{x-5} = x + 5,\ D : x \neq 5$

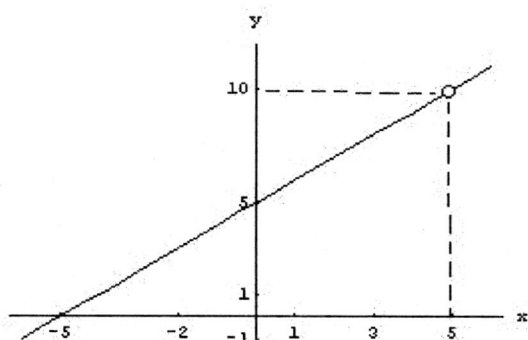

(c) $\lim\limits_{x \to -5} \dfrac{x^2 - 25}{x + 5} = \lim\limits_{x \to -5} \dfrac{(x - 5)(x + 5)}{x + 5} = \lim\limits_{x \to -5}(x - 5) = -10$

$f(x) = \frac{x^2 - 25}{x + 5} = x - 5,\ D : x \neq -5$

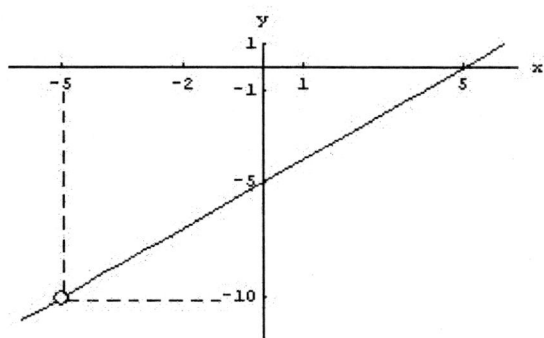

(d) $\lim\limits_{t \to 0} \dfrac{t^2 + \pi t}{t} = \lim\limits_{t \to 0} \dfrac{t(t + \pi)}{t} = \lim\limits_{t \to 0}(t + \pi) = \pi$

$f(t) = \frac{t^2 \pi t}{t} = t + \pi,\ D : t \neq 0$

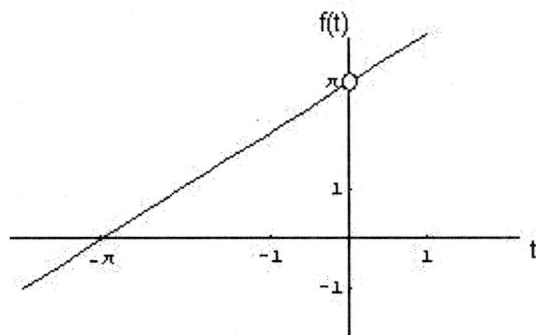

(e) $\lim\limits_{h \to 0} \dfrac{hk + h^2}{h} = \lim\limits_{h \to 0} \dfrac{h(k + h)}{h} = \lim\limits_{h \to 0}(k + h) = k$

$f(h) = \frac{hk^2 + h^2}{h} = h + k,\ k\text{-constant}\ D : h \neq 0$

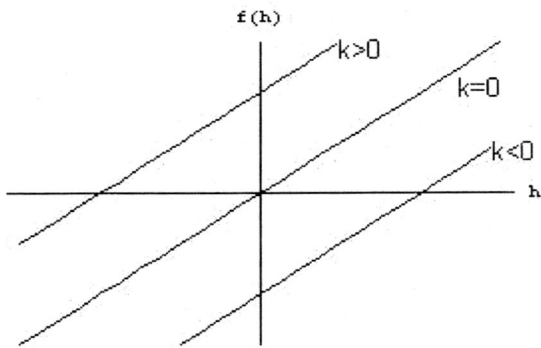

(f) $\lim\limits_{w\to 2} \dfrac{(w-3)(w+1)(w-2)}{3w-6} = \lim\limits_{w\to 0} \dfrac{(w-3)(w+1)(w-2)}{3(w-2)} = \lim\limits_{w\to 0} \dfrac{(w-3)(w+1)}{3} = \dfrac{-1\cdot 3}{3} = -1$

$f(w) = \dfrac{(w-3)(w+1)(w-2)}{3(w-2)} = \dfrac{1}{3}(w-3)(w+1),\ D: w \neq 2$

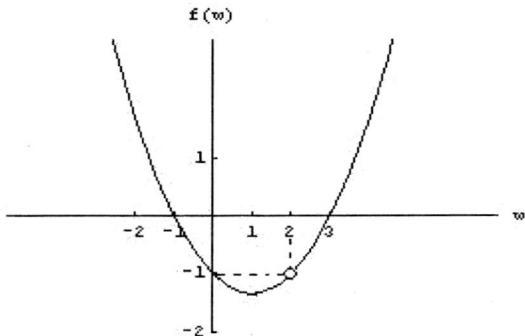

**Problem 6.**

$f(x) = \pi x - 4$
(a) $\lim\limits_{x\to 0} f(x) = -4$
(b) $\lim\limits_{x\to 1} f(x) = \pi - 4 \approx -.86$
(c) $\lim\limits_{x\to\infty} f(x) = \infty$

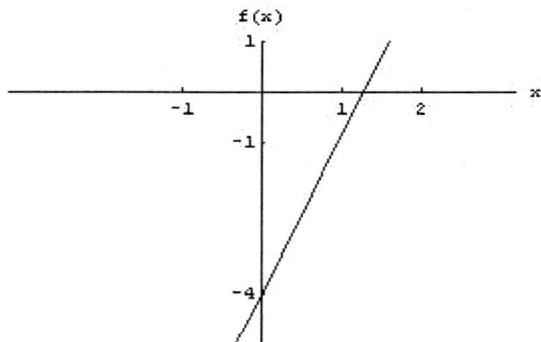

**Problem 8.**

$f(x) = \dfrac{x^2-2x}{x-2} = \dfrac{x(x-2)}{x-2} = x,\ D := x \neq 2$
(a) $\lim\limits_{x\to 0} f(x) = 0$
(b) $\lim\limits_{x\to 2} f(x) = 2$

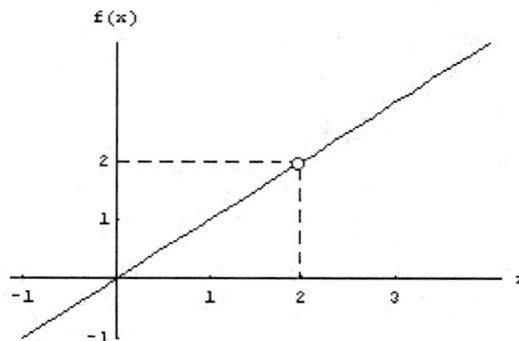

**Problem 10.**

$f(x) = \frac{(x+2)(x^2-x)}{x(x-1)}$

(a) $\lim_{x\to 0} \frac{(x+2)(x^2-x)}{x(x-1)} = 2$

(b) $\lim_{x\to 1} \frac{(x+2)(x^2-x)}{x(x-1)} = 3$

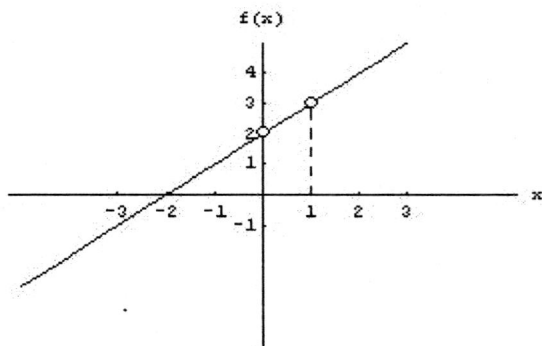

**Problem 12.**

$f(x) = \begin{cases} |x|, & x \neq 3 \\ 0, & x = 3 \end{cases}$

(a) $\lim_{x\to 0} f(x) = \lim_{x\to 0} |x| = 0$

(b) $\lim_{x\to 3} f(x) = \lim_{x\to 3} |x| = 3$

(c) $\lim_{x\to -\infty} f(x) = \lim_{x\to -\infty} |x| = \infty$

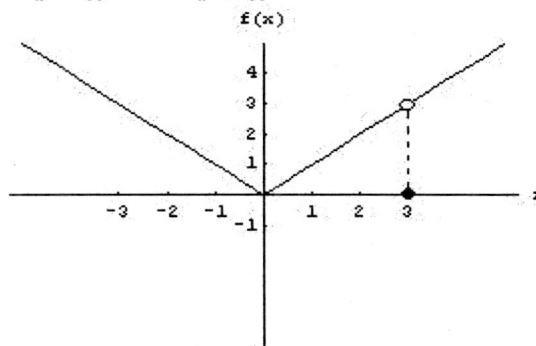

**Problem 14.**

$f(x) = e^x, \ a = 2$

$f'(2) = \lim_{x\to 2} \frac{e^x - e^2}{x-2} = e^2$

**Problem 17.**

$f(x) = \sqrt{7+x}, \ a = 0, \ f'(0) = \lim_{h\to 0} \frac{\sqrt{7+h}-\sqrt{7}}{h} = \lim_{h\to 0} \frac{(\sqrt{7+h}-\sqrt{7})(\sqrt{7+h}+\sqrt{7})}{h(\sqrt{7+h}+\sqrt{7})} = \lim_{h\to 0} \frac{7+h-7}{h(\sqrt{7+h}+\sqrt{7})} = \lim_{h\to 0} \frac{1}{\sqrt{7+h}+\sqrt{7}} = \frac{1}{2\sqrt{7}}$

**Problem 19.**

(b) $\lim_{x\to 5} g(x) = 7, \ g(5) = 8; \ g(x) = \begin{cases} 2x - 3 & , x \neq 5 \\ 8 & , x = 5 \end{cases}$

**Problem 20.**

(a) $\lim_{x\to\infty} f(x) = \infty, \ \lim_{x\to -\infty} f(x) = -\infty, \ \lim_{x\to 0} f(x) = 1; \quad f(x) = x + 1, \ f(x) = x^3 + 1.$

**Problem 22.**

$f(t) = \frac{t(3+t)}{t} = 3 + t; \ D : t \neq 0; \ Range : (-\infty, 3) \cup (3, +\infty)$

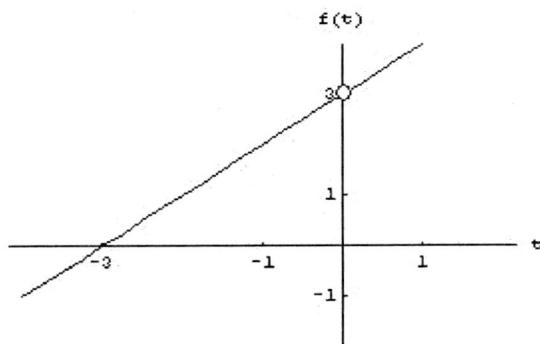

## Section 7.2    Left- and Right-Handed Limits

**Problem 2.**

(a) $\lim\limits_{x \to 2^-} \dfrac{1}{x-2} = -\infty$,

(b) $\lim\limits_{x \to 2^+} \dfrac{1}{x-2} = +\infty$

(c) $\lim\limits_{x \to 2} \dfrac{1}{x-2} = DNE$, because the left- and right-hand limits are not equal.

**Problem 4.**
$$f(x) = \frac{|x-3|}{x-3} = \begin{cases} \frac{x-3}{x-3} = 1 & \text{if } x > 3 \\ \frac{-(x-3)}{x-3} = -1 & \text{if } x < 3 \\ undefined & \text{if } x = 3 \end{cases}$$

(a) $\lim\limits_{x \to 0} f(x) = -1$

(b) $\lim\limits_{x \to 4} f(x) = 1$

(c) $\lim\limits_{x \to 3^+} f(x) = 1$

(d) $\lim\limits_{x \to 3^-} f(x) = -1$

(e) $\lim\limits_{x \to 3} f(x) = DNE$

**Problem 6.**

(a) $\lim\limits_{x \to \frac{1}{\pi}} f(x) = 2.$

(b) $\lim\limits_{x \to -\frac{1}{\pi}} f(x) = -2.$

(c) $\lim\limits_{x \to 0} f(x)$ does not exist because $\lim\limits_{x \to 0^-} f(x) = -1$ and $\lim\limits_{x \to 0^+} f(x) = 1$

**Problem 8.**
$$f(x) = \begin{cases} x+1 & , x \text{ not an integer} \\ 0 & , x \text{ an integer} \end{cases}$$

(a)

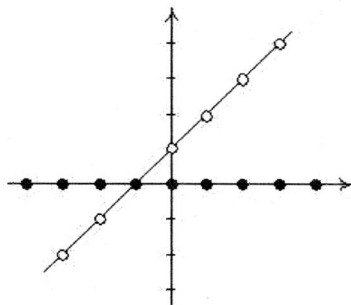

(b) $\lim\limits_{x \to 1.5} f(x) = 2.5,\ \lim\limits_{x \to 2} f(x) = 3,\ \lim\limits_{x \to 0} f(x) = 1$

(c) $\lim\limits_{x \to c} f(x) = c + 1$, for every $c$.

**Problem 9.**

$$g(x) = \begin{cases} x^2 & , x > 2 \\ x & , x \le 2 \end{cases}$$

(a) $\lim\limits_{x \to 2^+} g(x) = \lim\limits_{x \to 2^+} x^2 = 4$

$\lim\limits_{x \to 2^-} g(x) = \lim\limits_{x \to 2^-} x = 2$

So, $\lim\limits_{x \to 2} g(x) = DNE$

(b) $h(x) = \begin{cases} x^2 & , x > 2 \\ mx & , x \le 2 \end{cases}$

$mx = x^2$ when $x = 2$, so $2m = 4 \Rightarrow m = 2$.

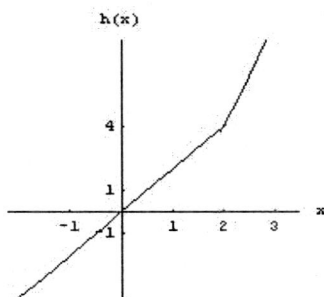

$$h(x) = \begin{cases} x^2 & , x > 2 \\ 2x & , x \le 2 \end{cases}$$

$\lim\limits_{x \to 2} h(x) = \begin{cases} \lim_{x \to 2^-} h(x) = \lim_{x \to 2^-} 2x = 4 \\ \lim_{x \to 2^+} h(x) = \lim_{x \to 2^+} x^2 = 4 \end{cases} \Rightarrow \lim\limits_{x \to 2} h(x) = 4$

$f(2) = 4 = \lim\limits_{x \to 2} h(x), \Rightarrow h$ is continuous at $x = 2$

**Problem 11.**

(a)

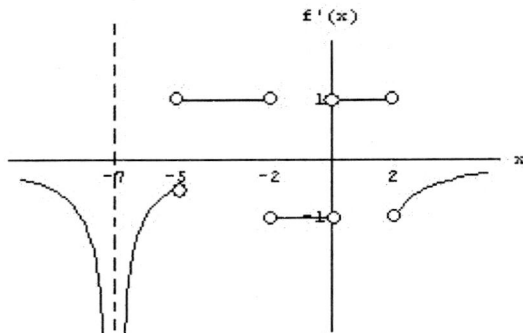

(b) $f'$ is undefined at $x = -7, -5, -2, 0, 2$.

(c)  (i)  $\lim\limits_{x \to \infty} f'(x) = 0$.

 (ii)  $\lim\limits_{x \to -2^-} f'(x) = 1$.

 (iii)  $\lim\limits_{x \to -2^+} f'(x) = -1$.

 (iv)  $\lim\limits_{x \to -2} f'(x)$ does not exist.

 (v)  $\lim\limits_{x \to -7^-} f'(x) = -\infty$.

 (vi)  $\lim\limits_{x \to -7^+} f'(x) = -\infty$.

**Problem 13.**

$$f(3) = 7 \implies \lim_{x \to 3} f(x) = 7$$

$$\text{Let } f(x) = \begin{cases} 2x + 1 & x = 3 \\ x - 4 & x \neq 3 \end{cases}$$

$$f(3) = 7 \text{ but } \lim_{x \to 3} f(x) = -1 \neq 7$$

**Problem 15.**

$$\text{Let } f(x) = \begin{cases} 3 + x & x \geq 0 \\ 2 - x & x < 0 \end{cases}$$

$$\lim_{x \to 0^-} f(x) = 2, \ \lim_{x \to 0^+} f(x) = 3, \text{ and } \lim_{x \to 0} f(x) = DNE$$

**Problem 17.**    **Problem 19.**    **Problem 21.**

(a) 0          (a) 0          (a) 0

(b) 0          (b) -2         (b) $-\infty$

(c) 0          (c) DNE        (c) DNE

## Section 7.3    A Streetwise Approach to Limits

**Problem 2.**

$$\lim_{x \to 3} \frac{2x^3 - 8x^2 + 5x + 3}{x - 3} = \lim_{x \to 3} \frac{(x-3)(2x^2 - 2x - 1)}{x - 3} = 2 \cdot 9 - 6 - 1 = 11$$

**Problem 3.**

$$\lim_{x \to 0} \frac{\sqrt{4+x} - 2}{x} = \lim_{x \to 0} \frac{(\sqrt{4+x} - 2)(\sqrt{4+x} + 2)}{x(\sqrt{4+x} + 2)} = \lim_{x \to 0} \frac{4 + x - 4}{x(\sqrt{4+x} + 2)} = \lim_{x \to 0} \frac{1}{\sqrt{4+x} + 2} = \frac{1}{4}$$

## Section 7.4    Continuity and the Intermediate and Extreme Value Theorems

**Problem 1.**

(a)

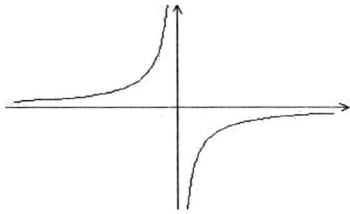

(i)  $\lim\limits_{x \to -\infty} \left(-\dfrac{3}{x}\right) = 0$

(ii)  $\lim\limits_{x \to \infty} \left(-\dfrac{3}{x}\right) = 0$

(iii)  $\lim\limits_{x \to \infty} \left(-\dfrac{3}{x} - 3\right) = -3$

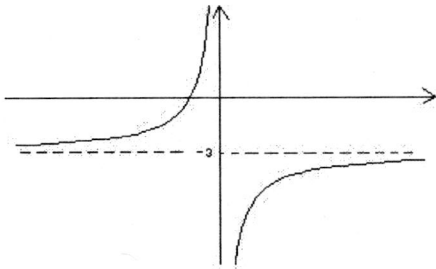

(iv)  $\lim\limits_{x \to \infty} \left(\dfrac{x+1}{x}\right) = \lim\limits_{x \to \infty} \left(1 + \dfrac{1}{x}\right) = 1$

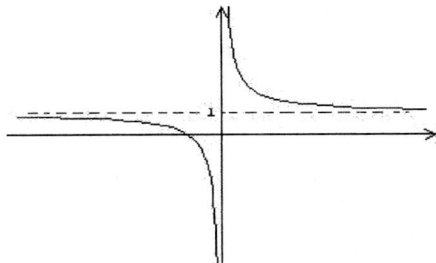

(v) $\lim\limits_{x\to\infty}\left(\dfrac{2x+3}{x}\right) = \lim\limits_{x\to\infty}\left(2+\dfrac{3}{x}\right) = 2$

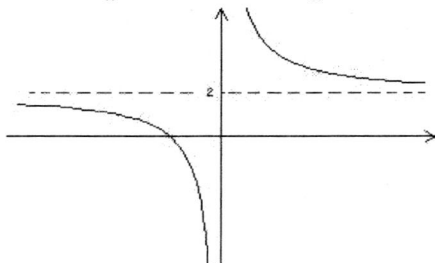

(b)  (i) $\lim\limits_{x\to-\infty}\left(-\dfrac{3}{x}\right) = \lim\limits_{x\to\infty}\left(\dfrac{3}{x}\right) = \lim\limits_{x\to\infty}(3)\cdot\lim\limits_{x\to\infty}\left(\dfrac{1}{x}\right) = 3\cdot 0 = 0$

(ii) $\lim\limits_{x\to\infty}\left(-\dfrac{3}{x}\right) = \lim\limits_{x\to\infty}(-3)\cdot\lim\limits_{x\to\infty}\left(\dfrac{1}{x}\right) = -3\cdot 0 = 0$

(iii) $\lim\limits_{x\to\infty}\left(-\dfrac{3}{x}-3\right) = \lim\limits_{x\to\infty}(-3)\cdot\lim\limits_{x\to\infty}\left(\dfrac{1}{x}\right) - \lim\limits_{x\to\infty}(3) = -3\cdot 0 - 3 = -3$

(iv) $\lim\limits_{x\to\infty}\left(\dfrac{x+1}{x}\right) = \lim\limits_{x\to\infty}\left(1+\dfrac{1}{x}\right) = \lim\limits_{x\to\infty}(1) + \lim\limits_{x\to\infty}\left(\dfrac{1}{x}\right) = 1 + 0 = 1$

(v) $\lim\limits_{x\to\infty}\left(\dfrac{2x+3}{x}\right) = \lim\limits_{x\to\infty}\left(2+\dfrac{3}{x}\right) = \lim\limits_{x\to\infty}(2) + \lim\limits_{x\to\infty}(3)\cdot\lim\limits_{x\to\infty}\left(\dfrac{1}{x}\right) = 2 + 3\cdot 0 = 2$

**Problem 2.**

(a) $\infty$

(b) $\infty$

(c) $\infty$

(d) $\infty$

(e) $-\infty$

(f) 8

**Problem 3.**

$$\lim_{x\to4}\frac{\sqrt{x}-2}{x-4} = \lim_{x\to4}\frac{\sqrt{x}-2}{(\sqrt{x}-2)(\sqrt{x}+2)} = \frac{1}{4}$$

**Problem 5.**

$f(x) = \dfrac{x^2-4}{x+2} = \dfrac{(x-2)(x+2)}{x+2} = x-2, \ D_f : x \neq -2.$

$x = -2$ is a removable point of discontinuity. We can define a function at $x = -2$ to make it continuous at -2. For example let $f(x) = \begin{cases} \frac{x^2-4}{x+2} & , x \neq -2 \\ -4 & , x = -2 \end{cases}$

**Problem 6.**

There is a removable point of discontinuity at $x = 0$; it can be removed by defining $f(0) = 0$.

**Problem 9.**

$f(x) = \begin{cases} -x^2 - 1 & , x > 0 \\ 5x - 1 & , x < 0 \end{cases}$ . $x = 0$ is a removable point of discontinuity, since we can define $f(x)$ to be $-1$ at $x = 0$, making our function continuous on $(-\infty, \infty)$

**Problem 11.**
$$f(x) = \begin{cases} -x^2 + 1 & , x > 0 \\ ax + b & , x \le 0 \end{cases}.$$
$ax + b = -x^2 + 1$ when $x = 0$, so $b = 1$. Hence, for $f$ to be continuous at $x = 0$ we need only constraint on $b$, $a$ can be arbitrary constant.

**Problem 13.**

(a) No.

(b) No. In fact, $f(x)$ is not continuous anywhere.

**Problem 15.**

(a) Yes

(b) Yes, $f'(0) = 0$

**Problem 17.**

(a) Yes

(b) No

**Problem 21.**

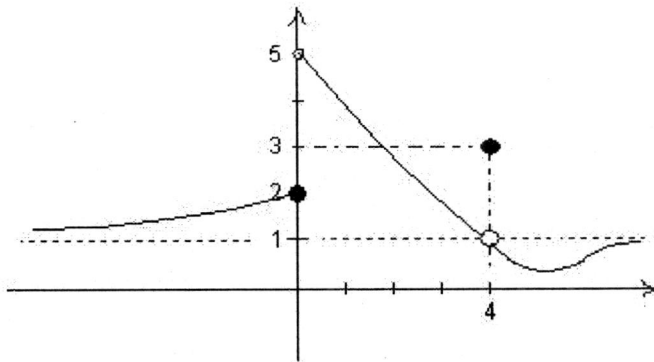

# CHAPTER 8
# Fruits of Our Labor

## Section 8.1    Local Linearity and the Derivative

**Problem 1.**

(a) $f(x) = \sqrt{x}$, $x = 25$.

Tangent line at $x = 25$ is $y - 5 = \frac{1}{2\sqrt{25}}(x - 25)$ or $y = 5 + 0.1(x - 25)$. So the linearization is $\sqrt{23} = f(23) \approx 5 + 0.1(23 - 25) \approx 4.8$. Using calculator we check that $\sqrt{23} \approx 4.795$

(c) $\sqrt{24.9} \approx f(24.9) = 0.1(24.9 - 25) + 5 = 4.99$; This is an overestimate, and using calculator we check that $|4.99 - \sqrt{24.9}| \approx |4.99 - 4.98999| = 0.00001$.

(f) $\sqrt{27} = f(27) \approx 5 + 0.1(27 - 25) \approx 5.2$. Using calculator we check that $\sqrt{27} \approx 5.196$

**Problem 2.**

(a) To approximate $\sqrt{102}$ use the graph of $\sqrt{x}$ and its tangent line at $x = 100$, then $\sqrt{102} \approx 10 + 0.05(102 - 100)$

(c) $f(x) = \sqrt{x}$ at $x = 16$; $\sqrt{18} \approx 4 + \frac{1}{8}(18 - 16)$

**Problem 4.**

The number $100 = 10^2$ is the closest perfect square to 98. Thus we use the tangent line approximation to $f(x) = \sqrt{x}$ at $x = 100$. Now $\frac{d}{dx}\sqrt{x}\big|_{x=100} = \frac{1}{2\sqrt{x}}\big|_{x=100} \frac{1}{2\sqrt{100}} = 0.05$, and hence the tangent line approximation to $f(x) = \sqrt{x}$ at $x = 100$ is $y = 0.05(x - 100) + 10$. Therefore, $\sqrt{98} \approx 0.05(98 - 100) + 10 = 0.05(-2) + 10 = 9.9$.

**Problem 7.**

As $V(r) = (10)(\pi r^2)$ in.$^3$, we obtain $\frac{dV}{dr} = 20\pi r$. When the radius is $r$, the volume changing with respect to radius at a rate of $20\pi r$ in.$^3$/in. Now $\frac{dV}{dr} = \lim\limits_{\Delta r \to 0} \frac{\Delta V}{\Delta r}$, and hence $\Delta V \approx (10)(2\pi r)(\Delta r)$. If the radius $r$ is increased by a small amount, $\Delta r$ inches, the volume will increase by about $\Delta V$ inches. To see this, the quantity $\Delta V$ is the volume of a thin shell of thickness $\Delta r$ that is added to the cylindrical side of the cylinder. The volume of this thin shell is roughly the volume of a rectangular prism of dimesions 10 in. by $2\pi r$ in. by $\Delta r$ in., which is $(10)(2\pi r)(\Delta r)$ in.$^3$.

## Section 8.2    The First and Second Derivatives in Context

**Problem 1.**

The first derivative $h'(t_*)$ is positive because the water level is rising at time $t_*$ as water is being poured into the bucket at time $t_*$ . The second deriviative $h''(t_*)$ is zero because rate at which water is being poured into the bucket is constant.

**Problem 3.**

(a) $D(t)$ - diameter as a length is positive

(b) $D'(t)$ - since $D(t)$ is a increasing function with respect to time its derivative is positive

(c) $D''(t)$ is negative, b/c $D'(t)$ is decreasing

## Section 8.3    Derivatives of Sums, Products, Quotients, and Power Functions

**Problem 1.**
$$f'(x) = 6x + 3 - 3x^{-2} - 6x^{-3}$$

**Problem 3.**
Product Rule. $f'(x) = \pi[(6x+7)(x-2) + (3x^2 + 7x + 1)] = \pi[6x^2 - 12x + 7x - 14 + 3x^2 + 7x + 1] = \pi[9x^2 + 2x - 13]$

**Problem 5.**
Quotient Rule. $f'(x) = \frac{(x+2)(1)-(x)(1)}{(x+2)^2} = \frac{2}{(x+2)^2}$

**Problem 7.**
$$f(x) = \left(\tfrac{5}{2}x^2 + 7x^5 - 5x\right)x = \tfrac{5}{2}x^3 + 7x^6 - 5x^2 \Rightarrow f'(x) = \tfrac{15}{2}x^2 + 42x^5 - 10x.$$

**Problem 9.**

(a) $f(x) = \frac{x^3 + 2x^2}{3} = \tfrac{1}{3}(x^3 + 2x^2), \quad f'(x) = \tfrac{1}{3}(3x^2 + 4x)$

(b) $f(x) = \frac{(x^2+1)^2}{x} = \frac{x^4 + 2x^2 + 1}{x} = x^3 + 2x + x^{-1}, \quad f'(x) = 3x^2 + 2 - x^{-2} = 3x^2 + 2 - \frac{1}{x^2}$

**Problem 11.**

(a) $f'(x) = 3x^2 - 1$

(b) $f'(x) = -\frac{2}{x^3}$

**Problem 12.**
$$f(x) = x + |x| = \begin{cases} 2x & , x \geq 0 \\ 0 & , x < 0 \end{cases}$$

(a)

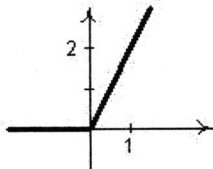

(b) $f'(x) = \begin{cases} 2 & , x > 0 \\ 0 & , x < 0 \\ undefined & , x = 0 \end{cases}$

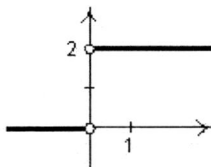

## Problem 15.

The function value $f(0)$ is defined, and $\lim\limits_{x \to 0^+} f(x) = \lim\limits_{x \to 0^-} = f(0) = 0$, so $f$ is continuous at 0.

As

$$f(x) = x + |x| = \begin{cases} 0, & x < 0 \\ 2x, & x \geq 0 \end{cases},$$

and we have

$$\lim_{h \to 0^-} \frac{f(h) - f(0)}{h} = \lim_{h \to 0^-} \frac{0 - 0}{h} = \lim_{h \to 0^-} 0 = 0,$$

and

$$\lim_{h \to 0^+} \frac{f(h) - f(0)}{h} = \lim_{h \to 0^+} \frac{2h - 0}{h} = \lim_{h \to 0^+} 2 = 2$$

Hence, $f'(0) = \lim\limits_{h \to 0} \dfrac{f(h) - f(0)}{h}$ does not exist.

## Problem 18.

Equation of tangent to $f(x)$ at the point $x = 1$ is given by $y - f(1) = f'(1)(x-1)$. We have $f(x) = x^3 + 2x$, $f'(x) = 3x^2 + 2$ and $f'(1) = 5$, $f(1) = 3$. Therefore $y - 3 = 5(x - 1)$ which is $y = 5x - 2$.

## Problem 21.

$$\frac{d}{dx}\left(\frac{\pi}{\pi x + \pi}\right) = \frac{d}{dx}\left(\frac{1}{x + 1}\right) = \frac{(x + 1)(0) - (1)(1)}{(x + 1)^2} = -\frac{1}{(x + 1)^2}.$$

## Problem 22.

$$\frac{d}{dx}\left(\frac{2x^2 + x + 1}{\sqrt{2x}}\right) = \frac{(\sqrt{2x})(4x + 1) - \left(\frac{1}{\sqrt{2x}}\right)(2x^2 + x + 1)}{2x} = \frac{(2x)(4x + 1) - (1)(2x^2 + x + 1)}{2\sqrt{2}x^{3/2}} =$$

$$\frac{8x^2 + 2x - 2x^2 - x - 1}{2\sqrt{2}x^{3/2}} = \frac{\sqrt{2}(6x^2 + x - 1)}{4x^{3/2}}.$$

## Problem 23.

$$\frac{d}{dx}\left(\frac{x^2 + 5x}{2x^{10}}\right) = \frac{d}{dx}\left(\frac{1}{2}x^{-8} + \frac{5}{2}x^{-9}\right) = -4x^{-9} - \frac{45}{2}x^{-10} = -\frac{4}{x^9} - \frac{45}{2x^{10}}$$

# CHAPTER 9
# Exponential Functions

## Section 9.1    Exponential Growth

**Problem 1.**

(a)

| t | B(t) |
|---|------|
| 1 | 600 |
| 2 | 1800 |
| 3 | 5400 |
| 4 | 16200 |
| 5 | 48600 |

(b)  At noon yesterday it was 67, at noon four days ago 2.5 (approximately)

(c)  $B(t) = B_0 \cdot 3^t$

(d)  $B(w) = B_0 \cdot 3^{7w}$

(e)  $B(1) = 200 \cdot 3^7 = 437400$

**Problem 3.**

(a)  $E(h) = 600 \cdot 2^{3h}$, 20 minutes $= \frac{1}{3}$ hour

(b)  $X(h) = 100 \cdot 2^{4h}$, 15 minutes $= \frac{1}{4}$ hour

(c)  $E(h) = X(h)$ when $600 \cdot 2^{3h} = 100 \cdot 2^{4h}$, which is $6 \cdot 8^h = 16^h$ or $6 = 2^h$. So $h \approx 2.584$.
Answer: approximately after 2 hours and 35 minutes.

## Section 9.2    Exponential: The Bare Bones

**Problem 1.**
$$\frac{x^{-1}+z^{-1}}{(z+x)x^{-2}} = \frac{x^2(x^{-1}+z^{-1})}{(z+x)} = \frac{x^2(\frac{1}{x}+\frac{1}{z})}{(z+x)} = \frac{x^2(\frac{z+x}{xz})}{(z+x)} = x^2(z+x)(zx)^{-1} \cdot (z+x)^{-1} = x^2z^{-1}x^{-1} = xz^{-1} = \frac{x}{z}$$

**Problem 3.**
$$\frac{\sqrt{2x^3}+\sqrt{12y^3x^4}}{\sqrt{6y^2x^5}} = \frac{\sqrt{x^2\cdot 2x}+\sqrt{4x^4y^2\cdot 3y}}{\sqrt{x^4y^2\cdot 6x}} = \frac{x\sqrt{2x}+2x^2y\sqrt{3y}}{x^2y\sqrt{6x}} = \frac{\sqrt{2x}+2xy\sqrt{3y}}{xy\sqrt{6x}}$$

**Problem 4.**
$$\frac{x^{-1}}{zx^{-1}+z^{-1}} = \frac{1}{x(zx^{-1}+z^{-1})} = \frac{1}{z+xz^{-1}} = \frac{1}{z+\frac{x}{z}} = \frac{1}{\frac{z^2+x}{z}} = \frac{z}{z^2+x}$$

**Problem 6.**
$$\frac{x^{n+1}y^{2n}}{(\frac{x}{y})^n} = \frac{x^{n+1}y^{2n}}{x^ny^{-n}} = x\cdot y^{3n}$$

**Problem 8.**
$$\frac{(ab)^{-x}}{a^{-x}+b^{-x}} = \frac{a^{-x}b^{-x}}{a^{-x}+b^{-x}} = \frac{1}{(ab)^x(a^{-x}+b^{-x})} = \frac{1}{a^xb^xa^{-x}+a^xb^xb^{-x}} = \frac{1}{b^x+a^x}$$

**Problem 10.**
$$b^{x+y} + b^x = b^x(b^y + 1)$$

**Problem 12.**
$$3b^{2x+1} - 4b^{2x-1} = 3b^{x+(x+1)} - 4b^{x+(x-1)} = b^x(3b^{x+1} - 4b^{x-1})$$

**Problem 14.**
$$b^{3x} - (2b)^{-1+2x} = b^{x+2x} - 2^{-1+2x} \cdot b^{x+x-1} = b^x(b^{2x} - 2^{2x-1} \cdot b^{x-1})$$

**Problem 17.**
$(a^x + b^x)^{\frac{1}{x}} = a+b$, FALSE. Let $a = 3, b = 4, x = 2$ then $LHS = (3^2 + 4^2)^{\frac{1}{2}} = 25^{\frac{1}{2}} = 5$;   $RHS = 3+4 = 7$. So $LHS \neq RHS$.

**Problem 19.**
FALSE. Take $a = 2, x = 1$

**Problem 21.**
TRUE

**Problem 23.**
$$\frac{a^{2x} - b^{4x}}{a^x + b^{2x}} = \frac{(a^x)^2 - (b^{2x})^2}{a^x + b^{2x}} = \frac{(a^x - b^{2x})((a^x + b^{2x}))}{a^x + b^{2x}} = a^x - b^{2x}$$

**Problem 26.**
Because $\lim\limits_{x \to -\infty} f(x) = 0$, $D = 0$. Now $3 = f(0) = C + D = C + 0 \Rightarrow C = 3$. Substituting the point $(1, 5)$ into the formula for $f(x)$ gives $f(1) = 3a = 5$, from which we obtain $a = \frac{5}{3}$. Thus, $f(x) = 3\left(\frac{5}{3}\right)^x$.

**Problem 28.**
Because $\lim\limits_{x \to -\infty} f(x) = -1$, $D = -1$. Now $-3 = f(0) = C + D = C - 1 \Rightarrow C = -2$. Substituting the point $(1, -4)$ into the formula for $f(x)$ gives $f(1) = -2 \cdot a - 1 = -4$, from which we obtain $a = \frac{3}{2}$. Thus, $f(x) = -2\left(\frac{3}{2}\right)^x - 1$.

**Problem 31.**
(a) (i); (b) (ii); (c) (vi); (d) (iii); (e) (iii); (f) (v)

**Problem 32.**

(a) $\frac{x^{2y} + x^{y+2}}{x^y} = \frac{x^y(x^y + x^2)}{x^y} = x^y + x^2$

(b) $-\frac{1}{x+y}$

(c) $\frac{A^{B+4} - A^{3B}}{A^B(A^2 - A^B)} = \frac{A^B(A^4 - A^{2B})}{A^B(A^2 - A^B)} = \frac{(A^2)^2 - (A^B)^2}{(A^2 - A^B)} = \frac{(A^2 - A^B)(A^2 + A^B)}{(A^2 - A^B)} = A^2 + A^B$

(d) $\frac{y^w(y^{2w} - y^4)}{y^w(y^w + y^2)} = \frac{(y^w - y^2)(y^w + y^2)}{(y^w + y^2)} = y^w - y^2$

**Problem 34.**

(a) 0

(b) 0

**Problem 36.**

(a) $\infty$

(b) 0

## Problem 38.

(a) $3b^x - b^{2x} = b^x(3 - b^x)$

(b) $(3b)^x - b^{x+2} = 3^x b^x - b^x b^2 = b^x(3^x - b^2)$

(c) $b^x(b^{\frac{x}{2}} - b^{x-1})$

## Section 9.3    Applications of the Exponential Function

## Problem 1.

(a) $C(t) = C_0 \left(\frac{1}{2}\right)^{t/5730}$.

(b) $C(1993 - 137) = C(1856) = C_0 \left(\frac{1}{2}\right)^{1856/5730} \approx (.7989)C_0$. In 1993, 79.98% of the original $C_{14}$ remained.

(c) $C(t) = C_0 \left(\frac{1}{2}\right)^{t/5730} = C_0 \left(\frac{1}{2^{1/5730}}\right)^t \approx C_0(0.99988)^t$. Each year the amount of $c_{14}$ in a deceased organism decreases by $100(1 - 0.99988)\% \approx 0.012\%$.

## Problem 4.

(a) If the population was increasing linearly, then population $P(t)$, where $t$ is measured in years after 1970, has the form $P(t) = \left(\dfrac{200,000 - 100,000}{20 - 0}\right) t + 100,000 = 5000t + 100,000$. In 1980, the population would be equal to $P(10) = 5000(10) + 1000,000 = 150,000$ people.

(b) If the population was increasing exponentially, then $P(t)$ has the form $P(t) = 100,000a^t$. Now $P(20) = 200,000 \Rightarrow 200,000 = 1000a^{20} \Rightarrow a = 2^{1/20}$. Hence $P(10) = 100,000(2^{10/20}) \approx 141,421 \leq 141,422$ people, which is less than 150,000 people.

## Problem 8.
Since we have an interval of 12 to 16 hours, we need to make two formulas: one for the slowest growth scenario, and one for the quickest.

Let $L(t)$ be the number of bacteria after $t$ hours, assuming a 12-hour doubling time. Then $L(t) == 10 \cdot 2^{t/12}$. Now $1000 = L(t) = 10 \cdot 2^{t/12} \Rightarrow 100 = 2^{t/12} \Rightarrow \ln 100 = \ln(2^{t/12}) \Rightarrow \ln 100 = \left(\frac{t}{12}\right) \ln 2$. $\Rightarrow t = \frac{12 \ln 100}{\ln 2} \approx 79.73$.

Let $M(t)$ be the number of bacteria after $t$ hours, assuming a 12-hour doubling time. Then $M(t) == 10 \cdot 2^{t/16}$. Now $1000 = L(t) = 10 \cdot 2^{t/16} \Rightarrow 100 = 2^{t/16} \Rightarrow \ln 100 = \ln(2^{t/16}) \Rightarrow \ln 100 = \left(\frac{t}{16}\right) \ln 2$. $\Rightarrow t = \frac{16 \ln 100}{\ln 2} \approx 106.30$.

If we begin with 10 bacteria that double in number every 12 to 16 hours, we can expect to see a 1000 bacteria in anytime from 79.7 to 106.3 hours. (Both values for $t$ above can be obtained using a graphing calculator.)

**Problem 10.**

(a) After 10 years, Harvard's total return was $((1.111)^{10} - 1) \cdot \% \approx 286.5\%$.

(b) Let $Y(t)$ be Yale's total return $t$ years after 1985. Then $2.873 = Y(10) = x^{10} \Rightarrow x^{10} = 2.873 \Rightarrow x = 1.1113$. Hence, Yale's average annual return was 11.13%.

(c) Yale got the higher return on its investments. This can be seen by comparing average annual returns or the total returns over the ten-year period.

(d) For Harvard, $1.111^t = 2 \Rightarrow t = \frac{\ln 2}{\ln 1.111} \approx 6.585$. Thus the doubling time is 6.585 years.

For Yale, $1.1113^t = 2 \Rightarrow t = \frac{\ln 2}{\ln 1.1113} \approx 6.568$. Thus the doubling time is 6.568 years.

**Problem 13.**

(a) Let $P(t)$ be the purchasing power of \$1 $t$ years from now. Then $P(t) = 0.98^t$.

(b) $0.5 = P(t) = (.98)^t \Rightarrow t = \frac{\ln 0.5}{\ln 0.98} \approx 34.3$. (This value of $t$ can be found using a graphing calculator.) It will take approximately 34.3 years for the purchasing power of the dollar to be cut in half.

**Problem 16.**

(a) For tuition: $x^{14} = 3.34 \Rightarrow x = 3.34^{1/14} \approx 1.0900$. $\Rightarrow$ The annual percent increase was 9%. For household income: $x^{1}4 = 1.82 \Rightarrow x = 1.82^{1/14} \approx= 1.0437$. $\Rightarrow$ The annual percent increase was 4.37%. For the CPI: $x^{1}4 = 1.74 \Rightarrow x = 1.74^{1/14} \approx 1.0404$. $\Rightarrow$ The annual percent increase was 4.04%.

(b) Let $T(t)$ denote the in-state tuition function $t$ years after the 1980-1981 school year. Then $2865 = T(14) = T_0 \cdot 3.34 \Rightarrow T(0) = T_0 = 857.78$. The average cost of tuition in 1980-1981 for in-state students was \$857.78.

(c)

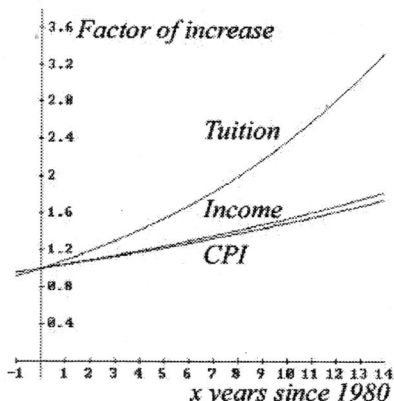

**Factor of increase** ... **Tuition** ... **Income** ... **CPI** ... *x years since 1980*

(d) Let $I(t)$ denote the family's income $t$ years after the 1980-1981 school year and $I_0 = I(0) = T_0/0.15$. We have $I(t) = I_0(1.82^{t/14})$. If $T(14) = I(14)r$, then $1.82I_0r = 3.34T_0 = 3.34(0.15I_0) \Rightarrow 1.82r = (3.34)(0.15)$ $r = \frac{(3.34)(0.15)}{1.82} \approx 0.2753$. Therefore, for the 1994-1995 school year, the family will need about 27.53% of their income to send the younger child to college.

## Section 9.4    The Derivative of an Exponential Function

**Problem 1.**

$$g(t) = 3^{5t}, \quad \frac{3^{5(t+h)} - 3^{5t}}{h} = \frac{3^{5t}3^{5h} - 3^{5t}}{h} = \frac{3^{5t}(3^{5h} - 1)}{h} = 3^{5t} \cdot \frac{3^{5t} - 1}{h} = g(t) \cdot \frac{g(h) - g(0)}{h}$$

**Problem 2.**

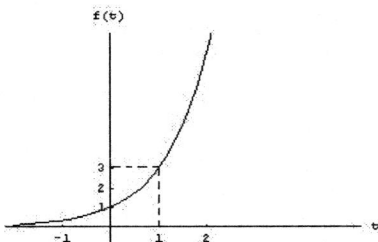

(a) Graph of $f(t) = 3^t$

(b) Slope of the secant line through $(1,3)$ and $(1.0001, 3.000329602)$ is 3.296, so $f'(1) \approx 3.296$

(c) Slope of the secant line through $(0,1)$ and $(0.0001, 1.000109867)$ is 1.099, so $f'(0) \approx 1.099$

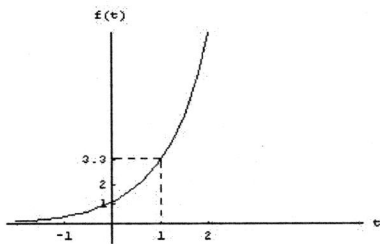

(d) Graph of $f'(t)$

**Problem 5.**

$$f'(x) = 3x^2 \cdot e^x + x^3 e^x$$

**Problem 6.**

$$f(x) = \frac{e^{2x}}{x} \Rightarrow f'(x) = \frac{2xe^{2x} - e^{2x}}{x^2} = \frac{e^{2x}}{x^2}(2x - 1)$$

**Problem 8.**

$$f'(x) = \frac{(2x+1)(e^x+1) - (x^2+x)(e^x)}{(e^x+1)^2}$$

**Problem 9.**

$$f'(x) = 2e^{2x}(x^2 + 2x + 2) + e^{2x}(2x + 2) = 2e^{2x}(x^2 + 2x + 2 + x + 1) = 2e^{2x}(x^2 + 3x + 3)$$

**Problem 11.**

As $\dfrac{d}{dx}e^x = e^x$, the slope of the tangent line to $f(x) = e^x$ at $x = 1$ is $e^1$. Hence the equation of the tangent line is $y - f(1) = e(x - 1) \Leftrightarrow (y - e) = e(x - 1) \Leftrightarrow y = ex$.

**Problem 12.**

(a) $f'(x) = \frac{2xe^x + x^2 e^x}{3} = \frac{e^x}{3}(x^2 + 2x)$

(b) $f'(x) = \frac{5}{3}\left(\frac{2x - x^2}{e^x}\right)$

(c) $f'(x) = \frac{-e^{5x} - 5xe^{5x}}{(xe^{5x})^2} = \frac{e^{5x}(-1 - 5x)}{x^2 e^{10x}} = \frac{-1 - 5x}{x^2 e^{5x}}$

**Problem 15.**

(a) Rewrite $f(x)$ as $x^2 e^{-x}$, then using the Product Rule $f'(x) = 2xe^{-x} - x^2 e^{-x} = \frac{2x - x^2}{e^x}$

(b) $f'(x) > 0$ if $\frac{2x - x^2}{e^x} > 0$. Since $e^x > 0$ for all $x's$ it is enough to solve $2x - x^2 > 0$ which is $x(2-x) > 0$, so $0 < x < 2$. Thus $f'(x)$ is positive if $0 < x < 2$ and $f'(x)$ is negative if $x < 0$ and $x > 2$.

(c) $f$ is increasing for $x \in (0,2)$ . $f(x)$ is decreasing for $x \in (-\infty, 0)$ and $x \in (2, \infty)$

(d) The smallest value of $f(x)$ is $f(0) = 0$, since $\lim\limits_{x \to \infty} f(x) = 0$ and $\lim\limits_{x \to -\infty} f(x) = \infty$.

# CHAPTER 10

# Optimization

## Section 10.1    Analysis of Extrema

**Problem 1.**

(a) $f'(x) = 3x^2 - 3$.  $f'(x) = 0 \Leftrightarrow 3x^2 - 3 = 0$, so $x^2 - 1 = 0$ and therefore $x = 1$ or $x = -1$ are critical points.

(b)

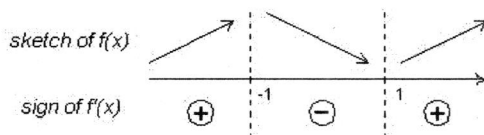

This tells us that at $x = -1$ $f$ has a local maximum and at $x = 1$ $f$ has a local minimum.

(c) There are no absolute minimum or maximum values.

**Problem 2.**

(a) From problem 1, we have that $x = -1$ and $x = 1$ are critical points. The endpoints of the domain $[-5, 5]$, $x = -5$ and $x = 5$, are also critical points.

(b) From problem 1, $x = -1$ is a local maximum point and $x = 1$ is a local minimum point. The absolute minimum occurs at $x = -5$, and the absolute maximum occurs at $x = 5$.

(c) The absolute minimum value is $f(-5) = -108$, and the absolute maximum value is $f(5) = 112$.

**Problem 3.**

(a) From problem 1, we have that $x = 1$ is a critical point. The endpoints of the domain, $x = 0$ and $x = 3$, are also critical points.

(b) The point $x = 0$ is neither a local maximum or minimum point nor an absolute minimum or maximum point. From problem 1, $x = 1$ is a local minimum point and the absolute minimum point. The point $x = 3$ is the absolute maximum point.

(c) The absolute minimum value is $f(1) = 0$, and the absolute maximum value is $f(3) = 20$.

**Problem 4.**

(a) From problem 1, we have that $x = 1$ is a critical point.

(b) As in problem 3, $x = 1$ is a local minimum point and the absolute minimum point.

(c) The absolute minimum value is $f(1) = 0$. There is no absolute maximum value.

**Problem 6.**

(a) From problem 5, we have that $x = -1$ and $x = 2$ are critical points. The endpoints of the domain, $x = -3$ and $x = 4$, are also critical points.

(b) From problem 5, $x = -1$ is a local minimum point and $x = 2$ is a local maximum point. The point $x = -3$ is the absolute maximum point, and $x = 4$ is the absolute minimum point.

(c) The absolute minimum value is $f(4) = -27$, and the absolute maximum value is $f(-3) = 50$.

**Problem 8.**

(a) From problem 7, $x = -\sqrt{2}$ is a critical point. The endpoints of the domain $[-2, 0]$, $x = -2$ and $x = 0$, are also critical points.

(b) The point $x = -2$ is neither a local maximum or minimum point nor an absolute minimum or maximum point. The point $x = -\sqrt{2}$ is a local and and absolute maximum point. The point $x = 0$ is the absolute minimum point.

(c) The absolute minimum value is $f(0) = 5$, and the absolute maximum value is $f(-\sqrt{2}) = 16\sqrt{2} + 5$.

**Problem 10.**

(a) $f'(x) = 12x^3 - 24x^2 = 12x^2(x - 2) \Rightarrow 0 = f'(x) = 12x^2(x - 2) \Rightarrow x = 0 \ x = 2$ are the critical points.

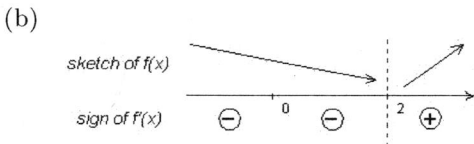

(b)

sketch of f(x)

sign of f'(x)   $\ominus$   0   $\ominus$   2   $\oplus$

The first derivative test implies that $x = 0$ is neither a local maximum nor a local minimum point and $x = 2$ is a local minimum point. Moreover, as $x = 0$ is not an endpoint of the domain, it is not an absolute maximum or minimum point As $f(x)$ increases without bound as $|x|$ increases without bound, $x = 2$ is also the absolute minimum point.

(c) The absolute minimum value is $f(2) = -13$, and there is no absolute maximum value.

**Problem 13.**

Note that 0 is not in the natural domain of $f(x)$ since $f(0)$ is undefined.

(a) $f'(x) = x^2 + 2 - \dfrac{3}{x^2} = \dfrac{x^4 + 2x^2 - 3}{x^2} = \dfrac{(x^2 + 3)(x + 1)(x - 1)}{x^2} \Rightarrow 0 = f'(x) \Rightarrow x = -1$ are critical points $x = 1$. The point $x = 0$ is not a critical point, because 0 is not in the domain of $f$.

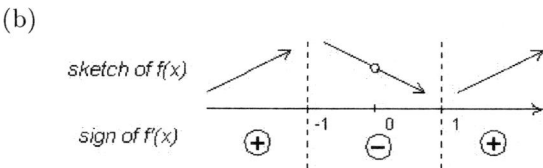

(b)

sketch of f(x)

sign of f'(x)   $\oplus$   -1   $\ominus$   0   1   $\oplus$

The first derivative test implies that $x = -1$ is a local maximum point and $x = 1$ is a local minimum point. As $f(x)$ increases without bound as as $x$ increases without bound, $x = -1$ is not an absolute maximum point. As $f(x)$ decreases without bound as as $x$ decreases without bound, $x = 1$ is not an absolute minimum point.

(c) There are no absolute minimum or maximum values.

**Problem 16.**

(a) $f'(x) = \dfrac{(x^2 + 4)(0) - (2x)(1)}{(x^2 + 4)^2} = -\dfrac{2x}{(x^2 + 4)^2} \Rightarrow 0 = f'(x) \Rightarrow x = 0$ is the only critical point.

(b)

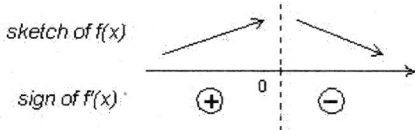

*sketch of f(x)*

*sign of f'(x)*   $\oplus$   0   $\ominus$

The first derivative test implies $x = 0$ is a local and absolute maximum point.

(c) The absolute maximum value is $f(0) = \frac{1}{4}$ and there is no absolute minimum value.

**Problem 18.**

(a) $f'(x) = 2xe^{-x} + x^2(-e^{-x}) = -x(x - 2)e^{-x}; \; 0 = f'(x) \Rightarrow x = 2$ and $x = 0$ are the critical points.

(b) $f'(x)$ is positive on $(0, 2)$ and is negative on $(-\infty, 0) \cup (2, \infty)$. The first derivative test implies that $x = 0$ is a local minimum point and $x = 2$ is a local maximum point.

(c) No. The function values of $f$ increase without bound as $x$ decreases without bound.

(d) Yes, the absolute minimum value of $f(0) = 0$ is achieved at $x = 0$.

**Problem 20.**

$f'(x) = \dfrac{10(x^2+1) - 20x^2}{(x^2+1)^2} = \dfrac{-10x^2 + 10}{(x^2+1)^2} = 0 \iff -10(x^2 - 1) = 0$, so $x = \pm 1$ are critical points.
At $x = -1$ $f$ has a local minimum, at $x = 1$ $f$ has a local maximum. No absolute extremum.

**Problem 22.**

$f'(x) = \dfrac{e^x(1) - e^x(x)}{e^{2x}} = \dfrac{(1 - x)}{e^x}; \; 0 = f'(x) \Rightarrow x = 1$ is a critical point. $f'(x)$ is positive on $(-\infty, 1)$ and is negative on $(1, \infty)$, and hence $x = 1$ is both a local and an absolute maximum point. The absolute maximum value is $f(1) = \frac{1}{e}$. As $f(x)$ decreases without bound as $x$ decreases without bound, $f$ has not absolute minimum value.

**Problem 24.**

$f'(x) = \dfrac{-(x^2+9) - (4-x)2x}{(x^2+9)^2} = \dfrac{-x^2 - 9 - 8x + 2x^2}{(x^2+9)^2} = \dfrac{x^2 - 8x - 9}{(x^2+9)^2} = 0 \iff x^2 - 8x - 9 = 0$ or $(x - 9)(x + 1) = 0$.
Thus at $x = -1$ $f$ has a local maximum, at $x = 9$ $f$ has a local minimum.
No absolute extremum.

**Problem 26.**

$f'(x) = \dfrac{e^x 2x - e^x 2}{4x^2} = \dfrac{e^x(2x - 2)}{4x^2}$.
$f'(x) = 0$ if $2x - 2 = 0 \Rightarrow x = 1$ is a critical point.
At $x = 1$ has a local and absolute minimum. The absolute value minimum is $f(1) = \frac{e}{2}$.

## Section 10.2    Concavity and The Second Derivative

**Problem 1.**

(a) $f'(x) = 3x^2 - 6 = 0$, $3(x^2 - 2) = 0$, so $x^2 = 0 \Rightarrow x = \sqrt{2}$ and $x = -\sqrt{2}$ are critical points.

(b) Second derivative $f''(x) = 6x$. Since $f''(-\sqrt{2}) = -6\sqrt{2} < 0$, $f$ has a local maximum at $x = -\sqrt{2}$. Since $f''(\sqrt{2}) = 6\sqrt{2} > 0$, $f$ has a local minimum at $x = \sqrt{2}$

**Problem 3.**

(a) $f'(x) = 3x^2 + 9x - 12 = 3(x+4)(x-1)$; $0 = f'(x) \Rightarrow x = -4$ and $x = 1$ are the critical points.

(b) $f''(x) = 6x + 9$. As $f''(-4) < 0$, $x = -4$ is a local maximum point; as $f''(1) > 0$, $x = 1$ is a local minimum point.

**Problem 5.**

(a) $f'(x) = 8x^3 + 64$; $0 = f'(x) \Rightarrow x = -2$ is the only critical point.

(b) $f''(x) = 24x^2$. As $f''(-2) > 0$, $x = -2$ is a local minimum point.

**Problem 7.**

(a) $f'(x) = 4x^3 + 12x^2 = 4x^2(x+3)$; $0 = f'(x) \Rightarrow x = -3$ and $x = 0$ are the critical points.

(b) $f''(x) = 12x^2 + 24x$. As, $f''(-3) > 0$, $x = -3$ is a local minimum point. As $f''(0) = 0$, the second derivative test does not apply. Moreover, as $f'(x)$ is positive on $(-3, 0) \cup (0, \infty)$, $x = 0$ is neither a local maximum nor a local minimum point.

**Problem 9.**

(a) $f'(x) = e^x - 1 = 0$, so $e^x = 1$ and $x = 0$.

(b) Second Derivative Test $f''(x) = e^x$. Since $f''(0) = e^0 = 1 > 0$, $f$ has a local minimum at $x = 0$.

**Problem 11.**

(a) $f'(x) = x^4 - 4x^3 + 4x^2$; $f'(x) = 0 \Leftrightarrow x^2(x^2 - 4x + 4) = 0$. So, the critical points are $x = 0$ and $x = 2$.

(b) Second Derivative Test fails ($f''(0) = 0$). By First Derivative Test there is no local extremum ($f'$ is always non-negative)

SECTION 10.2 *Concavity and The Second Derivative* 73

## Problem 14.

(a) Local minimum: $f(x) = (x-4)^4 + 1$; (b) Local maximum: $f(x) = -(x-4)^4 + 1$; (c) Neither: $f(x) = (x-4)^3 + 1$.

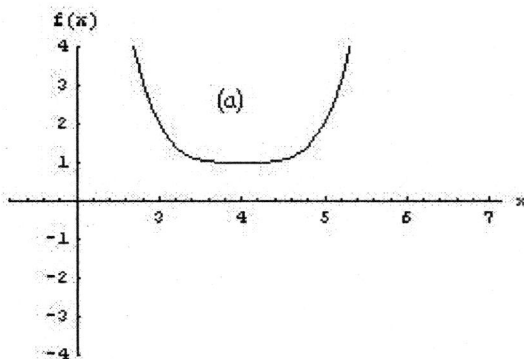

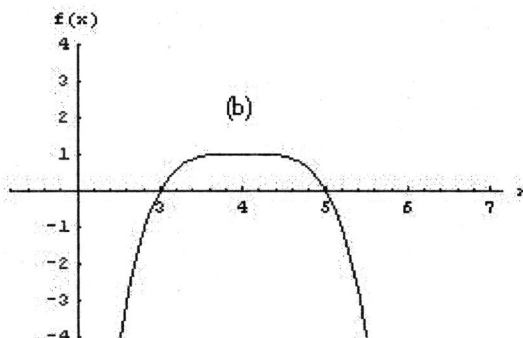

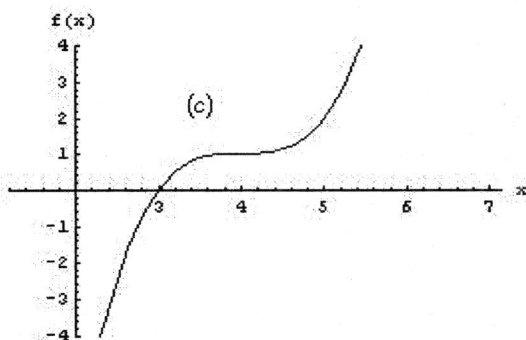

## Problem 16.

(a) (iii) and (vi)

(b) The critical points are $x = -8$, $x = -4$, and $x = -3$. The given information about the sign of $h'$ implies that $h$ has no local maxima and $x = -4$ is a local minimum point. The absolute maximum must occur at one (or both) endpoints, and the absolute minimum must occur at $x = -4$.

## Problem 18.

$f(x) = x^5 - 2x^4 - 7 \quad [-1, 1]$

$f'(x) = 5x^4 - 8x^3 = x^3(5x - 8) = 0$, so $x = 0$ or $x = \frac{8}{5}$, but since $f$ is restricted to $[-1, 1]$ we have one point $x$ such that $f'(x) = 0$ which is $x = 0$.

(a) At $x = 0$ $f$ has an absolute maximum, which is $f(0) = -7$.

(b) $f(-1) = -10$, $f(1) = -8$, therefore absolute minimum is $-10$ at $x = -1$

(c) no absolute minimum on $(-1, 1)$

## Section 10.3    Principles in Action

**Problem 1.**

The critical points occur at the values of $x$ for which $f'(x) = \frac{e^x(x-1)^2}{(x^2+1)^2}$ is zero or undefined; hence $x = 1$ is the only critical point. As $f'(x) \geq 0$, for all $x's$, $x = 1$ is neither a local maximum nor a local minimum point. $f$ has no absolute extrema.

**Problem 3.**

Again let $l$ feet and $w$ feet be the dimensions of the 90-square foot garden. Now $lw = 90 \Rightarrow l = \frac{90}{w}$, and we proceed to minimize the perimeter function $P(w) = 2l + 2w = 2(\frac{90}{w}) + 2w$ on $(0, \infty)$. We calculate $P'(w) = -\frac{180}{w^2} + 2$; hence, $P'(w) = 0$ when $w = \sqrt{90} = 3\sqrt{10}$. As $P''(w) = \frac{360}{w^3} > 0$ for all $w > 0$, $P(3\sqrt{10})$ is indeed the minimum value of $P$ on $(0, \infty)$. Now $w = 3\sqrt{10} \Rightarrow l = \frac{90}{3\sqrt{10}} = 3\sqrt{10}$, and therefore, the plot is a square.

**Problem 5.**

From the Pythagorean Theorem, we have $h^2 + w^2 = 14^2$, and hence $h = \sqrt{196 - w^2}$. The strength of the beam is given by $S = kh^2w = k(196 - w^2)w = -kw^3 + 196kw$, where $k > 0$ is the proportionality constant. Now $S'(w) = -3kw^2 + 196k$. $S'(w) = 0 \Rightarrow w = \sqrt{\frac{196}{3}}$, and $S''(W) = -6kw < 0$ for all positive values of $w$. Hence the absolute maximum value of $S(w)$ is achieved when $w = \sqrt{\frac{196}{3}} \approx 8.08$ inches. At this width, the height $h = \sqrt{196 - \frac{196}{3}} = 14\sqrt{\frac{2}{3}} \approx 11.43$ inches.

**Problem 7.**

(i) $g(x) = 2f(x)$

    (a) stretching vertically by a factor of 2

    (b) we can obtain the new derivative from the old one by stretching it by a factor of 2.

    (c) local minimum at $x = -1$; all points in $(0, 1]$

(ii)  (a) shifting down three units to obtain the graph of $j$.

    (b) the graph of $j'$ is the same as the graph of $f'$.

    (c) local minimum at $x = -1$, and $(0, 1]$

(iii) (a) Where the graph of $f$ lies above or on the $x$-axis the graph of $m$ is identical. Where the graph of $f$ lies below the $x$-axis the graph of $m$ is obtained by flipping the graph of $f$ over the $x$-axis.

    (b) On interval $(0, 4)$ the graph of $m'$ is the same as $f'$. On interval $(-2, 0)$ the graph of $m'$ is obtained by flipping the graph of $f'$ over the $x$-axis.

    (c) $[0, 1]$ - all points in.

(iv) (a) The graph of $f$ is shifted two units to the right to obtain the graph of $k$.

    (b) The graph of $f'$ is shifted two units to the right to obtain the graph of $k'$.

    (c) $x = 1$, and all points in $(2, 3]$

(v)  (a) The graph of $f$ is horizontally compressed by a factor of 2 to obtain the graph of $h$.

    (b) The graph of $f$ is horizontally compressed by a factor of 2 and then vertically stretched by a factor of 2 to obtain the graph of $h'$.

    (c) $x = -\frac{1}{2}$, and all points in $(0, \frac{1}{2}]$

**Problem 9.**

(a)

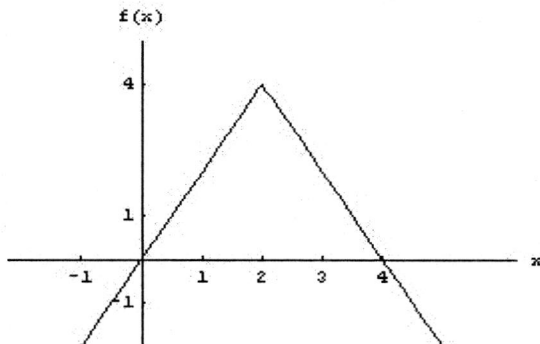

(b) $f$ attains its maximum value at $x = 2$; $f'(2)$ does not exist

(c) No

(d) Yes, maximum value at $x = 3$, minimum value at $x = 8$.

**Problem 13.**

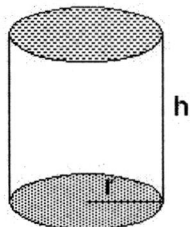

(a) $\pi r^2 \cdot h = 300 cm^3$

(b) $2 \cdot \pi r^2 + 2 \cdot \pi r^2 + 2\pi rh$

(c) $\pi r^2 \cdot h = 300 \;\Rightarrow\; h = \frac{300}{\pi r^2}$, therefore if we plug in $\frac{300}{\pi r^2}$ instead of $h$ into the equation from part $(b)$ we get: $4\pi r^2 + 2\pi r \cdot \frac{300}{\pi r^2}$. So, cost of material is $C(r) = 4\pi r^2 + \frac{600}{r}$

(d) $r = \sqrt[3]{\frac{75}{\pi}}$

(e) $h = \dfrac{300}{\pi \cdot \frac{75^{2/3}}{\pi^{2/3}}} = \dfrac{300}{\sqrt[3]{\pi} \cdot \sqrt[3]{5625}} = \dfrac{300}{\sqrt[3]{5625\pi}}$

**Problem 15.**

Let $W$ square centimeters be the the surface area of the box, $l$ cm be the side length of the square base, and $h$ cm be the height. We now have $W = l^2 + 4(lh)$, $V = l^2 h$, and hence $h = \frac{V}{l^2}$. As a function of $l$, we can express $W$ as $W(l) = l^2 + 4\frac{V}{l}$. Now $W'(l) = 2l - \frac{4V}{l^2}$; $W'(l) = 0 \Rightarrow l = \sqrt[3]{2V}$. As $W''(l) = 2 + \frac{8V}{l^3} > 0$ for all $l > 0$, $W(l)$ is minimized when $l = \sqrt[3]{2V}$. The optimal ratio of height to length is $\dfrac{h}{l} = \dfrac{\frac{V}{l^2}}{l} = \dfrac{V}{l^3} = \dfrac{V}{2V} = \dfrac{1}{2}$.

**Problem 17.**

Let $w$ inches and $h$ inches be the dimensions of the printed area and $A$ be the area of an entire page including margins. We have that $A = (w + 2(0.5))(h + 0.5 + 1) = wh + h + 1.5w + 1.5$, $wh = 33.75$, and $h = \frac{33.75}{w}$. As a function of $w$, we express $A$ by $A(w) = 33.75 + \frac{33.75}{w} + 1.5w + 1.5$. Now $A'(w) = -\frac{33.75}{w^2} + 1.5$. $A'(w) = 0 \Rightarrow w = \sqrt{\frac{45}{2}} \approx 4.74$ in. and $A''(w) = \frac{135}{2w^3} > 0$ for all $w > 0$. Thus $A(w)$ is minimized when $w = \sqrt{\frac{45}{2}} \approx 4.74$ inches. If $w = \sqrt{\frac{45}{2}}$, then $h = \dfrac{33.75}{\sqrt{\frac{45}{2}}} \approx 7.12$ inches. Therefore, the dimensions that would minimize the page area are (approximately) 4.74 by 7.12 inches.

**Problem 19.**

(a) $f'(x) = x^2 - 2 + \frac{1}{x^2}$

(b) $f''(x) = 2x - \frac{2}{x^3}$

(c) Domain of $f$: $x \neq 0$; $f'(x) = 0 \Leftrightarrow x^2 - 2 + \frac{1}{x^2} = 0$, multiplying both sides by $x^2$ gives $x^4 - 2x^2 + 1 = 0$, that is $(x^2 - 1)^2 = 0$. So $x^2 - 1 = 0$ and $x = \pm 1$. Critical points: $x = 1$ and $x = -1$ (they are also stationary points).

(d) $f''(x) = 2x - \frac{2}{x^3} = 0$, multiplying both sides by $x^2$ gives $2x^4 - 2 = 2(x^4 - 2) = 2(x^2 - 1)(x^2 + 1) = 0$, so $x \pm 1$. Now, $f$ changes its concavity at $x = -1$ and $x = 1$, therefore both these points are inflection points.

(e) $x = 0$ is not the $x$-value of a critical point because it is not in the domain of function $f$

(f) No absolute minimum, no absolute maximum.

# CHAPTER 11

# A Portrait of Polynomials and Rational Functions

## Section 11.1    A Portrait of Cubics from a Calculus Perspective

**Problem 1.**

Answers will vary; one possibility is: $f(x) = x(x+2)(x-3)$.

**Problem 3.**

Answers will vary; one possibility is: $f(x) = -(x-1)^3$.

**Problem 5.**

Answers will vary; one possibility is: $f(x) = (x-1)^3 = x^3 - 3x^2 + 3x - 1$.

**Problem 7.**

Answers will vary; one possibility is: $f(x) = -\frac{2}{27}(x-3)^3$. As $f$ is always decreasing, $f$ must have exactly one zero. Hence, $f$ is of the form $f(x) = k(x-3)^3$, where $k < 0$ is a constant. Using $f(0) = 2$, we have that $2 = k(0-3)^3$, from which we obtain $k = -\frac{2}{27}$.

**Problem 10.**

The dimensions (in inches) of the box are $(21 - 2x) \times (16 - 2x) \times x$. We wish to maximize its volume, $V(x) = (21 - 2x)(16 - 2x)x = 4x^3 - 74x^2 + 336x$ on the interval $[0, 8]$. (Note that $x > 0$ and if $x > 8$, then at least one of the dimensions would be negative.) Now $V'(x) = 12x^2 - 148x + 336 = 0$, and $V'(x) = 0 \Rightarrow 3x^2 - 37x + 84 = 0 \Rightarrow (3x - 28)(x - 3) = 0 \Rightarrow x = 3$ or $x = \frac{28}{3}$. As $x = \frac{28}{3} > 8$, it is not in the domain of $V(x)$. To show that $x = 3$ is maximizes the volume, we consider the second derivative $V''(x) = 24x - 148$. Now $V''(3) = -76 < 0$ and hence $x = 3$ inches maximizes the volume.

**Problem 11.**

(a) $f(x) = x(x-3)(x+5) = x^3 + 2x^2 - 15x$. $\Rightarrow$ The $x$-intercepts are at $x = -5, 0, 3$, and $f'(x) = 3x^2 + 4x - 15$. Now $f'(x) = 0 \Rightarrow x = \frac{-4 \pm \sqrt{4^2 - 4(3)(-15)}}{6}$ Applying the quadratic formula reveals a local maximum at $x = -3$ and a local minimum at $x = \frac{5}{3}$.

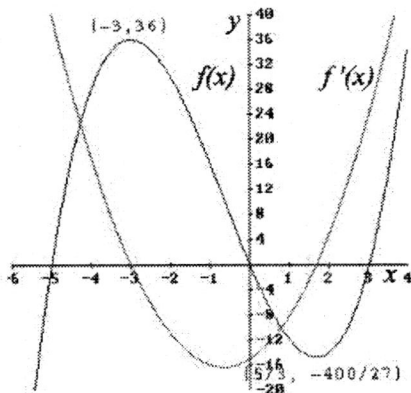

(b) Factoring out the $x$ and applying the quadratic formula shows that the $x$-intercepts are located at $x = 0, \frac{-1 \pm \sqrt{5}}{2}$. Differentiating yields $f'(x) = 3x^2 + 6x - 9$, giving a local maximum at $x = -3$ and a local minimum at $x = 1$.

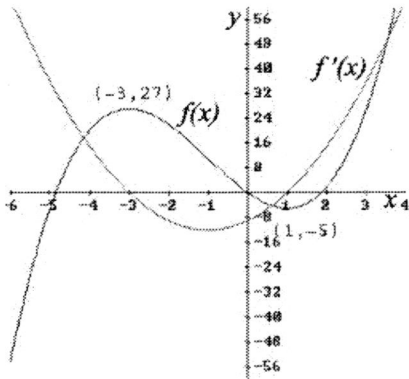

**Problem 12.**

(a) $f(x) = x^3 - 3x + 1 \Rightarrow f'(x) = 3x^2 - 3 = 3(x^2 - 1) = 3(x + 1)(x - 1)$. $0 = f'(x) \Rightarrow x = -1$ and $x = 1$ are the critical points. Now $f''(x) = 6x$ gives $f''(-1) = -6 < 0$ and $f''(1) = 6 > 0$. Therefore, $x = -1$ is a local maximum point and $x = 1$ is a local minimum point. Moreover neither of these critical points are absolute extreme points because $\lim\limits_{x \to \infty} f(x) = \infty$ and $\lim\limits_{x \to -\infty} f(x) = -\infty$.

(b) $f(x) = x^3 + 3x + 1 \Rightarrow f'(x) = 3x^2 + 3 = 3(x^2 + 1)$. $0 = f'(x) \Rightarrow x^2 = -1$. Hence there are no real values of $x$ for which $0 = f'(x)$ and consequently no critical points.

**Problem 14.**

$f(x) = x^3 + x^2 + x + 1 \Rightarrow f'(x) = 3x^2 + 2x + 1 = 0$. Now $0 = f'(x) \Rightarrow x = \frac{-2 \pm \sqrt{-8}}{6}$. Hence there are no real values of $x$ for which $0 = f'(x)$ and consequently no critical points.

**Problem 17.**                                       **Problem 19.**

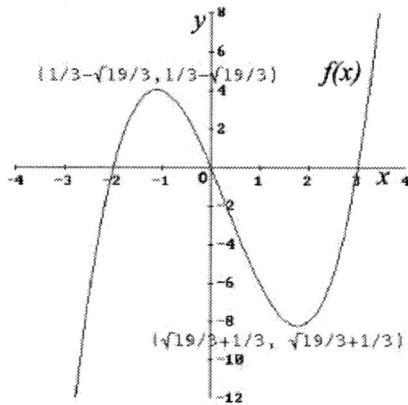                                 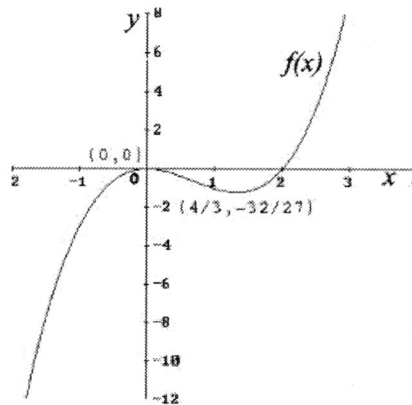

**Problem 21.**

$f(x) = -2x^3 + 3x^2 + 6x - 2 \Rightarrow f'(x) = -6x^2 + 6x + 6 \Rightarrow f''(x) = -12x + 6 \Rightarrow$ The point of inflection is $\left(\frac{1}{2}, f\left(\frac{1}{2}\right)\right) = \left(\frac{1}{2}, \frac{3}{2}\right)$. Now $f'\left(\frac{1}{2}\right) = \frac{15}{2}$. The equation of the tangent line is $\left(y - \frac{3}{2}\right) = \frac{15}{2}\left(x - \frac{1}{2}\right) \Leftrightarrow$ $y = \frac{15}{2}x - \frac{9}{4}$.

## Section 11.2    Polynomial Functions and Their Graphs

**Problem 4.**

Answers will vary; one possibility: $P(x) = x^4 + 5$.

**Problem 6.**

Answers will vary; one possibility: $P(x) = x(x-9)^2(x-3)(x+e)$.

**Problem 8.**

Answers will vary; one possibility: $P(x) = (x+3)^4 + 2$.

**Problem 9.**

$P(x) = k(x-2)^2(x+3)^2$, for some $k \neq 0$. Now $-2 = P(0) = k(4)(9) = 36k \Rightarrow k = -\frac{1}{18}$. This answer is unique.

**Problem 11.**

$P(x) = k(x+1)^3$, for some $k \neq 0$. As $\lim\limits_{x \to \infty} P(x) = \infty$, $k$ can be any positive number, and hence the answer is not unique.

**Problem 13.**

$P(x) = kx^3(x-1)(x+2)$, for some $k \neq 0$. Now $2 = P(-1) = k(-1)^3(-1-1)(-1+2) = 2k \Rightarrow k = 1$. This answer is unique.

**Problem 15.**

(a) $f(x) = 2x^3 + 2x^2 - 12x = 2x(x+3)(x-2)$. The zeros are $x = -3, 0$, and 2.

(b) $g(x) = 2x^3 + 2x^2 + 12x = 2x(x^2 + x + 6)$. Now if $x^2 + x + 6 = 0$, then $x = \frac{-1 \pm \sqrt{1-4(6)}}{2} = \frac{-1 \pm \sqrt{-23}}{2}$, which are not real solutions. The only zero is $x = 0$.

**Problem 17.**

(a) $P(x) = x^3 - x^2 - 4x + 4 = (x^2 - 4)(x - 1) = (x + 2)(x - 2)(x - 1)$. The zeros are $x = -2, 1$, and 2.

(b) $Q(x) = x^3 - x^2 + 4x - 4 = (x - 1)(x^2 + 4)$. Now $x^2 + 4 = 0$ has no real solutions; hence, the zero is $x = 1$.

**Problem 19.**

$P(x) = x^4 - 2x^3 - 6x^2 + 12x = x(x-2)(x^2 - 6) = x(x-2)(x - \sqrt{6})(x + \sqrt{6})$. The zeros are $x = 0, 2, \sqrt{6}$, and $-\sqrt{6}$.

**Problem 22.**

(a) Definitely false. A polynomial that is symmetric about the origin has odd degree.

(b) Definitely true. A polynomial that is symmetric about the origin has odd degree.

(c) Definitely false. A polynomial that is symmetric about the origin has odd degree, and odd-degree polynomials have at least one zero.

(d) Definitely true. A polynomial that is symmetric about the origin has odd degree, and odd-degree polynomials have at least one zero.

(e) Definitely true. Because $P(x)$ is symmetric about the origin, $(0,0)$ cannot be a turning point, and if $(a, P(a))$ is a turning point, $(-a, -P(a))$ is also a turning point. Hence the number of turning points is a (nonegative) even number.

**Problem 24.**

(a) Possibly true. True if $P(x) = x(x-1)(x-2)(x-3)(x-4)$; false if $P(x) = x^5$.

(b) Definitely true. A polynomial of degree $n$ can have no more than $n$ real roots.

(c) Definitely false. A polynomial of degree $n$ can have at most $n-1$ turning points.

(d) Possibly true. True if $P(x) = x(x-1)(x-2)(x-3)(x-4)$; false if $P(x) = x^5$.

(e) Definitely true. A polynomial of degree $n$ can have at most $n-1$ turning points.

(f) Definitely true. Odd-degree polynomials have at least one real root.

(g) Definitely false. Since $\lim_{x \to \infty} P(x) = \infty$, it must be that $\lim_{x \to -\infty} P(x) = -\infty$.

**Problem 26.**

(a) (i) Even multiplicity; (ii) Simple; (iii) Odd multiplicity; (iv) Simple; (v) Even multiplicity

(b) (i) Zero; (ii) Positive; (iii) Zero; (iv) Negative; (v) Zero

## Section 11.3    Polynomial Functions and Their Graphs

**Problem 2.**

(a) Since $P'(x)$ is degree 5, $P$ is degree 6.

(b) The critical points of $P$ occur whenever $P'(x) = 0$, that is at $x = 0, -2$.

(c) Because $\lim_{x \to -\infty} P'(x) = -\infty$, $\lim_{x \to \infty} P'(x) = \infty$, and $P'(x)$ is continuous everywhere, $P(x)$ has an absolute minimum. We know it must be obtained at one of the critical points, so we calculate $P''(x) = 2x(x + 2)^3 + 3x^2(x+2)^2$; unfortunately, $P''(0) = 0$ and $P''(-2) = 0$, so the second derivative test lends no information. (How could you have predicted that?) Instead we look at the sign of $P'(x)$; across $x = 0$ the sign remains positive, but across $x = -2$ the sign changes from negative to positive. Thus, there is an absolute minimum at $x = -2$. However, we cannot find the value of the function here since we don't have an expression for the function; we can determine only the value of the function relative to some other value of the function.

**Problem 4.**

(a) Note that $P'(x)$ has two zeros, at $x = 0, 1$; these are the critical points of $P(x)$.

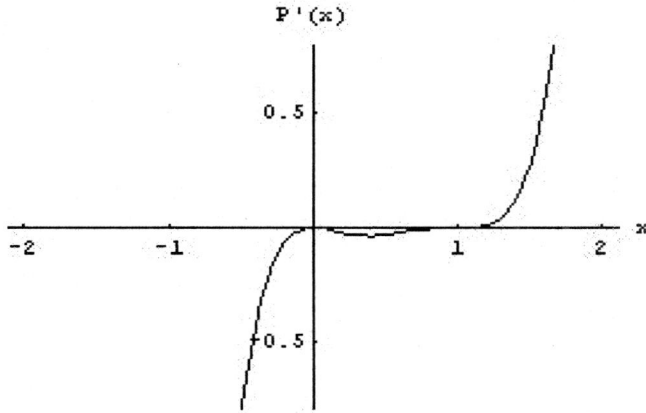

(b)

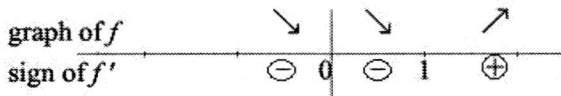

Note that $P' > 0$ on $(1, \infty)$ and $P' < 0$ on $(-\infty, 0) \cup (0, 1)$.

(c) Because $P(x) < 0$ at $x = 0$, $P(x) < 0$ at $x = 1$ since $P$ is decreasing on the interval. Since $P'(x)$ is positive and increasing for $x > 1$, $P(x)$ must be increasing at an increasing rate for $x > 1$; because $P(x) < 0$ at $x = 1$, $P(x)$ must cross the $x$-axis.

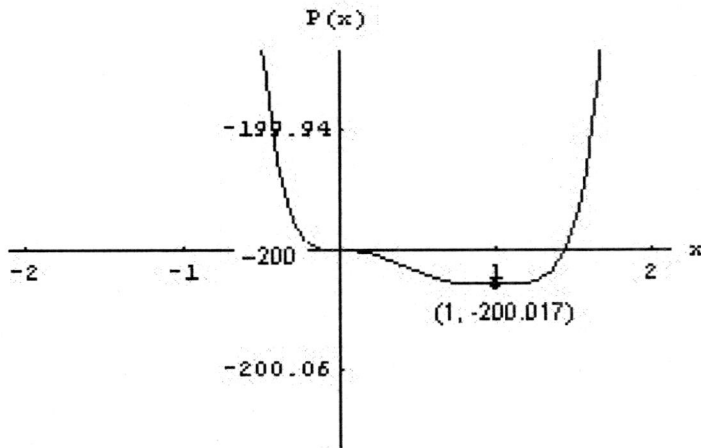

**Problem 6.**

(a) True.

(b) Possible; for $P(x) = 2x^7 + x$ we determine $P'(x)$ has no zeros, but for $P(x) = 2x^7 - x$ we see $P'(x)$ has multiple zeros.

(c) True.

(d) Possible; for $P(x) = 2x^7$ we have $P''(x) = 84x^5$ and thus only 1 point of inflection, while for $P(x) = 2x^7 - 28x^6 + 147x^5 - 350x^4 + 336x^3$ we have $P''(x) = 84x^5 - 840x^4 + 2940x^3 - 4200x^2 + 2016x = 84x(x-1)(x-2)(x-3)(x-4)$ and thus 5 points of inflection. (This example is provided for reference, not as an exercise you are expected to perform; it serves to show that thinking about this graphically is often significantly easier than writing out the precise equation describing the graph!)

**Problem 8.**

The strategy here should be to use the $x-$intercepts to establish the factors of the appropriate polynomial and then to use the additional point provided to determine the multiplicative factor.

(a) The zeros of the function are $x = -2, 1$, and 2. As the graph has the shape of a graph of a cubic polynomial, the function could have an equation of the form $P(x) = k(x+2)(x-1)(x-2)$, where $k$ is a constant. Because the graph has a $y-$intercept of 3, we have $3 = P(0) = k(0+2)(0-1)(0-2) \Rightarrow k = \frac{3}{4}$. Therefore, $P(x) = \frac{3}{4}(x+2)(x-1)(x-2)$.

(b) This graph is the reflection of the graph in part (a) in the $x-$axis. Thus, $P(x) = -\frac{3}{4}(x+2)(x-1)(x-2)$.

(c) This function has an even order zero at $x = -2$ and an odd order zero at $x = 0$. As the graph has the shape of the graph of a cubic polynomial, the function could have an equation of the form $P(x) = kx(x+2)^2$. Because the graph contains the point $(1, 2)$, we have $2 = P(1) = k(1)(1+2)^2 = 9k \Rightarrow k = \frac{2}{9}$. Therefore, $P(x) = \frac{2}{9}x(x+2)^2$

**Problem 10.**

Let $P(x)$ be the polynomial function graphed in the figure. We see that $\lim_{x \to \infty} P(x) = \infty$ and $\lim_{x \to -\infty} P(x) = -\infty$. Hence $P(x)$ is an odd-degree polynomial function. From the graph, we see that there are three points of inflection, which are approximately located at $x = -3, -1.5$, and 0. Thus, $P(x)$ must have degree at least five. Now $P(x)$ has an odd order zero at $x = -3$ and an even order zero at $x = 1$. Thus, $P(x)$ can have the form $P(x) = k(x+3)^3(x-1)^2$, for some constant $k$. As the $y-$intercept is 1, we have $1 = P(0) = k(0+3)^3(0-1)^2 = 27k \Rightarrow k = \frac{1}{27}$. Therefore, $P(x) = \frac{1}{27}(x+3)^3(x-1)^2$.

**Problem 12.**

(a)  (i) True.

(ii) False; $P(x) = (x-1)(x-2)(x-3)(x-4)(x-5)$ has 5 zeros.

(iii) False; $P(x) = x^5 + x$ has derivative $P'(x) = x^4 + 1$ which is positive for all $x$, forcing $P(x)$ to have no turning points.

(iv) True.

(b) Statement (i) is true. Because $P'(\pi) = 0$ and $P''(\pi) > 0$, $P(x)$ has a local minimum at $x = \pi$; because $P(x)$ has degree 5, it cannot have an absolute minimum.

**Problem 14.**

(a) Graphs (i) and (iv) could be the graphs of polynomials. Graph (i) has even degree greater than or equal to 6 because it has 4 points of inflection and $\lim_{|x| \to \infty} P(x) = \infty$. Graph (iv) has odd degree greater than or equal to 5 because $\lim_{x \to \infty} P(x) = -\infty$, $\lim_{x \to -\infty} P(x) = \infty$, and it has 3 points of inflection.

(b) Graph (iii) could be the graph of such a function.

(c) Graph (ii) has turning points and a horizontal asymptote, making it incompatible with both categories; Graphs (v) and (vi) have vertical asymptotes, which are not characteristics of either class of function.

# Section 11.4    Rational Functions and Their Graphs

**Problem 1.**

(a) The graph has a single simple zero at $x = 0$ and a vertical asymptote at $x = -1$. As the sign of $y$ changes across the vertical asymptote, there is an odd power of $(x + 1)$ in the denominator of the function. As there is a horizontal asymptote at $y = 2$, the degrees of the numerator and denominator of the function are equal, and the lead coefficient of the numerator is 2. Hence $y = \frac{2x}{x+1} = 2 - \frac{2}{x+1}$.

(b) The graph has simple zeros at $x = -2$ and $x = 0$ and a vertical asymptote at $x = -1$. As the sign of $y$ does not change across the vertical asymptote, there is an even power of $(x+1)$ in the denominator of the function. As there is a horizontal asymptote at $y = 2$, the degrees of the numerator and denominator of the function are equal, and the lead the coefficient of the numerator is 2. Hence $y = \frac{2x(x+2)}{(x+1)^2} = -\frac{2}{(x+1)^2} + 2$.

(c) The graph has no zeros and always lies above the $x$−axis. There is a vertical asymptote at $x = -1$, which, in this case, implies that there is an even power of $(x + 1)$ in the denominator of the function. As there is a horizontal asymptote at $y = 0$, the degree of the numerator of the function is less than the degree of the denominator of the function. Hence $y = \frac{2}{(x+1)^2}$.

(d) The graph has no zeros and vertical asymptotes at $x = -1$ and $x = 2$. The sign of $y$ changes across both of these vertical asymptotes, which implies that there are odd powers of $(x + 1)$ and $(x - 2)$ in the denominator of the function. As there is a horizontal asymptote at $y = 0$, the degree of the numerator of the function is less than the degree of the denominator of the function. Hence the equation has the form $y = \frac{k}{(x+1)(x-2)}$, where $k$ is a nonzero constant. As $y < 0$ for $|x| > 2$, $k < 0$. For simplicity, we choose $k = -1$. Therefore, $y = -\frac{1}{(x+1)(x-2)}$.

(e) The graph has no zeros and vertical asymptotes at $x = -1$ and $x = 2$. The sign of $y$ changes across $x = -1$ but does not change across $x = 2$. Thus, there is an odd power $(x + 1)$ and an even power of $(x - 2)$ in the denominator of the function. As there is a horizontal asymptote at $y = 0$, the degree of the numerator of the function is less than the degree of the denominator of the function. Hence the equation has the form $y = \frac{k}{(x+1)(x-2)^2}$, where $k$ is a nonzero constant. As $y < 0$ for $x > 2$, $k < 0$. For simplicity, we choose $k = -1$. Therefore, $y = -\frac{1}{(x+1)(x-2)^2}$

(f) The graph has no zeros, lies above the $x$−axis and has vertical asymptotes at $x = -3$ and $x = 1$. Thus, there are even powers of $(x + 3)$ and $(x - 1)$ in the denominator of the function. As there is a horizontal asymptote at $y = 1$, the equation has the form $y = \frac{1}{(x+3)^2(x-1)^2} + 1 = \frac{1+(x+3)^2(x-1)^2}{(x+3)^2(x-1)^2} = \frac{x^4+4x^3-2x^2-12x+10}{x^4+4x^3-2x^2-12x+9}$.

(g) The graph has no zeros, lies above the $x$−axis and has vertical asymptotes at $x = -3$ and $x = 1$. Thus, there are even powers of $(x + 3)$ and $(x - 1)$ in the denominator of the function. As there is a horizontal asymptote at $y = 0$, the degree of the numerator of the function is less than the degree of the denominator of the function. Hence the equation has the form $y = \frac{k}{(x+3)^2(x-1)^2}$. As $y > 0$ for $x > 1$, $k > 0$, and, for simplicity, we choose $k = 1$. Therefore, the equation is $y = \frac{1}{(x+3)^2(x-1)^2}$.

(h) The graph has no zeros and a vertical asymptote at $x = 0$. As the $y$ changes sign across $x = 0$, there is an odd power $x$ in the denominator of the function. As there are no vertical asymptotes, the degree of the numerator of the function is greater than the degree of the denominator. Hence the equation $y = \frac{x^2+1}{x}$ suffices.

**Problem 3.**

(a) The graph has no zeros and vertical asymptotes at $x = -1$ and $x == 2$. The sign of $y$ changes across both of these vertical asymptotes, which implies that there are odd powers of $(x+1)$ and $(x-2)$ in the denominator of the function. As there is a horizontal asymptote at $y = 0$, the degree of the numerator of the function is less than the degree of the denominator of the function. Hence the equation has the form $y = \frac{k}{(x+1)(x-2)}$, where $k$ is a nonzero constant. As $y < 0$ for $|x| > 2$, $k < 0$. For simplicity, we choose $k = -1$. Therefore, $y = -\frac{1}{(x+1)(x-2)}$.

(b) The graph has simple zeros at $x = -2$ and $x = 0$ and a vertical asymptote at $x = -1$. As the sign of $y$ does not change across the vertical asymptote, there is an even power of $(x+1)$ in the denominator of the function. As there is a horizontal asymptote at $y = 2$, the degrees of the numerator and denominator of the function are equal, and the lead the coefficient of the numerator is 2. Hence $y = \frac{2x(x+2)}{(x+1)^2}$.

(c) The graph has a simple zero at $x = -2$ and an even-ordered zero at $x = 0$. Thus, the numerator has factors of $(x+2)$ and $x^2$. There are vertical asymptotes at $x = -1$ and $x = 1$. The sign of $y$ changes across the vertical asymptote $x = 1$ but not across $x = -1$. Thus, the denominator has factors of $(x+1)^2$ and $(x-1)$. As there is a horizontal asymptote at $y = 2$, the degrees of the numerator and denominator of the function are equal, and the lead the coefficient of the numerator is 2. Therefore, the equation has the form $y = \frac{2x^2(x+2)}{(x+1)^2(x-1)}$.

**Problem 5.**

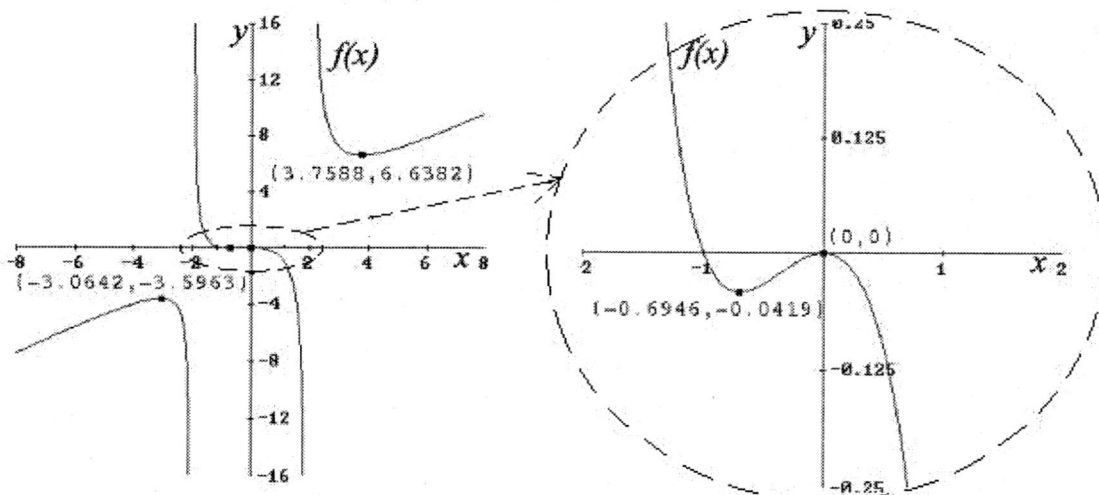

Calculating the derivative of $f(x) = \frac{x^3 + x^2}{x^2 - 4}$ via the quotient rule gives:

$$f'(x) = \frac{(3x^2 + 2x)(x^2 - 4) - (x^3 + x^2)(2x)}{(x^2 - 4)^2} = \frac{(3x^4 + 2x^3 - 12x^2 - 8x) - (2x^4 + 2x^3)}{(x^2 - 4)^2} =$$

$$\frac{x^4 - 12x^2 - 8x}{(x^2 - 4)^2} = \frac{x(x^3 - 12x - 8)}{(x^2 - 4)^2} = \frac{x^4 - 12x^2 - 8x}{(x^2 - 4)^2}.$$

Now, as $f'(0) = 0$, $x = 0$ is a critical point. A graphing calculator will verify that there is a local minimum at $x = 0$. Furthermore, a graphing calculator will find a local maximum at $x \approx -3.0642$, a local minimum at $x \approx -0.6946$, and a local minimum at $x \approx 3.7588$.

**Problem 7.**

(a) Note that $f(x) = \frac{x^2-4}{x^2-3x-4} = \frac{(x+2)(x-2)}{(x+1)(x-4)}$. The $x$-intercepts are $x = \pm 2$; the $y$-intercept is $y = 1$; the vertical asymptotes are $x = -1, 4$; the horizontal asymptote is $y = 1$.

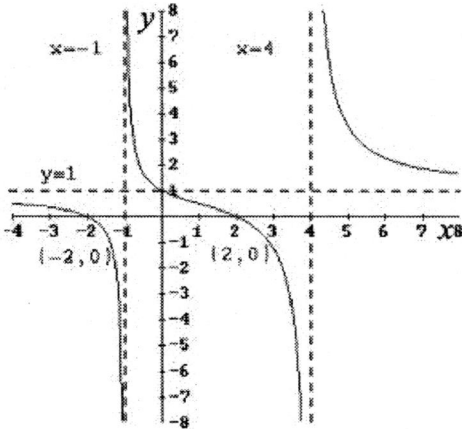

## Problem 8.

(a) $f(x) = \frac{4}{x^2+1} - 4 = -\frac{4x^2}{x^2+1}$.

(b) $f(x) = \frac{-2x(x+3)(x-3)}{(x+2)(x-1)^2}$.

## Problem 11.

(a) $f(x) = \frac{4}{x} + x \Rightarrow f'(x) = \frac{-4}{x^2} + 1 = \frac{(x+2)(x-2)}{x^2}$.

$f'(x) > 0$ on $(-\infty, -2) \cup (2, \infty)$, and hence the function is increasing here; $f'(x) < 0$ on $(-2, 0) \cup (0, 2)$, and hence the function is decreasing here.

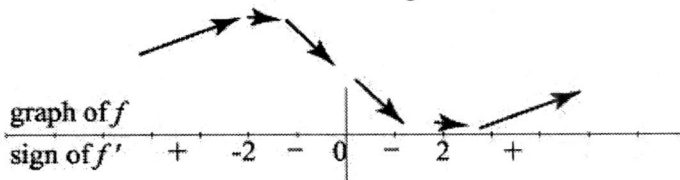

| graph of $f$ | | | | | |
|---|---|---|---|---|---|
| sign of $f'$ | + | -2 | − | 0 | − | 2 | + |

(b) $f'(x) = \frac{-4}{x^2} + 1 \Rightarrow f''(x) = \frac{8}{x^3}$ On $(-\infty, 0)$ $f''(x) < 0$, and hence the function is concave down; on $(0, \infty)$ $f''(x) > 0$, and hence the function is concave up.

| concavity of graph of $f$ | down | | up |
|---|---|---|---|
| sign of $f''$ | − | 0 | + |

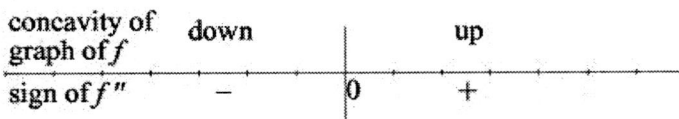

(c) There is a local maximum at $(-2, -4)$ and a local minimum at $(2, 4)$.

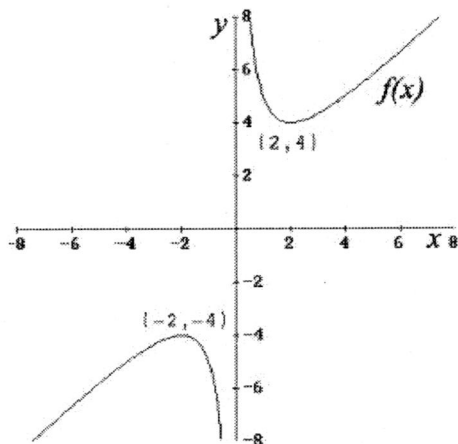

(d) Now $\lim\limits_{x \to 0^-} f(x) = -\infty$ and $\lim\limits_{x \to 0^+} f(x) = \infty$ imply that $f$ has no absolute extrema.

**Problem 12.**

(i) The $x$-intercept is the origin; the $y$-intercept is the origin; the vertical asymptotes are $x = \pm 1$; the horizontal asymptote is the $x$-axis.

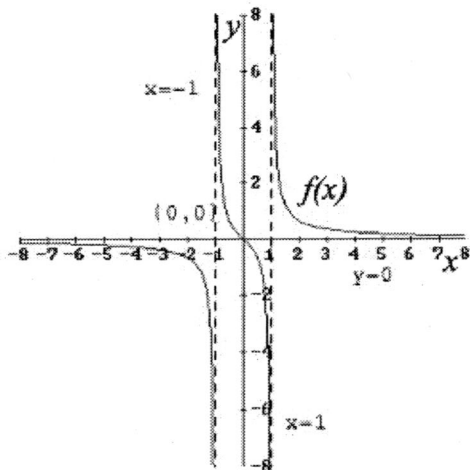

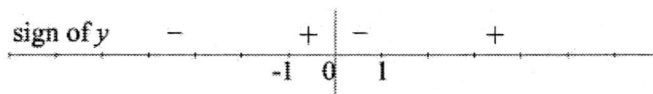

(ii),(iii) The $x$-intercepts are $x = 0, 2$; the $y$-intercept is the origin; the vertical asymptotes are $x = \pm 1$; there are no horizontal asymptotes.

(v) The $x$-intercepts are $x = 0, 2$; the $y$-intercept is the origin; the vertical asymptotes are $x = \pm 1$; the horizontal asymptote is the $x$-axis.

(viii) The $x$-intercept is the origin; the $y$-intercept is the origin; there are no vertical asymptotes; the horizontal asymptote is $y = -1$. Note that $y < 0$ for all values of $x$.

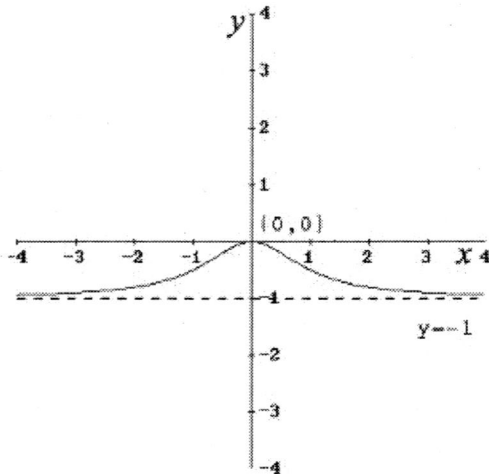

**Problem 14.**

(a) i. $x = 3, 7$; ii. Insufficient information; iii. $x = 0, \frac{4}{3}$; iv. $x = -2, 0, 2$.

(b) i. $x = 1$, $x = 6$; ii. $x = -2$, $x = 3$; iii. $x = -\frac{2}{3}$ $x = 1$; iv. $x = \pm\sqrt{3}$.

(c) i. $y = 5$; ii. $y = 2$; iii. $y = 10$; iv. $y = 5$.

# CHAPTER 12

# Inverse Functions: Can What Is Done Be Undone?

## Section 12.1    What Does It Mean for $f$ and $g$ to Be Inverse Functions?

**Problem 1.**

(a) 1-to-1 and invertible (as long as social security numbers are not reassigned after someone dies)

(b) Not 1-to-1 and hence not invertible; five people are put into each group.

(c) Not 1-to-1 and not invertible; many sites have the same altitude.

**Problem 3.**

    (a) $f^{-1}(x) = \frac{x-1}{2}$                        (b) $f^{-1}(x) = \sqrt{x+2}$, $x > -2$

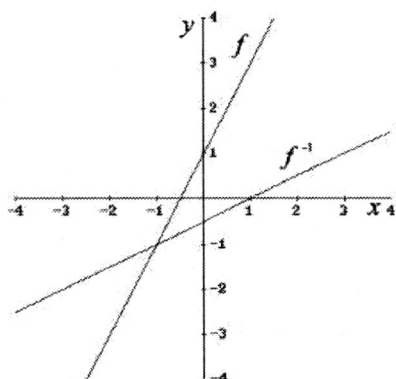
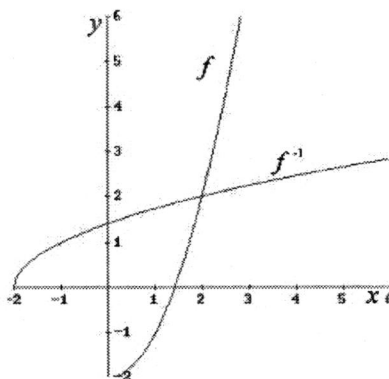

  (c)                                    (d)

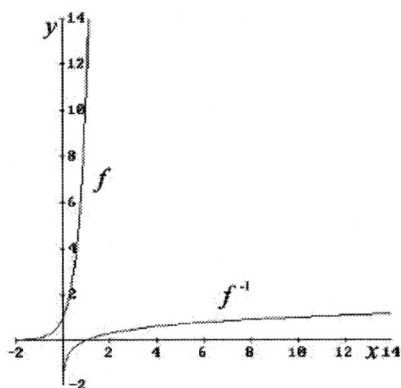
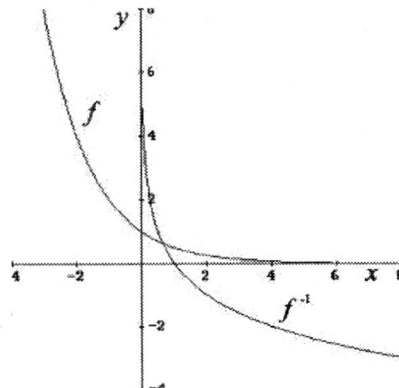

**Problem 6.**

(a) Consider $f'(x) = 3x^2 + 6x + 6$, which is defined and continuous for all real values of $x$. The discrimanant of $3x^2 + 6x + 6 = 0$ is $6^2 - 4(3)(6) = -36 < 0$; hence $f'(x)$ is never equal to zero. Moreover, as $f'(0) = 6 > 0$, $f'$ is increasing everywhere; that is, if $a < b$, then $f(a) < f(b)$. Hence $f$ is 1-to-1 and, therefore, invertible.

(b) Answers will vary. $(12, 0)$, $(22, 1)$, $(44, 2)$ are on the graph of $f^{-1}(x)$, because $(0, 12)$, $(1, 22)$, and $(2, 44)$ are points on the graph of $f(x)$.

**Problem 7.**

(a) Invertible.

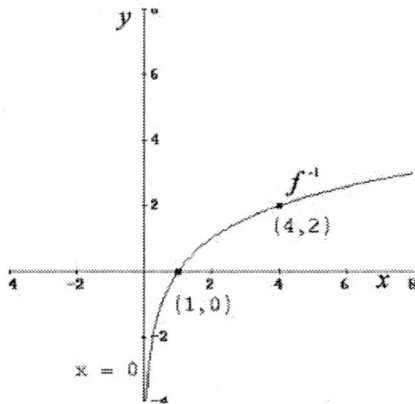

(b) Not invertible; to make it invertible restrict the domain to $\left[-\frac{\pi}{2}, \frac{\pi}{2}\right]$.

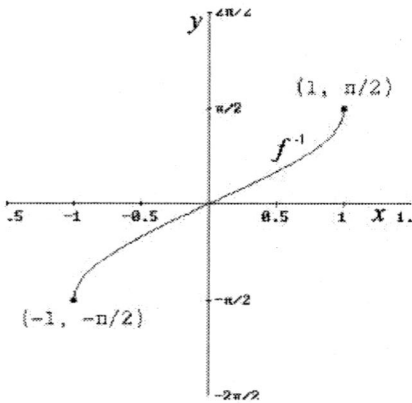

(c) Invertible.

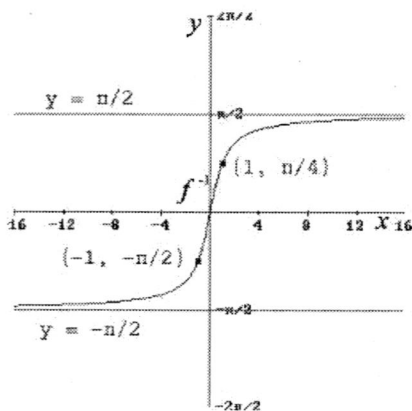

## Section 12.2    Finding the Inverse of a Function

**Problem 2.**

(a) Given $y = 2 - \frac{x+1}{x}$, interchange $x$ and $y$ to obtain $x = 2 - \frac{y+1}{y} \Leftrightarrow xy = 2y - (y+1) \Leftrightarrow y = \frac{1}{1-x}$. Hence $f^{-1}(x) = \frac{1}{1-x}$.

(b) Given $y = \frac{x^5}{10} + 7$, interchange $x$ and $y$ to obtain $x = \frac{y^5}{10} + 7 \Leftrightarrow 10(x-7) = y^5 \Leftrightarrow y = \sqrt[5]{10x - 70}$. Hence $f^{-1}(x) = \sqrt[5]{10x - 70}$.

**Problem 4.**

Given $y = \frac{2x-1}{3x+4}$, interchange $x$ and $y$ and solve for $y$. Now $x = \frac{2y-1}{3y+4} \Rightarrow 3xy + 4x = 2y - 1 \Rightarrow 3xy - 2y = -4x - 1 \Rightarrow y = \frac{-4x-1}{3x-2}$. Hence $f^{-1}(x) = \frac{-4x-1}{3x+2}$.

**Problem 6.**

Given $y = \frac{x}{x+3}$, interchange $x$ and $y$ and solve for $y$. Now $x = \frac{y}{y+3} \Rightarrow xy + 3x = y \Rightarrow y(1-x) = 3x \Rightarrow y = \frac{3x}{1-x}$. Hence $f^{-1}(x) = \frac{3x}{1-x}$. The domain of $f^{-1}$ is $(-\infty, 1) \cup (1, \infty)$.

**Problem 9.**

Given $y = 2\sqrt{x-6}$, interchange $x$ and $y$ and solve for $y$. Now $x = 2\sqrt{y-6} \Rightarrow x^2 = 4(y-6) \Rightarrow y = \frac{1}{4}x^2 + 6$. Hence $f^{-1}(x)\frac{1}{4}x^2 + 6$. The domain of $f^{-1}$ is the range of $f$, which is $[0, \infty)$.

**Problem 10.**

Given $y = x^3 + 1$, interchange $x$ and $y$ and solve for $y$. Now $x = y^3 + 1 \Rightarrow x - 1 = y^3 \Rightarrow y = \sqrt[3]{x-1}$. Hence $f^{-1} = \sqrt[3]{x-1}$. The domain of $f^{-1}$ is $(-\infty, \infty)$.

**Problem 12.**

$f(x) = x^3 - 2x + 3 \Rightarrow f'(x) = 3x^2 - 2$. Now $f'(x)$ is zero at $x = \pm\sqrt{\frac{2}{3}}$, negative on $(-\sqrt{\frac{2}{3}}, \sqrt{\frac{2}{3}})$, a nd positive elsewhere. The first derivative test implies that there are local extrema at $x = \pm\sqrt{\frac{2}{3}}$, and hence $f$ cannot be 1-to-1.

**Problem 14.**

$f(x) = x^2 + 2x - 1 \Rightarrow f'(x) = 2x + 2$. Now $f'(-1) = 0$, $f'(x) > 0$ for $x > -1$ and $f'(x) < 0$ for $x < -1$. The first derivative test implies that there is a local minimum at $x = -1$, and hence $f$ cannot be 1-to-1.

**Problem 15.**

$f(x) = 3 \cdot 2^x \Rightarrow f'(x) = 3 \ln 2 \cdot 2^x$. Now $f'(x) > 0$ for all $x$. Thus $f$ is increasing and hence 1-to-1.

To determine $f^{-1}$, interchange the roles of $x$ and $y$ in the equation $y = 3 \cdot 2^x$ and solve for $y$. Now $x = 3 \cdot 2^y \Rightarrow \frac{x}{3} = 2^y \Rightarrow \ln\left(\frac{x}{3}\right) = y \ln 2 \Rightarrow y = \frac{\ln\left(\frac{x}{3}\right)}{\ln 2}$. Hence $f^{-1}(x) = \frac{\ln\left(\frac{x}{3}\right)}{\ln 2}$. (Note that logarithmic calculations are discussed in Chapter 13.)

## Section 12.3    Interpreting the Meaning of Inverse Functions

**Problem 2.**

(a)   (i) The cost of $3A$ pounds of apricots is $C(3A) = \frac{3}{A}(3A) = 9$ dollars.

    (ii) The amount of apricots that can be purchased for \$6 is $C^{-1}(6) = \frac{A}{3}(6) = 2A$ pounds.

(iii) The amount of apricots that can be purchased for \$1 is $C^{-1}(1) = \frac{A}{3}(1) = \frac{A}{3}$ pounds.

(b)   (i) True.

    (ii) True.

    (iii) True.

    (iv) True.

(c) Statement iv.

## Problem 4.

(a) To earn \$70 on a given day, the typist must type 50 words per minute.

(b) If the typist types five more words per minute today than he did yesterday, he will earn 10% more than he did yesterday.

(c) The will need to type $D^{-1}(B+10)$ words per minute to earn \$10 more today than he earned yesterday.

## Problem 5.

(a) Spending half as much on advertising would result in \$80,000 less in revenue.

(b) Increasing advertising expenditures from \$30,000 to \$30,001 would result in roughly a \$2.80 increase in revenue.

(c) To generate twice as much revenue as last year, the company would need to spend $R^{-1}(2C)$ on advertising.

## Problem 7.

(a)

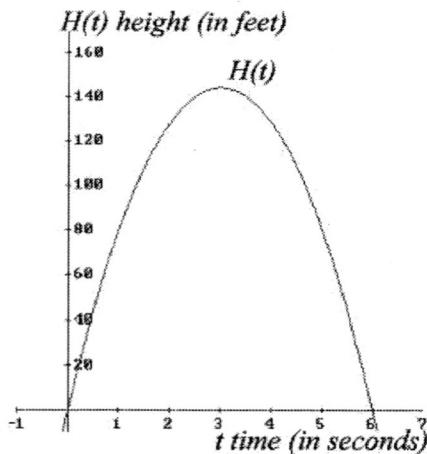

(b) Domain of H: $[0, 6]$. Note that values of $t$ less than zero are meaningless because the ball not yet been thrown and that the values of $t$ greater than 6 are meaningless because the ball hits the ground at $t = 6$ ($H(6) = 0$). Range of H: $[0, 144]$ (See explanation in part (c).)

(c) As $H$ is a quadratic function with a negative lead coefficient, $H$ will achieve its maximum value at the value of $t$ for which $H'(t) = -32t + 96 = 0$. Thus $H$ achieves its maximum value of $H(3) = 144$ at $t = 3$. Therefore the ball's maximum height is 144 feet, which is achieved 3 seconds after the ball is thrown.

(d) The inverse relation for $H(t)$ is not a function because its graph contains the points $(0,0)$ and $(0,6)$.

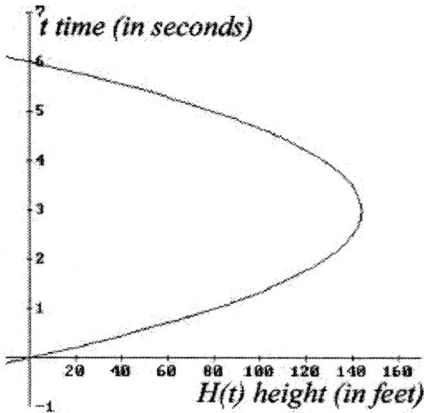

(e) Let domain be $[0,3]$; the ball's fall is no longer represented.

(f) $H^{-1}(80) = 1$ because $H(1) = 80$. The ball reaches a height of 80 feet 1 second after it is thrown.

## Exploratory Problems for Chapter 12

**Problem 1.**

(a) The derivative of $f(x) = x^2$ at $(3,9)$ is $f'(3) = 2(3) = 6$. The derivative of $f^{-1}(x) = \sqrt{x}$ at $(9,3)$ is $(f^{-1})'(9) = \frac{1}{2\sqrt{9}} = \frac{1}{6}$. These two derivatives are reciprocals.

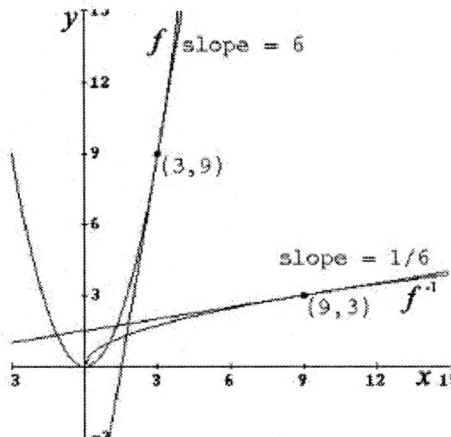

(b) The derivative of $f(x) = x^2$ at $(a,b)$ is $f'(a) = 2a$.

(c) The derivative of $f^{-1}(x) = \sqrt{x}$ at $(b,a)$ is $(f^{-1})'(b) = \frac{1}{2\sqrt{b}}$.

(d) As $b = f(a) = a^2$, $(f^{-1})'(b) = \frac{1}{2\sqrt{a^2}} = \frac{1}{2a}$. These two derivatives are reciprocals.

# CHAPTER 13
# Logarithmic Functions

## Section 13.1    The Logarithmic Function Defined

**Problem 2.**

   (a) "log base 5 of 14"; 5; 14

   (b) $\log_4(8)$; 4; 8

   (c) $\ln(5)$; "natural log of 5"

**Problem 4.**

   (a) It is between -2 and -1, because $\log_{10}(0.05)$ is the number to which 10 must be raised to get 0.05, and $10^{-2} = 0.01$ while $10^{-1} = 0.1$.

   (b) It is between 3 and 4, because $\log_3(29)$ is the number to which 3 must be raised to get 29, and $3^3 = 27$ while $3^4 = 81$.

**Problem 6.**

   (a)  $\log_2 \sqrt{8} = \log_2((2^3)^{\frac{1}{2}}) = \log_2(2^{\frac{3}{2}}) = \frac{3}{2}$.

   (b)  $\log_{10} 0.001 = \log_{10} 10^{-3} = -3$.

   (c)  $\log_2(\frac{4}{\sqrt{8}}) = \log_2(\sqrt{2}) = \frac{1}{2}$.

   (d)  $\log_3 \frac{1}{9} = \log_3 3^{-2} = -2$.

   (e)  $\log_k k^{3x} = 3x$.

   (f)  $\log_k 1 = 0$.

   (g)  $\log_k(k^x k^y) = \log_k(k^{x+y}) = x + y$.

## Section 13.2    The Properties of Logarithms

**Problem 2.**

   (a)  $\log_2(u^2 w) = \log_2(u^2) + \log_2 w = 2\log_2 u + \log_2 w = 2A + B$

   (b)  $\log_2(u^3/w^2) = \log_2(u^3) - \log_2(w^2) = 3\log_2 u - 2\log_2 w = 3A - 2B$

   (c)  $\log_2(1/\sqrt{w}) = \log_2 1 - \log_2 \sqrt{w} = 0 - \log_2 w^{\frac{1}{2}} = -\frac{1}{2}\log_2 w = -\frac{1}{2}B$

   (d)  $\log_2(\frac{2}{\sqrt{uw}}) = \log_2 2 - \log_2(uw)^{\frac{1}{2}} = 1 - \frac{1}{2}(\log_2(uw)) = 1 - \frac{1}{2}(\log_2 u + \log_2 w) = 1 - \frac{1}{2}(A + B)$

**Problem 4.**

(a) $3^2 10^{2\log 5} = 9(10^{\log 25}) = 9(25) = 225.$

(b) $5e^{-3\ln 2} = 5e^{\ln 2^{-3}} = 5(2^{-3}) = \frac{5}{8}.$

**Problem 5.**

(a) $10^{\log 2 + 1} = 10^{\log 2} 10^1 = 2(10) = 20.$

(b) $e^{3 - \ln 2} = e^3 e^{-\ln 2} = e^3 e^{\ln 2^{-1}} = e^3(2^{-1}) = \frac{e^3}{2}.$

**Problem 7.**

(a) $10^{\log 2 - \log 3} = 10^{\log 2} 10^{-\log 3} = 2(10^{\log 3^{-1}}) = 2(3^{-1}) = \frac{2}{3}.$

(b) $e^{2\ln 5 - \ln 2} = e^{2\ln 5} e^{-\ln 2} = e^{\ln 5^2} e^{\ln 2^{-1}} = (5^2)(2^{-1}) = \frac{25}{2}.$

**Problem 9.**

(a) $10^{\frac{\log 8 + 1}{2}} = (10^{\log 8 + 1})^{\frac{1}{2}} = \sqrt{10^{\log 8} 10^1} = \sqrt{8(10)} = \sqrt{80} = 4\sqrt{5}.$

(b) $e^{-\frac{\ln 8}{3} + 2} = e^{-\frac{\ln 8}{3}} e^2 = e^{-\frac{1}{3}\ln 8} e^2 = (e^{\ln 8})^{-\frac{1}{3}} e^2 = 8^{-\frac{1}{3}} e^2 = \frac{e^2}{\sqrt[3]{8}} = \frac{e^2}{2}.$

**Problem 11.**
$\frac{\log 12}{2} = \frac{1}{2}\log(2^2 3) = \frac{1}{2}(\log 2^2 + \log 3) = \frac{1}{2}(2\log 2 + \log 3) = \frac{1}{2}(2a + b) = a + \frac{b}{2}.$

**Problem 12.**
$5\log \sqrt[3]{6} = 5\log(6^{\frac{1}{3}}) = \frac{5}{3}\log 6 = \frac{5}{3}(\log 2 + \log 3) = \frac{5}{3}(a + b).$

**Problem 14.**

(a)  (i) $\lim\limits_{x \to \infty} \frac{\sqrt{x}}{\ln x} = \lim\limits_{x \to \infty} \frac{\frac{1}{2\sqrt{x}}}{\frac{1}{x}} = \lim\limits_{x \to \infty} \frac{x}{2\sqrt{x}} = \lim\limits_{x \to \infty} \frac{\sqrt{x}}{2} = \infty$

(ii) $\lim\limits_{x \to \infty} \frac{\ln x}{\sqrt{x}} = \lim\limits_{x \to \infty} \frac{\frac{1}{x}}{\frac{1}{2\sqrt{x}}} = \lim\limits_{x \to \infty} \frac{2}{\sqrt{x}} = 0$

(b) $\sqrt{x}$ grows faster than $\ln x$ as $x \to \infty$.

**Problem 16.**
$a\ln(x + 3) - b\ln(\frac{1}{x}) - c\ln(x + 1) = \ln(x + 3)^a + \ln(\frac{1}{x})^{-b} + \ln(x + 1)^{-c} = \ln(x + 3)^a(\frac{1}{x})^{-b}(x + 1)^{-c} = \ln\frac{(x+3)^a x^b}{(x+1)^c}.$

## Exploratory Problems for Chapter 13

**Problem 1.**
$\log C(t) = mt + b$, where $m$ and $b$ are constants. Now $\log C(0) = 3 \Rightarrow b = 3$, and consequently, $\log C(7) = 5 \Rightarrow m = \frac{2}{7}$. Hence $\log C(t) = \frac{2}{7}t + 3 \Rightarrow C(t) = 10^3 \cdot 10^{2t/7}.$

## Section 13.3    Using Logarithms and Exponentiation to Solve Equations

**Problem 2.**

(a) $\log(10^{2x}) = \log 93 \Rightarrow 2x = \log 93 \Rightarrow x = \frac{\log 93}{2} (\approx 0.984)$.

(b) $\log 10^{3x+2} = \log 1,000,000 \Rightarrow 3x + 2 = 6 \Rightarrow x = \frac{4}{3}$

(c) $x + 1 = \log_2 7 \Rightarrow x = \log_2(7) - 1 (\approx 1.807)$.

(d) $3^{x+x^2} = 3^1 \Rightarrow \log_3(3^{x+x^2}) = \log_3(3^1) \Rightarrow x + x^2 = 1 \Rightarrow x = \frac{-1 \pm \sqrt{5}}{2} (\approx 0.618, -1.618)$.

(e) $\ln(5B^x) = \ln((2C)^{x+1}) \Rightarrow x \ln(5B) = (x+1) \ln(2C) \Rightarrow x(\ln(5B) - \ln(2C)) = \ln(2C) \Rightarrow x = \frac{\ln(2C)}{\ln(5B) - \ln(2C)}$.

(f) $\ln x = 3 \Rightarrow e^{\ln x} = e^3 \Rightarrow x = e^3 (\approx 20.086)$.

(g) $10^{\log_{10} x} = 10^{17} \Rightarrow x = 10^{17}$

(h) $e^{\ln(5x-40)} = e^3 \Rightarrow 5x - 40 = e^3 \Rightarrow x = \frac{e^3+40}{5} (\approx 12.017)$.

(i) $10^{\log_{10}(2x^2+4)} = 10^2 \Rightarrow 2x^2 + 4 = 100 \Rightarrow x = \pm 4\sqrt{3} (\approx \pm 6.928)$.

(j) $2^{x/7} = \frac{16}{3} \Rightarrow \log_2(2^{x/7}) = \log_2(\frac{16}{3}) \Rightarrow \frac{x}{7} = \log_2(\frac{16}{3}) \Rightarrow x = 7 \log_2(\frac{16}{3}) (\approx 16.905)$.

**Problem 5.**

(a) $2^{x^2+x} = 3^x \Rightarrow \ln(2^{x^2+x}) = \ln(3^x) \Rightarrow (x^2 + x) \ln 2 = x \ln 3 \Rightarrow (\ln 2)x^2 + (\ln 2 - \ln 3)x = 0 \Rightarrow$
$x(x \ln x + \ln 2 - \ln 3) = 0 \Rightarrow x = 0$ or $x = \frac{\ln 3}{\ln 2} - 1$.

(b) $\log_3(3^{x^2+2x}) = \log_3 1 \Rightarrow x^2 + 2x = 0 \Rightarrow x(x+2) = 0 \Rightarrow x = 0$ or $x = -2$.

(c) $12 \ln x - 2(\ln 2 + \ln x) = 10 \Rightarrow 10 \ln x = 10 + 2 \ln 2 \Rightarrow \ln x = \frac{10+2\ln 2}{10} \Rightarrow x = e^{\frac{10+2\ln 2}{10}}$.

(d) $(e^x + 3)(e^x - 2) = 0 \Rightarrow e^x = -3$ (not possible) or $e^x = 2 \Rightarrow x = \ln 2$.

(e) $e^{2x} + 8 = 6e^x \Rightarrow e^{2x} - 6e^x + 8 = 0 \Rightarrow (e^x - 2)(e^x - 4) = 0 \Rightarrow e^x = 4$ or $e^x = 2 \Rightarrow x = \ln 4$ or $x = \ln 2$.

(f) $(\ln x)(\ln 5) = \ln 4 + \ln x \Rightarrow (\ln x)(\ln 5 - 1) = \ln 4 \Rightarrow \ln x = \frac{\ln 4}{\ln 5 - 1} \Rightarrow x = e^{\frac{\ln 4}{\ln 5 - 1}}$.

**Problem 6.**

(a) $M_{CR} - M_M = \log(\frac{I_{CR}}{I_M}) \Rightarrow 10^{M_{CR}-M_M} = 10^{6.7-5} = \frac{I_{CR}}{I_M} \Rightarrow I_{CR} = 10^{1.7} I_M$. The Costa Rican earthquake was about $10^{1.7} \approx 50.1$ times more powerful.

(b) As above, $10^{M_T-M_{SF}} = 10^{7.4-7.1} = \frac{I_T}{I_{SF}}$, from which we obtain $I_T = 10^{0.3} I_{SF}$. The earthquake in Turkey was about $10^{0.3} \approx 2.0$ times as intense.

**Problem 8.**
$$\ln\left(\frac{3}{2^{x-3}}\right) = \ln(7^{2x+1}) \Rightarrow \ln(3) - \ln(2^{x-3}) = (2x+1)\ln(7) \Rightarrow \ln(3) - (x-3)\ln(2) = (2x)\ln(7) + \ln(7)$$
$$\Rightarrow \ln(3) + 3\ln(2) - \ln(7) = x(2\ln(7) + \ln(2)) \Rightarrow x = \frac{\ln(3)+3\ln(2)-\ln(7)}{2\ln(7)+\ln(2)} (\approx 0.269).$$

**Problem 9.**

$\ln(\sqrt{\pi}3^{1+2x}) = \ln(5^x) \Rightarrow \ln(\sqrt{\pi}) + \ln(3^{1+2x}) = x\ln(5) \Rightarrow \ln(\sqrt{\pi}) + (1+2x)\ln(3) = x\ln(5) \Rightarrow \ln(\sqrt{\pi}) + \ln(3) = x(\ln(5) - 2\ln(3)) \Rightarrow x = \frac{\ln(\sqrt{\pi}) + \ln(3)}{\ln(5) - 2\ln(3)} (\approx -2.843).$

**Problem 11.**

$\ln(e^{2+x}) = \ln(\pi^{3x+3}) \Rightarrow 2 + x = (3x+3)\ln(\pi) \Rightarrow 2 - 3\ln(\pi) = 3x(\ln(\pi)) - x \Rightarrow 2 - 3\ln(\pi) = x(3\ln(\pi) - 1)$
$\Rightarrow x = \frac{2 - 3\ln(\pi)}{3\ln(\pi) - 1} (\approx -0.589).$

**Problem 13.**

$7 + \pi 3^{x+2} = 6\pi \Rightarrow 3^{x+2} = \frac{6\pi - 7}{\pi} \Rightarrow \ln(3^{x+2}) = \ln(\frac{6\pi - 7}{\pi}) \Rightarrow (x+2)\ln(3) = \ln(6\pi - 7) - \ln(\pi) \Rightarrow$
$x + 2 = \frac{\ln(6\pi - 7) - \ln(\pi)}{\ln(3)} \Rightarrow x = \frac{\ln(6\pi - 7) - \ln(\pi)}{\ln(3)} - 2 (\approx -0.792).$

**Problem 14.**

$\log x - \log(x+1) = \log(\frac{x}{x+1}) = 2 \Rightarrow \frac{x}{x+1} = 10^2 \Rightarrow x = 100(x+1) \Rightarrow x = -\frac{100}{99}$, which cannot be a solution of the original equation because $x < 0$. $\Rightarrow$ No solution.

**Problem 16.**

$\ln x^{1/2} + \ln x^2 = 1 - 2\ln x \Rightarrow \frac{1}{2}\ln x + 2\ln x = 1 - 2\ln x \Rightarrow \frac{9}{2}\ln x = 1 \Rightarrow \ln x = \frac{2}{9} \Rightarrow x = e^{\frac{2}{9}} (\approx 1.2494).$

**Problem 18.**

$[\ln(2x+3)]^2 = 9 \Rightarrow \ln(2x+3) = 3$ or $\ln(2x+3) = -3 \Rightarrow 2x+3 = e^3$ or $2x+3 = e^{-3} \Rightarrow x = \frac{e^3-3}{2} (\approx 8.543)$ or $x = \frac{e^{-3}-3}{2} (\approx -1.475).$

**Problem 20.**

$e^x(e^x - 5) = 6 \Rightarrow (e^x)^2 - 5e^x - 6 = 0$
$(e^x - 6)(e^x + 1) = 0$
$e^x = 6$ or $e^x = -1$ (not possible) $\Rightarrow x = \ln 6 (\approx 1.792).$

**Problem 21.**

$e^{2x} - 4e^x + 3 = 0 \Rightarrow (e^x - 3)(e^x - 1) = 0 \Rightarrow e^x = 3$ or $e^x = 1 \Rightarrow x = \ln 3 (\approx 1.099)$ or $x = \ln(1) = 0.$

**Problem 23.**

$e^{-2x} - e^{-x} = 6 \Rightarrow (e^{-x})^2 - e^{-x} - 6 = 0 \Rightarrow (e^{-x} - 3)(e^{-x} + 2) = 0 \Rightarrow e^{-x} = 3$ or $e^{-x} = -2$ (not possible) $\Rightarrow x = -\ln(3) = \ln\left(\frac{1}{3}\right) (\approx -1.099).$

**Problem 25.**

$e^x - 1 = e^{-x} \Rightarrow e^x(e^x - 1) = e^x(e^{-x}) \Rightarrow (e^x)^2 - (e^x) - 1 = 0 \Rightarrow e^x = \frac{1 \pm \sqrt{5}}{2}$ (only positive solution is possible) $\Rightarrow e^x = \frac{1 + \sqrt{5}}{2} \Rightarrow x = \ln(\frac{1 + \sqrt{5}}{2})(\approx 0.481).$

**Problem 27.**

$3^{\ln x} = 5x \Rightarrow \ln(3^{\ln x}) = \ln 5x \Rightarrow (\ln x)(\ln 3) = \ln 5 + \ln x \Rightarrow (\ln x)((\ln 3) - 1) = \ln 5 \Rightarrow \ln x = \frac{\ln 5}{(\ln 3) - 1} \Rightarrow$
$x = e^{\frac{\ln 5}{(\ln 3) - 1}} (\approx 1.225 \times 10^7).$

**Problem 28.**

$\frac{4}{\ln(x+1)} + 5 = 13 \Rightarrow \frac{4}{\ln(x+1)} = 8 \Rightarrow 4 = 8\ln(x+1) \Rightarrow \frac{1}{2} = \ln(x+1) \Rightarrow x+1 = e^{\frac{1}{2}} \Rightarrow x = \sqrt{e} - 1 (\approx 0.649).$

**Problem 30.**

$\frac{3}{(e^x+1)^2} = 27 \Rightarrow 3 = 27(e^x + 1)^2 \Rightarrow \frac{1}{9} = (e^x + 1)^2 \Rightarrow e^x + 1 = \pm\frac{1}{3} \Rightarrow e^x = -\frac{4}{3}$ (not possible) or $e^x = -\frac{2}{3}$
(not possible).

**Problem 32.**

$\ln(x-3) - \ln(2x+1) = 1 \Rightarrow \ln(\frac{x-3}{2x+1}) = 1 \Rightarrow \frac{x-3}{2x+1} = e \Rightarrow x - 3 = e(2x+1) \Rightarrow x(1-2e) = e+3 \Rightarrow$
$x = \frac{e+3}{1-2e} \approx -1.289$. As values of $x$ must be greater than 3 in the original equation, there is no solution.

**Problem 34.**

(a) $2Q^{x+5} = R \Rightarrow \ln(2Q^{x+5}) = \ln(R) \Rightarrow \ln(2) + (x+5)\ln(Q) = \ln(R) \Rightarrow (x+5)\ln(Q) = \ln(R) - \ln(2) = \ln(\frac{R}{2})$
$\Rightarrow x = \frac{\ln(\frac{R}{2})}{\ln(Q)} - 5.$

(b) $(2Q)^{x+5} = R \Rightarrow \ln((2Q)^{x+5}) = \ln(R) \Rightarrow (x+5)\ln(2Q) = \ln(R) \Rightarrow x + 5 = \frac{\ln(R)}{\ln(2Q)} \Rightarrow x = \frac{\ln(R)}{\ln(2Q)} - 5.$

**Problem 35.**

(a) $(Q+R)^x = S \Rightarrow \ln((Q+R)^x) = \ln(S) \Rightarrow x\ln(Q+R) = \ln(S) \Rightarrow x = \frac{\ln(S)}{\ln(Q+R)}.$

(b) $(QR)^x = S \Rightarrow \ln((QR)^x) = \ln(S) \Rightarrow x\ln(QR) = \ln(S) \Rightarrow x = \frac{\ln(S)}{\ln(Q)+\ln(R)}.$

**Problem 37.**

(a) If $[H^+]$ increases tenfold, pH decreases by 1.
$\text{pH}_{new} = -\log(10[H^+]_{old}) = -(\log(10) + \log([H^+]_{old})) = -1 - \log([H^+]_{old}) = \text{pH}_{old} - 1.$

(b) $\text{pH} = -\log(3.15 \times 10^{-8}) = -(\log(3.15) + \log(10^{-8})) = -(\log(3.15) - 8) \approx 7.5.$

(c) Something neutral has a higher $[H^+]$.
$\text{pH} = -\log([H^+])$, so $[H^+] = 10^{-\text{pH}}.$
$\frac{[H^+]_{neutral}}{[H^+]_{blood}} = \frac{10^{-7}}{10^{-7.5}} = 10^{0.5} \approx 3.162$ times greater.

**Problem 39.**
$e^{x^3} = (e^x)^3 \Rightarrow \ln(e^{x^3}) = \ln((e^x)^3) \Rightarrow x^3 = 3\ln(e^x) \Rightarrow x^3 = 3x \Rightarrow x(x^2 - 3) = 0 \Rightarrow x = 0$ or $x = \pm\sqrt{3}.$

**Problem 41.**
$\frac{\ln x}{\ln 2} = \ln x - \ln 2 \Rightarrow \ln x = (\ln 2)(\ln x - \ln 2) \Rightarrow (\ln x)(1 - \ln 2) = -(\ln 2)^2 \Rightarrow \ln x = \frac{-(\ln 2)^2}{1 - \ln 2} \Rightarrow x =$
$e^{\frac{(\ln 2)^2}{(\ln 2)-1}} (\approx 0.209).$

**Problem 43.**
$10^{2x} = 10^2 10^x \Rightarrow \log(10^{2x}) = \log(10^2 10^x) \Rightarrow 2x = \log(10^2) + \log(10^x) = 2 + x \Rightarrow x = 2.$

**Problem 44.**
$\ln x = (\frac{1}{2}\ln x)^2 \Rightarrow \ln x = \frac{(\ln x)^2}{4} \Rightarrow 4\ln x = (\ln x)^2 \Rightarrow (\ln x)^2 - 4\ln x = (\ln x - 4)(\ln x) = 0 \Rightarrow \ln x = 4$ or
$\ln x = 0 \Rightarrow x = e^4 (\approx 54.6)$ or $x = 1.$

**Problem 45.**

(a) Solve $M(t) = M_0 + \frac{1}{2}M_0 = \frac{3}{2}M_0$. Now $\frac{3}{2}M_0 = M_0(1 + \frac{0.05}{12})^{12t} \Rightarrow \frac{3}{2} = (1 + \frac{0.05}{12})^{12t} \Rightarrow \ln(\frac{3}{2}) =$
$\ln((1 + \frac{0.05}{12})^{12t}) \Rightarrow \ln(\frac{3}{2}) = (12t)\ln(1 + \frac{0.05}{12}) \Rightarrow 12t = \frac{\ln(\frac{3}{2})}{\ln(1+\frac{0.05}{12})} \Rightarrow t \approx 8.13$. It will take approximately
8.13 years for the amount of money in the account to increase by 50%.

(b) Solve $M(8) = 2M_0$ for $r$. Now $M_0(1 + \frac{r}{12})^{12(8)} = 2M_0 \Rightarrow (1 + \frac{r}{12})^{96} = 2 \Rightarrow \ln((1 + \frac{r}{12})^{96}) = \ln(2) \Rightarrow$
$96\ln(1 + \frac{r}{12}) = \ln(2) \Rightarrow \ln(1 + \frac{r}{12}) = \frac{\ln(2)}{96} \Rightarrow 1 + \frac{r}{12} = e^{\frac{\ln(2)}{96}} \Rightarrow r = 12(e^{\frac{\ln(2)}{96}} - 1)(\approx 0.087) = 8.7\%.$

**Problem 47.**
Slope: $\frac{\ln(2+\epsilon)-\ln 2}{(2+\epsilon)-2} = \frac{\ln(2+\epsilon)-\ln 2}{\epsilon} \Rightarrow y - \ln 2 = \frac{\ln(2+\epsilon)-\ln 2}{\epsilon}(x-2) \Leftrightarrow y = \frac{\ln(2+\epsilon)-\ln 2}{\epsilon}x - 2\frac{\ln(2+\epsilon)-\ln 2}{\epsilon} + \ln 2.$

## Section 13.4    Graphs of Logarithmic Functions: Theme and Variations

**Problem 1.**                                                    **Problem 3.**

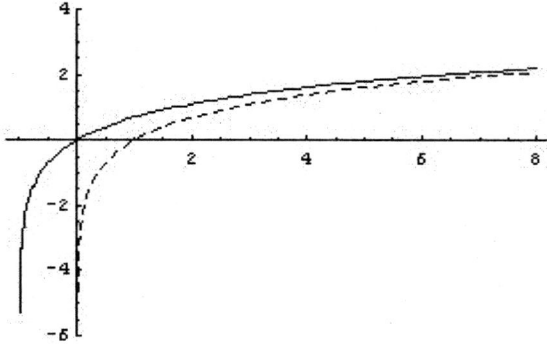    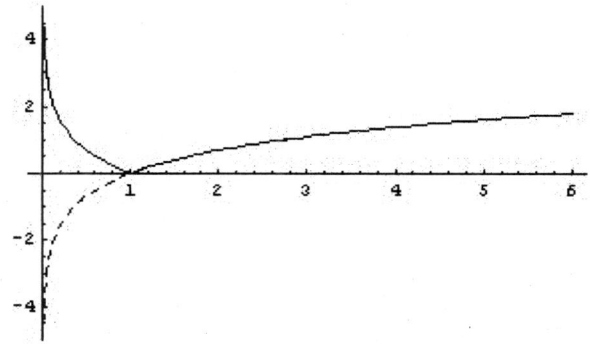

**Problem 4.**                                                    **Problem 6.**

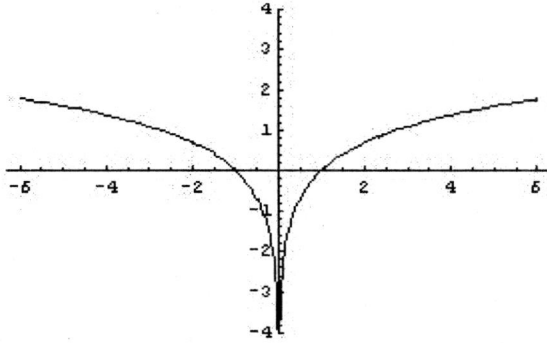    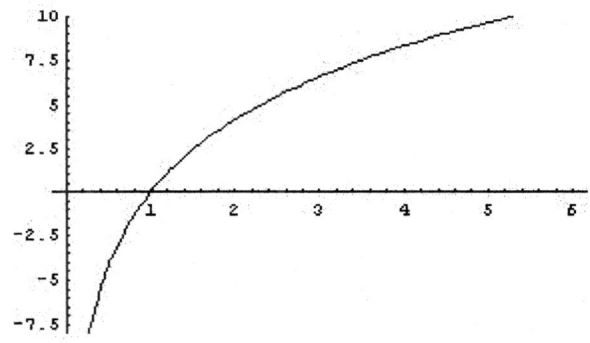

$$y = \ln(x) - \ln(x^3) + 4\ln(x^2) =$$
$$\ln(x) - 3\ln(x) + 8\ln(x) = 6\ln(x)$$

# CHAPTER 14

# Differentiating Logarithmic and Exponential Functions

## Exploratory Problems for Chapter 14

**Problem 1.**

- 

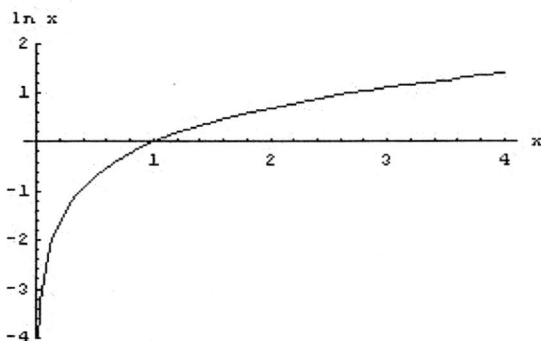

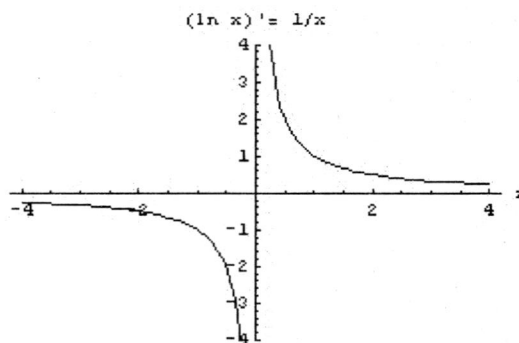

- In the table below, $f'(x)$ is approximated by $\frac{f(x+h)-f(x)}{h}$, where $h = 0.000001$ and values are rounded to the nearest ten thousandth.

| $x$ | $f'(x)$ (approximated) |
|-----|------------------------|
| 0.1 | 10.0000 |
| 0.5 | 2.0000 |
| 1 | 1.0000 |
| 2 | 0.5000 |
| 3 | 0.3333 |
| 4 | 0.2500 |
| 5 | 0.2000 |
| 6 | 0.1667 |

Conjecture: The derivative of $\ln x$ is $\frac{1}{x}$.

## Section 14.1     The Derivative of Logarithmic Functions

**Problem 2.**
$$y = \pi \ln \sqrt{x} \Rightarrow y = \pi(\tfrac{1}{2}) \ln(x) \Rightarrow y' = (\tfrac{\pi}{2})(\tfrac{1}{x}) = \tfrac{\pi}{2x}.$$

**Problem 4.**
$$y = x \ln x \Rightarrow y' = (1)(\ln(x)) + (x)(\tfrac{1}{x}) = \ln(x) + 1.$$

**Problem 5.**
$$y = \frac{\ln \sqrt{2x}}{x} \Rightarrow y = \frac{\tfrac{1}{2}\ln(2x)}{x} = \frac{\ln(2)+\ln(x)}{2x} \Rightarrow y' = \frac{(\tfrac{1}{x})(2x)-(2)(\ln(2)+\ln(x))}{(2x)^2} = \frac{1-(\ln(2)+\ln(x))}{2x^2}.$$

**Problem 7.**

$$y = \frac{\log_2 x}{3} \Rightarrow y = \frac{\ln x}{3\ln 2} \Rightarrow y' = \frac{1}{3\ln 2}\left(\frac{1}{x}\right) = \frac{1}{3x\ln 2}.$$

**Problem 9.**

$g(x) = \pi \log_2(\pi x) - \pi^2 = \pi \log_2(x) + \pi \log_2(\pi) - \pi^2 \Rightarrow g'(x) = \pi\left(\frac{1}{\ln(2)}\frac{1}{x}\right) = \frac{\pi}{x\ln(2)}$. We see that $g'(x) > 0$ on $(0, \infty)$, which is the domain of $g$. Thus $g$ is increasing and hence invertible.

To solve for $g^{-1}(x)$, we have : $x = \pi \log_2(\pi g^{-1}(x)) - \pi^2 \Rightarrow \frac{x+\pi^2}{\pi} = \log_2(\pi g^{-1}(x)) \Rightarrow 2^{\frac{x+\pi^2}{\pi}} = \pi g^{-1}(x)$
$\Rightarrow g^{-1}(x) = \frac{1}{\pi}2^{\frac{x+\pi^2}{\pi}}.$

**Problem 11.**

$g'(x) = 2x \ln(x) + x^2\frac{1}{x} = 2x \ln(x) + x = x(2\ln(x) + 1)$. Now $g'(x) = 0$ when $x = 0$, which is not in the domain, or when $\ln(x) = -\frac{1}{2} \Rightarrow x = \frac{1}{\sqrt{e}}$. Hence the only critical point is $x = \frac{1}{\sqrt{e}}$. Moreover, $g'(x) < 0$ for $0 < x < \frac{1}{\sqrt{e}}$ and $g'(x) > 0$ for $x > \frac{1}{\sqrt{e}}$, and hence the absolute minimum occurs at $x = \frac{1}{\sqrt{e}}$. The absolute minimum value is $g(\frac{1}{\sqrt{e}}) = -\frac{1}{2e}$. There is no absolute maximum because $\lim_{x\to\infty} g(x) = \infty$.

**Problem 13.**

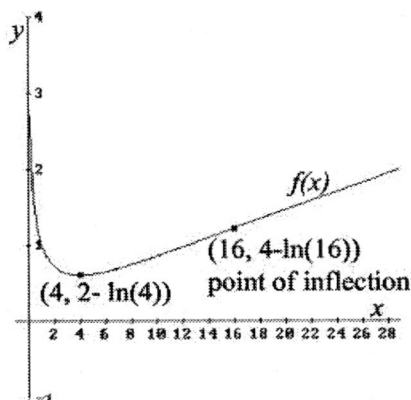

(16, 4-ln(16))
point of inflection

(4, 2- ln(4))

$f(x)$

(a) $f'(x) = \frac{1}{2\sqrt{x}} - \frac{1}{x}$. Now $f'(x) = 0$ when $x = 4$. Hence $x = 4$ is the only critical point. Note that $f'(x) < 0$ for $0 < x < 4$ and $f'(x) > 0$ for $x > 4$. Hence there is a local minimum point at $x = 4$. Critical point at $x = 4$

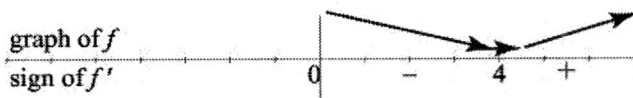

(b) $f''(x) = -\frac{1}{4x^{3/2}} + \frac{1}{x^2}$. Now $f''(x) = 0$ when $x = 16$, $f''(x) > 0$ for $0 < x < 16$, and $f''(x) < 0$ for $x > 16$. Hence $(16, f(16)) = (16, 4 - \ln 16)$ is the only point of inflection.

(c) $\lim_{x\to 0^+} (\sqrt{x} - \ln x) = (\lim_{x\to 0^+} \sqrt{x}) - (\lim_{x\to 0^+} \ln x) = 0 - (-\infty) = \infty$. $\lim_{x\to\infty} (\sqrt{x} - \ln x) = \lim_{x\to\infty} (\ln(e^{\sqrt{x}}) - \ln x) =$
$\lim_{x\to\infty} \ln(\frac{e^{\sqrt{x}}}{x}) = \ln(\lim_{x\to\infty} \frac{e^{\sqrt{x}}}{x}) = \ln(\infty) = \infty.$

## Section 14.2    The Derivative of $b^x$ Revisited

**Problem 1.**

$$y = x^2 \cdot 2^x \Rightarrow \frac{dy}{dx} = (2x)(2^x) + (\ln(2)2^x)(x^2) = x(2^x)(2 + \ln(2)x).$$

**Problem 3.**

$$y = \tfrac{x^5 5^x}{5} \Rightarrow \frac{dy}{dx} = \frac{1}{5}((5x^4)(5^x) + (\ln(5)5^x)(x^5)) = \frac{1}{5}(x^4)(5^x)(5 + \ln(5)x).$$

**Problem 5.**

(a) $f(x) = x^2 + e^x + x^e + e^2 \Rightarrow f'(x) = 2x + e^x + ex^{e-1}$.

(b) $f(x) = (\pi - \frac{6}{\sqrt{29}})e^x \Rightarrow f'(x) = (\pi - \frac{6}{\sqrt{29}})e^x$

(c) $f(x) = (3e^3)e^x \Rightarrow f'(x) = (3e^3)e^x = 3e^{x+3}$

**Problem 7.**

The slope of the tangent to $y = (2.7)^x$ at $x = 0$ is approximately $\dfrac{(2.7)^{0.000001} - (2.7)^0}{0.000001} \approx 0.993$, hence $\dfrac{d}{dx}(2.7)^x \approx (0.993)(2.7)^x$. The slope of the tangent to $y = (2.8)^x$ at $x = 0$ is approximately $\dfrac{(2.8)^{0.000001} - (2.8)^0}{0.000001} \approx 1.03$; hence $\dfrac{d}{dx}(2.8)^x \approx (1.03)(2.8)^x$.

**Problem 8.**

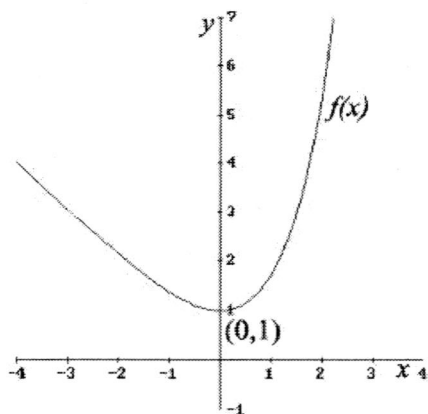

(a) $f'(x) = e^x - 1$. $f'(x) = 0$ when $x = 1$. $f'(x) < 0$ for $x < 0$ and $f'(x) > 0$ for $x > 0$.

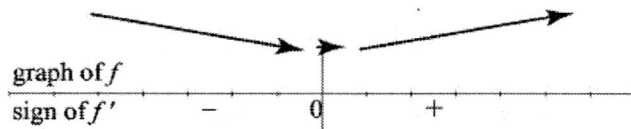

(b) By the first derivative test and part (a), $x = 0$ is the only local minimum point. The corresponding local minimum value is $f(0) = e^0 - 0 = 1$. See the labeled graph above.

(c)

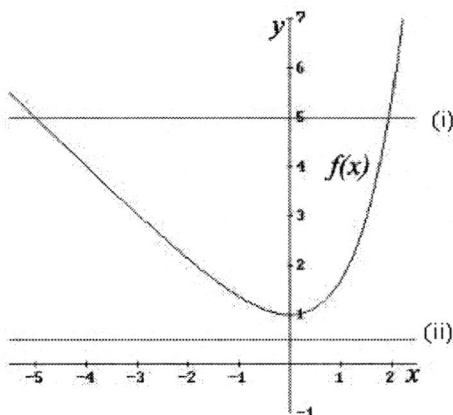

(i)  2 solutions

(ii)  No solution

**Problem 10.**

$y = e^{3x} \ln\left(\dfrac{1}{\sqrt{5x}}\right) \Rightarrow y = (e^3)^x \ln((5x)^{-1/2}) = -\dfrac{1}{2}((e^3)^x)(\ln(5) + \ln(x)) \Rightarrow$

$\dfrac{dy}{dx} = -\dfrac{1}{2}((\ln e^3)e^{3x})(\ln(5)+\ln(x)) + (\dfrac{1}{x})(e^{3x})) = -\dfrac{1}{2}((3e^{3x})(\ln(5)+\ln(x)) + (\dfrac{1}{x})(e^{3x})) = -\dfrac{1}{2}e^{3x}(3(\ln(5) + \ln(x)) + \dfrac{1}{x}).$

## Section 14.3   Worked Examples Involving Differentiation

**Problem 1.**

$f(x) = \dfrac{\frac{3}{2}\ln(x)}{x} \Rightarrow f'(x) = \dfrac{3}{2}\dfrac{(\frac{1}{x})x - (1)\ln(x)}{x^2} = \dfrac{3(1 - \ln(x))}{2x^2} \Rightarrow. \ f'(e) = \dfrac{3(1-\ln(e))}{2e^2} = 0$

**Problem 3.**

$y = \dfrac{\ln x^2}{5e^{3x}} = \dfrac{2}{5}\dfrac{\ln(x)}{(e^{3x})} \Rightarrow y' = \dfrac{2}{5}\dfrac{(\frac{1}{x})(e^{3x}) - (3e^{3x})(\ln x)}{(e^{3x})^2} = \dfrac{2(\frac{1}{x} - 3\ln x)}{5e^{3x}}.$

**Problem 5.**

$y = x\ln\left(\dfrac{1}{x}\right) = x\ln(x^{-1}) = -x\ln x \Rightarrow y' = -\left((1)\ln x + \left(\dfrac{1}{x}\right)(x)\right) = -\ln x - 1.$

**Problem 6.**

$f(x) = x\ln\left(\dfrac{1}{x}\right) = x\ln(x^{-1}) = -x\ln x \Rightarrow f'(x) = -\left((1)\ln x + \left(\dfrac{1}{x}\right)(x)\right) = -\ln x - 1.$

**Problem 8.**

$f(x) = e^{5x}\ln\left(\dfrac{\pi}{\sqrt{x}}\right) = e^{5x}(\ln\pi - \ln\sqrt{x}) = e^{5x}\left(\ln\pi - \dfrac{1}{2}\ln x\right) = (e^{5x})\left(\ln\pi - \dfrac{1}{2}\ln x\right)$

$\Rightarrow f'(x) = (5e^{5x})\left(\ln\pi - \dfrac{1}{2}\ln(x)\right) + \left(-\dfrac{1}{2x}\right)(e^{5x}) = e^{5x}\left(5\left(\ln\pi - \dfrac{\ln x}{2}\right) - \dfrac{1}{2x}\right).$

**Problem 10.**

$f(x) = \dfrac{\ln(2x^3)}{3e^x} = \dfrac{\ln 2 + 3\ln x}{3e^x} \Rightarrow f'(x) = \dfrac{\left(\frac{3}{x}\right)(3e^x) - (\ln 2 + 3\ln x)(3e^x)}{9e^{2x}} = \dfrac{3 - x\ln 2 - 3x\ln x}{3xe^x}.$

**Problem 11.**

$f(x) = \frac{x+\ln\left(\frac{1}{x}\right)}{x^2} = \frac{x+\ln(x^{-1})}{x^2} = \frac{x-\ln x}{x^2} \Rightarrow f'(x) = \frac{\left(1-\frac{1}{x}\right)(x^2) - (2x)(x-\ln(x))}{x^4} = \frac{x^2 - x - 2x^2 + 2x\ln(x)}{x^4} = \frac{2\ln(x) - x - 1}{x^3}.$

**Problem 13.**

$f(x) = x^2 \left(\ln x + \ln\left(\sqrt{\frac{61}{2x}}\right)\right) = x^2 \left(\ln x + \left(\frac{1}{2}\right)\left(\ln\left(\frac{61}{2x}\right)\right)\right) = x^2 \left(\ln x + \left(\frac{1}{2}\right)(\ln 61 - \ln(2x))\right) = x^2 \left(\ln x + \left(\frac{1}{2}\right)(\ln 61 - \ln 2 - \ln x)\right) = \left(\frac{1}{2}\right)x^2(\ln 61 - \ln 2 + \ln x) \Rightarrow$
$f'(x) = \left(\frac{1}{2}\right)\left[(2x)(\ln 61 - \ln 2 + \ln x) + x^2\left(\frac{1}{x}\right)\right] = x(\ln 61 - \ln 2 + \ln x) + \frac{x}{2}.$

**Problem 14.**

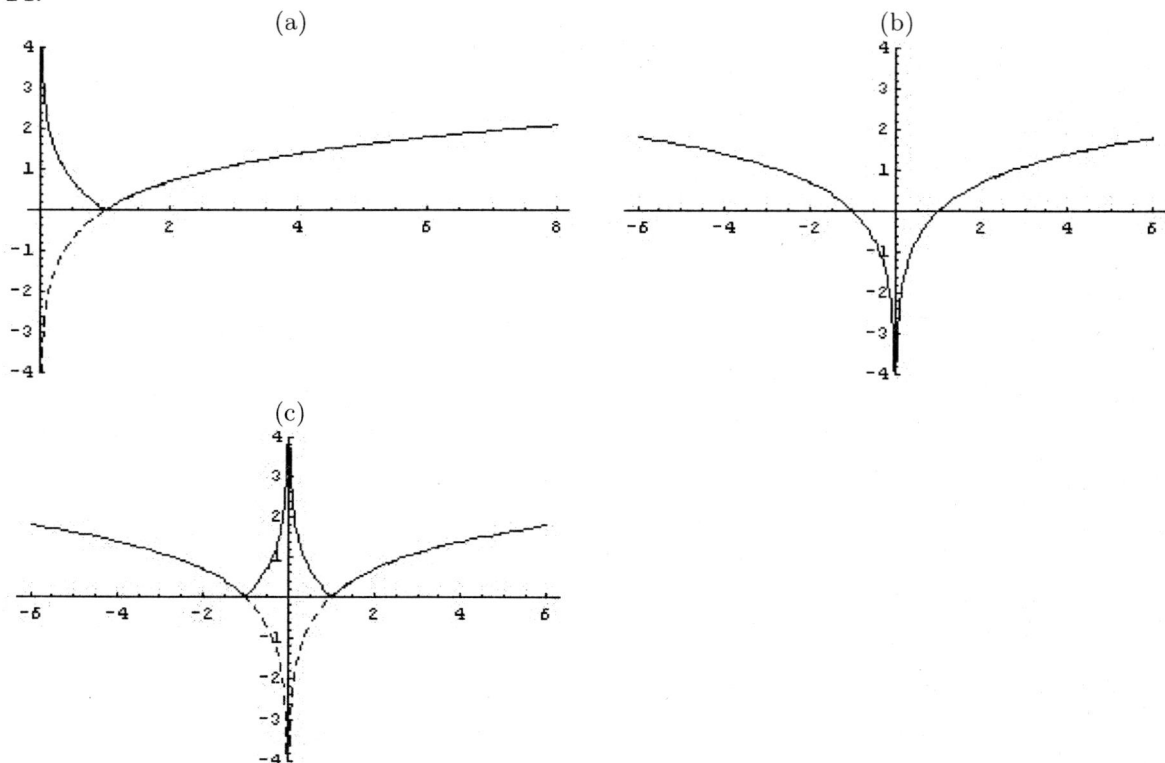

(a)

(b)

(c)

(d)  (a) $y'(x) = \frac{|\ln x|}{x\ln x} \Rightarrow x = 1$ is the only critical point. (Note that $y'(x)$ is never zero and that $y'(1)$ is undefined.) For $0 < x < 1$, $\ln x < 0$, and thus $y' < 0$, for $0 < x < 1$. Similarly, for $x > 1$, $\ln x > 0$, and hence $y' > 0$ for $x > 1$. The first derivative test implies that there is a local minimum at at $x = 1$. The local minimum value at $x = 1$ is $y(1) = |\ln 1| = 0$.

(b) $y' = \left(\frac{1}{|x|}\right)\left(\frac{|x|}{x}\right) = \frac{1}{x}$. Now $y' < 0$ for $x < 0$ and $y' > 0$ for $x > 0$, and hence the function has no critical points. (Note that $x = 0$ is not a critical point because it is not in the domain of the function.)

(c) $y' = \left(\frac{|\ln|x||}{\ln|x|}\right)(\ln|x|)' = \frac{|\ln|x||}{x\ln|x|}$. Now $\ln|x| = 0$ for $x = 1$ and $x = -1$, and thus $y'$ is not defined for $x = 1$ and $x = -1$. These two values of $x$ are the critical points. (As in (b), note that $x = 0$ is not a critical point because it is not in the domain of the function.) As $\ln|x| > 0$ for $|x| > 1$ and $\ln|x| < 0$ for $0 < |x| < 1$, both $x = 1$ and $x = -1$ are local minimum points. The corresponding minimum values are $y(-1) = 0$ and $y(1) = 0$.

**Problem 16.**

(a) $y = 3\ln 5x + 6\ln\left(\frac{3}{x}\right) = 3\ln 5 + 3\ln x + 6\ln 3 - 6\ln x = 3\ln 5 + 6\ln 3 - 3\ln x \Rightarrow y' = -\frac{3}{x}$.

(b) $y = 20\log\left(\frac{x}{100}\right) = 20\log x - 20\log(100) \Rightarrow y' = \frac{20}{x\ln 10}$

(c) $y = \frac{ke^{2kx}}{\sqrt{k+1}} = \frac{k}{\sqrt{k+1}}e^{2kx} \Rightarrow y' = \frac{k}{\sqrt{k+1}}(2k)e^{2kx} = \frac{2k^2}{\sqrt{k+1}}e^{2kx}$.

(d) $y = \frac{(\ln 2)e^{5x}}{\ln 4} + \frac{(\ln 2)e^{\ln 2}}{\ln 3} = \frac{(\ln 2)e^{5x}}{\ln 4} + \frac{2\ln 2}{\ln 3} \Rightarrow y' = \frac{(5\ln 2)e^{5x}}{\ln 4}$.

**Problem 18.**

(a) $\frac{9.8 \text{ billion}}{5.7 \text{ billion}} \approx 1.719$. The world's population will increase by 71.9% .

(b) Let $P(t)$ be the total population of the world in billions $t$ years since 1995. As $P(t)$ is exponential, $P(t)$ has the form $P(t) = P_0e^{kt}$, where $P_0$ and $k$ are constants. Now $P(0) = 5.7 = P_0$, and thus $P(t) = 5.7e^{kt}$. As $P(55) = 9.8$, we have that $P(55) = 9.8 = 5.7e^{55k} \Rightarrow e^{55k} = \frac{9.8}{5.7} \Rightarrow 55k = \ln\left(\frac{9.8}{5.7}\right) \Rightarrow k = \frac{1}{55}\ln\left(\frac{9.8}{5.7}\right) \approx 0.00985$. Therefore, $P(t) = 5.7e^{0.00985t}$. Now $\frac{P(t+1)}{P(t)} = \frac{e^{0.00985(t+1)}}{e^{0.00985t}} = e^{0.00985} \approx 1.0099$. Therefore the annual percentage growth rate is about 0.99%.

(c) $P'(t) = (0.00985)5.7e^{0.00985t} = 0.056145e^{0.00985t} \Rightarrow P'(0) = 0.056145e^{0.00985(0)} = 0.056145$ and $P'(55) = 0.056145e^{0.00985(55)} \approx 0.096513957$. At $t = 0$, the population is growing at a rate of 56.145 million people per year, and at $t = 55$, the population is growing at 96.513957 million people per year.

(d) The average rate of growth is $\frac{P(55)-P(0)}{55-0} = \frac{9.8-5.7}{55} \approx 0.074545454$ billion people per year $= 74,545,455$ people per year.

(e) Solve $\frac{P(55)-P(0)}{55-0} = \frac{4.1}{55} = 0.056145e^{0.00985t}$. Now $e^{0.00985t} = \frac{4.1}{(55)(0.056145)} \Rightarrow 0.00985t = \ln\left(\frac{4.1}{(55)(0.056145)}\right) \Rightarrow t = \left(\frac{1}{0.00985}\right)\ln\left(\frac{4.1}{(55)(0.056145)}\right) \approx 28.8$ years. In 2024, the instantaneous rate of growth and the average rate of growth over the 55-year period are equal.

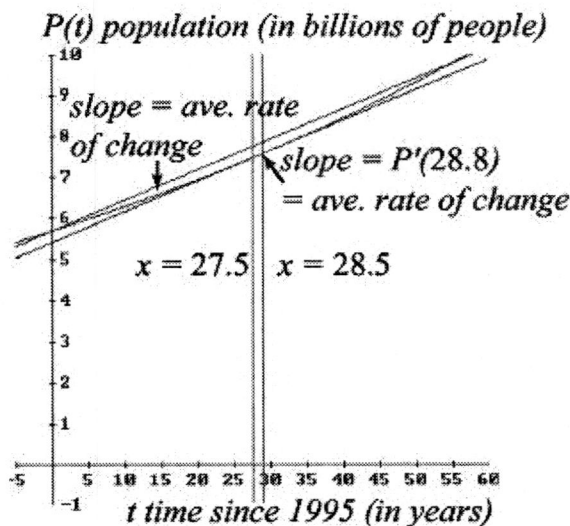

(f) For North America, $N(t) = N_0(1 + (0.30)\frac{t}{55}) = \frac{0.3N_0t}{55} + N_0$. For Africa, $A(t) = A_0(1 + (3.00)\frac{t}{55}) = \frac{3A_0t}{55} + A_0$.

**Problem 20.**

(a) $R(5000) = R_0 e^{k(5000)} = \frac{1}{2} R_0 \Rightarrow e^{5000k} = \frac{1}{2} \Rightarrow k = \frac{1}{5000} \ln\left(\frac{1}{2}\right) \approx -1.39 \times 10^{-4}$. Therefore, $R(t) = R_0 e^{(-1.39 \times 10^{-4})t}$.

(b) Now $R(t) = 3000 e^{(-1.39 \times 10^{-4})t}$, and hence $R'(t) = (-1.39 \times 10^{-4})3000 e^{(-1.39 \times 10^{-4})t}$. Thus, $R'(0) = (-1.39 \times 10^{-4})3000 = -0.416$. The radioactive substance is decaying at a rate of $-0.416$ mg per year at $t = 0$.

**Problem 22.**

(a) $M(t) = 500(e^{0.08})^t = 500(1.083)^t$.

(b) $\frac{M(t+1)}{M(t)} = \frac{500(1.083)^{t+1}}{500(1.083)^t} = 1.083$
Annual growth is 8.3%.

(c) $\frac{dM}{dt} = 500(0.08)e^{0.08t} = 40e^{0.08t}$.

(d) We need to find the value of $t$ for which $M(t) = 1599$. This equation implies $500e^{0.08t} = 1599 \Rightarrow e^{0.08t} = \frac{1599}{500} \Rightarrow t = 12.5 \ln\left(\frac{1599}{500}\right) \approx 14.5$. It will take about 14.5 years to have enough money for the ticket.

**Problem 24.**

(a) $P(0) = 200 = P_0 e^{k(0)} = P_0$. $P(30) = 900 = 200e^{k(30)} \Rightarrow e^{30k} = 4.5 \Rightarrow k = \frac{1}{30} \ln 4.5 \approx 0.05$.

(b) As in part (a), $M_0 = 100$. Since the population doubles every 10 days, we have $M(t) = (100)(2^{t/10})$.

(c) The populations will be equal when $M(t) = P(t)$. From this equation, we have $200e^{0.05t} = 100(2^{t/10}) \Rightarrow e^{0.05t} = 2^{(t-10)/10} \Rightarrow 0.05t = \frac{(t-10)\ln 2}{10} \Rightarrow t = 2(t-10)\ln 2 \Rightarrow t = \frac{-20\ln 2}{1-2\ln 2} \approx 35.9$. The populations will be equal after 35.9 days.

(d) $P'(t) = 200(0.05)e^{0.05t} = 10e^{0.05t}$.

(e) $M'(t) = 10(\ln 2)(2^{t/10})$.

(f) $P'(35.9) = 10e^{0.05(35.9)} \approx 60.2$ flies/day. $M'(35.9) = 10(\ln 2)(2^{35.9/10}) \approx 83.5$ mosquitoes/day. The mosquito population is growing more rapidly.

**Problem 26.**
$f(x) = \frac{2\ln x}{x} \Rightarrow f'(x) = \frac{(\frac{2}{x})(x) - (1)(2\ln x)}{x^2} = \frac{2 - 2\ln x}{x^2}$. Now $f'(x)$ is defined for all values of $x$ in its domain, and $f'(x) = 0$ when $x = e$. Thus $x = e$ is the only critical point. Moreover, $f'(x) > 0$ for $0 < x < e$ and $f'(x) < 0$ for $x > e$. Hence, the global maximum of $f$ occurs at $x = e$, and its value is $f(e) = \frac{2\ln e}{e} = \frac{2}{e}$. Now $\lim_{x \to 0^+} f'(x) = \lim_{x \to 0^+} \frac{2 - 2\ln x}{x^2} = \left(\lim_{x \to 0^+}(2 - 2\ln x)\right)\left(\lim_{x \to 0^+}\left(\frac{1}{x^2}\right)\right) = \infty \cdot \infty = \infty$. As $f'(x)$ increases without bound as $x$ approaches 0 from the right, $f(x)$ decreases without bound as $x$ approaches 0 from the right. Therefore, there is no global minimum.

**Problem 28.**
$h(x) = x^3 \cdot 3^x \Rightarrow h'(x) = (3x^2)(3^x) + (\ln(3)3^x)(x^3) = x^2 3^x(3 + \ln(3)x)$. $h'(x) = 0$ at $x = 0$ and $x = -\frac{3}{\ln(3)} \approx -2.731$; these two values of $x$ are the critical points. Note that $h'(x) < 0$ on $\left(-\infty, -\frac{3}{\ln 3}\right)$ and $h'(x) > 0$ on $\left(-\frac{3}{\ln 3}, 0\right) \cup (0, \infty)$. Hence $x = -\frac{3}{\ln 3}$ is a local minimum point and $x = 0$ is not a local extreme point. Moreover, the information about $h'(x)$ implies that the absolute minimum value of $h$

occurs at $x = -\frac{3}{ln3}$; this minimum value is $h\left(-\frac{3}{ln3}\right) \approx -1.014$. There is no absolute maximum because $h(x)$ grows without bound as $x$ increases without bound.

# CHAPTER 15
## Take It to the Limit

### Section 15.1    An Interesting Limit

**Problem 2.**

Inflation is compounded continuously. To determine the doubling time $t$ under the inflation rate $r$, We solve $2M_0 = M_0 e^{rt}$ for $t$. This equation implies $2 = e^{rt}$. Solving for $t$ we have $\ln 2 = rt$ or $t = \frac{\ln 2}{r}$.

For 22%, $r = 0.22$: $t = \frac{\ln(2)}{0.22} \approx 3.15$ years.

For 131%, $r = 1.31$: $t = \frac{\ln(2)}{1.31} \approx 0.53$ years.

For 2600%, $r = 26.00$: $t = \frac{\ln(2)}{26} \approx 0.027$ years.

For 3%, $r = 0.03$: $t = \frac{\ln(2)}{0.03} \approx 23.1$ years.

**Problem 4.**

(a) $\lim\limits_{x \to 3} \dfrac{3}{x-3}$ does not exist.

(b) $\lim\limits_{x \to 3} \dfrac{3}{(x-3)^2} = \infty$.

(c) $\lim\limits_{x \to \infty} \left(1 - \dfrac{1}{3x}\right)^{7x} = \lim\limits_{x \to \infty} \left(\left(1 - \dfrac{1/3}{x}\right)^x\right)^7 = \left(\lim\limits_{x \to \infty}\left(1 + \dfrac{-1/3}{x}\right)^x\right)^7 = (e^{-\frac{1}{3}})^7 = e^{-\frac{7}{3}}$.

(d) Substitute $h = \frac{1}{t}$. We have $\lim\limits_{t \to 0^+} (1 - 2t)^{1/t} = \lim\limits_{h \to \infty}\left(1 - \dfrac{2}{h}\right)^h = e^{-2}$.

**Problem 6.**

Use the compound interest formula $M(t) = M_0(1 + \frac{r}{n})^{nt}$, where $r$ is the rate (expressed as a decimal) and $n$ is the number of compounding periods per year. For this problem $r = 5\% = 0.05$, and $t$ is the number of years since the money was deposited.

(a) Quarterly $\Rightarrow n = 4$: $M(1) = (10,000)(1 + \frac{0.05}{4})^{(4)(1)} = \$10,509.45$.

(b) Daily $\Rightarrow n = 365$: $M(1) = (10,000)(1 + \frac{0.05}{365})^{(365)(1)} = \$10,512.67$.

(c) Continuously: $M(1) = (10,000)e^{0.05(1)} = \$10,512.71$.

**Problem 8.**

(a) $M(1) = M_0(1 + \frac{0.05}{4})^{4(1)} \approx (1.0509)M_0$. The effective rate is 5.09%.

(b) $M(1) = (1.05)M_0 = M_0(1 + \frac{r}{4})^{4(1)}$

$1.05 = (1 + \frac{r}{4})^{4(1)}$

$\ln(1.05) = 4\ln(1 + \frac{r}{4})$

$1 + \frac{r}{4} = e^{\frac{\ln(1.05)}{4}} \Rightarrow r = 4e^{\frac{\ln(1.05)}{4}} - 4 \approx 4.909\%$. The nominal interest rate is 4.909%.

**Problem 10.**

(a) Substitute $h = 1/x$. We have $\lim_{x \to 0^+} (1+x)^{1/x} = \lim_{h \to \infty} \left(1 + \frac{1}{h}\right)^h = e^1$.

(b) $\lim_{w \to \infty} \left(\frac{w+2}{w}\right)^w = \lim_{w \to \infty} \left(1 + \frac{2}{w}\right)^w = e^2$.

(c) $\lim_{x \to \infty} \left(\frac{x-1}{x}\right)^{2x} = \lim_{x \to \infty} \left(1 - \frac{1}{x}\right)^{2x} = \lim_{x \to \infty} \left(\left(1 - \frac{1}{x}\right)^x\right)^2 = (e^{-1})^2 = e^{-2}$.

(d) $\lim_{n \to \infty} \left(\frac{n}{n+1}\right)^n = \lim_{n \to \infty} \left(\frac{n+1}{n}\right)^{-n} = \lim_{n \to \infty} \left(1 + \frac{1}{n}\right)^{-n} = \lim_{n \to \infty} \left(\left(1 + \frac{1}{n}\right)^n\right)^{-1} = e^{-1}$.

(e) Substitute $h = \frac{1}{2x}$. We have $\lim_{x \to 0^+} (1+2x)^{3/(2x)} = \lim_{h \to \infty} \left(1 + \frac{1}{h}\right)^{3h} = \lim_{h \to \infty} \left(\left(1 + \frac{1}{h}\right)^h\right)^3 = e^3$.

**Problem 11.**

(a) $M(7) = 6000e^{0.05(7)} \approx \$8,514.41$.

(b) $M(t) = 6000e^{0.05t}$ dollars.

(c) $\frac{dM}{dt} = 6000(0.05e^{0.05t}) = 300e^{0.05t}$ dollars per year.

(d) True: $\frac{dM}{dt} = 6000(0.05e^{0.05t}) = 0.05M(t)$.

(e) $M(t) = 6000e^{0.05t} = 6000(e^{0.05})^t \approx 6000(1.051)^t$.

(f) Solve $M(t+1) = (1+p)M(t)$ for $p$. Now $6000(1.051)^{t+1} = (1+p)(6000(1.051)^t) \Rightarrow 1.051 = 1 + p \Rightarrow p = 5.1\%$ The money grows 5.1% annually.

**Problem 13.**

For the account that compounds interest at 4% continuously, we have $M(t) = M_0 e^{0.04t}$. To find the effective yield, we solve solve $M(t+1) = (1+p)M(t)$ for $p$. Now $M_0 e^{0.04(t+1)} = (1+p)M_0 e^{0.04t} \Rightarrow e^{0.04} = 1 + p \Rightarrow p = e^{0.04} - 1 \approx 4.081\%$. Hence the effective annual yield is 4.081% for the continuously compounding account. The account that offers 4.2% interest, compounded annually is a better deal because its effective yield is 4.2%.

# Exploratory Problems for Chapter 15

**Problem 2.**

(a) $\frac{dQ}{dt} = 2Q - 6 = 2(Q-3)$. Let $y = Q - 3$. Then $\frac{dy}{dt} = \frac{dQ}{dt}$ and the differential equation becomes $\frac{dy}{dt} = 2y$. The general solution to the differential equation in terms of $y$ is $y = Ce^{2t}$. Substituting $Q = y + 3$, the general solution becomes $Q(t) - 3 = Ce^{2t} \Rightarrow Q(t) = Ce^{2t} + 3$.

(b) $\frac{dM}{dt} = 0.1M - 200 = 0.1(M - 2000)$. Let $y = M - 2000$. Then $\frac{dy}{dt} = \frac{dM}{dt}$ and the differential equation becomes $\frac{dy}{dt} = 0.1y$. The general solution to the differential equation in terms of $y$ is $y = Ce^{0.1t}$. Substituting $M = y + 2000$, the general solution becomes $M(t) - 2000 = Ce^{0.1t} \Rightarrow M(t) = Ce^{0.1t} + 2000$.

## Section 15.2    Introducing Differential Equations

**Problem 2.**

(a) $F(t) = F_0(1 - 0.1)^t = F_0(0.9)^t$. To find the number of years it will take for half of the forestland to be destroyed, we find the value of $t$ for which $0.5F_0 = F_0(0.9)^t$. Solving this equation, we have $0.5 = (0.9)^t$ $\Rightarrow \ln 0.5 = t \ln 0.9 \Rightarrow t = \frac{\ln 0.5}{\ln 0.9} \approx 6.58$ years.

(b) $F(t) = F_0 e^{-0.1t}$. To find the number of years it will take for half of the forestland to be destroyed, we find the value of $t$ for which $0.5F_0 = F_0 e^{-0.1t}$. Solving this equation, we have $0.5 = e^{-0.1t} \Rightarrow \ln 0.5 = -0.1t$ $\Rightarrow t = -\frac{\ln 0.5}{0.1} \approx 6.93$ years.

(c) $\frac{dF}{dt} = -0.1F_0 e^{-0.1t} = -0.1F$. The sign of the proportionality constant is negative because the amount of forestland is decreasing.

**Problem 4.**

(a) $y(t) = Ce^{-2t}$. $\sqrt{2} = Ce^{-2(0)} \Rightarrow C = \sqrt{2} \Rightarrow y(t) = \sqrt{2}e^{-2t}$.

(b) $y(t) = -t^2 + C$. $\sqrt{2} = -(0)^2 + C \Rightarrow C = \sqrt{2} \Rightarrow y(t) = -t^2 + \sqrt{2}$.

(c) $y(t) = -2t + C$. $\sqrt{2} = -2(0) + C \Rightarrow C = \sqrt{2} \Rightarrow y(t) = -2t + \sqrt{2}$.

**Problem 5.**

(a) $\frac{dP}{dt} = C(2e^{2t}) = 2(Ce^{2t}) = 2P$.

(b) $\frac{dP}{dt} = 2e^{2t} \neq 2(e^{2t} + C)$.

**Problem 7.**

(a) No: $\dfrac{dy}{dt} = e^t + \dfrac{1}{t}$, and $y - \dfrac{y}{t} = e^t + \ln t - \dfrac{e^t + \ln t}{t}$.

(b) No: $\dfrac{dy}{dy} = te^t + (1)e^t = (t+1)e^t$, and $y - \dfrac{y}{t} = te^t - e^t = (t-1)e^t$.

**Problem 9.**

(a) We have $\frac{dM}{dt} = kM$ and $250 = k(5000)$. Hence $k = 0.05$, and $\frac{dM}{dt} = 0.05M$.

(b) We have $\frac{dB}{dt} = kB$. The general solution is $B(t) = Ce^{kt}$. We know that $600 = Ce^{k(0)}$, and hence $C = 600$. We also know that $800 = 600e^{k(10)}$, and hence $k = \frac{\ln(\frac{4}{3})}{10} \approx 0.0288$. Therefore $\frac{dB}{dt} = 600e^{(\ln(4/3))/10} = 600 \left(\frac{4}{3}\right)^{1/10}$.

**Problem 10.**

(a) $D(t) = Ce^{kt}$.

(b) $\dfrac{dT}{dt} = k(T - R)$. If $D(t) = T(t) - R$, then $\dfrac{dD}{dt} = \dfrac{dT}{dt}$. If we substitute these two expressions into the second differential equation, we obtain the differential equation in part (a).

(c) The sign of the proportionality constant $k$ is negative. To see this, observe that $D$ is decreasing and hence $\frac{dD}{dt} = kD < 0$. As $D > 0$, we see that $k < 0$.

(d) (i)

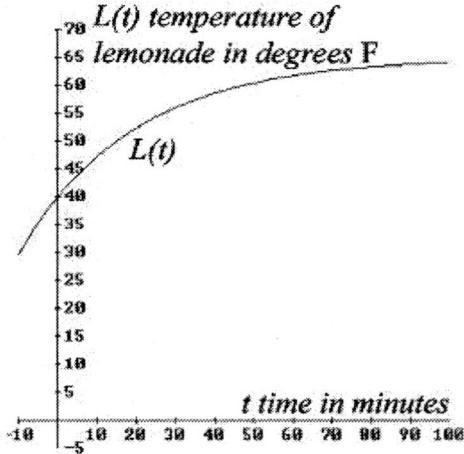

(ii) The temperature $L(t)$ is increasing at a decreasing rate.

(iii) $\frac{dL}{dt} = k(L - 65)$, where $k$ is a negative constant.

(iv) Let $D(t)$ be the difference in temperature; that is, $D(t) = L(t) - 65$. Now $\frac{dD}{dt} = \frac{dL}{dt}$. Substituting for $L$ into the differential equation, we obtain the differential equation $\frac{dD}{dt} = kD$. The general solution to this equation is $D(t) = Ce^{kt}$. Rewriting this equation in terms of $L$, we have $L(t) = Ce^{kt} + 65$. As $L(0) = 40$, we have $40 = Ce^{k(0)} + 65 \Rightarrow C = -25$. As $L(15) = 50$, we have $50 = -25e^{15k} + 65$ $\Rightarrow \frac{3}{5} = e^{15k} \Rightarrow \ln(3/5) = 15k \Rightarrow k = \frac{\ln(3/5)}{15} \approx -0.034$. Therefore $L(t) = -25e^{\frac{t\ln(3/5)}{15}} + 65 = -25\left(\frac{3}{5}\right)^{t/15} + 65$.

(v) The temperature will reach 55 degrees when $55 = L(t)$. Solving this equation, we obtain $55 = -25e^{\frac{t\ln(3/5)}{15}} + 65 \Rightarrow \frac{2}{5} = e^{\frac{t\ln(3/5)}{15}} \Rightarrow \ln\left(\frac{2}{5}\right) = \frac{t\ln(3/5)}{15} \Rightarrow t = \frac{15\ln 0.4}{\ln 0.6} \approx 26.9$ minutes.

**Problem 12.**
$\frac{dM}{dt} = 0.05M - 2000$ and $M(0) = 30,000$.

**Problem 14.**
$\frac{dy}{dt} = 2y - 6 = 2(y - 3)$. Let $x = y - 3$. Then $\frac{dy}{dt} = \frac{dx}{dt}$, and hence $\frac{dx}{dt} = 2x$. Now $x(t) = Ce^{2x}$. In terms of $y$, this last equation becomes $y(t) = x(t) + 3 = Ce^{2x} + 3$. Using the initial condition $y(0) = 2000$, we have $y(0) = 2000 = Ce^{2(0)} + 3 = C + 3 \Rightarrow C = 1997$. Therefore, the particular solution to the initial value problem is $y(t) = 1997e^{2x} + 3$.

**Problem 17.**

(a) $\dfrac{dN}{dt} = kN(800 - N)$ where $k$ is a negative constant of proportionality.

(b) We minimize the quadratic function $R(N) = kN(800 - N) = -kN^2 + 800kN$. Now $\dfrac{dR}{dN} = -2kN + 800k = -2k(N - 400)$. $R(N)$ is minimized when $\dfrac{dR}{dN} = 0$, which occurs when $N = 400$. At this time, there are 400 moles of substance A.

# CHAPTER 16
## Taking the Derivative of Composite Functions

### Section 16.1    The Chain Rule

**Problem 2.**

(a) $f'(x) = 3(-5)(x+2)^{-6} \left( \frac{d}{dx}(x+2) \right) = -15(x+2)^{-6}$

(b) $f'(x) = 2(-8)(3x+7)^{-9} \left( \frac{d}{dx}(3x+7) \right) = -16(3x+7)^{-9}(3) = -48(3x+7)^{-9}$

**Problem 3.**

$f(x) = \ln\sqrt{\pi x + 1} + \sqrt{\pi x} + (\pi x + \pi)^5 + \frac{1}{(\pi x^2 + 1)^3} = \frac{1}{2}\ln(\pi x + 1) + (\pi x)^{1/2} + (\pi x + \pi)^5 + (\pi x^2 + 1)^{-3} \Rightarrow$

$f'(x) = \left( \frac{1}{2} \right) \left( \frac{1}{\pi x + 1} \right) \left( \frac{d}{dx}(\pi x + 1) \right) + \frac{1}{2}(\pi x)^{-1/2} \left( \frac{d}{dx}(\pi x) \right) + 5(\pi x + \pi)^4 \left( \frac{d}{dx}(\pi x + \pi) \right) - 3(\pi x^2 + 1)^{-4} \left( \frac{d}{dx}(\pi x^2 + 1) \right) =$

$\frac{\pi}{2\pi x + 2} + \frac{1}{2}\pi(\pi x)^{-1/2} + 5\pi(\pi x + \pi)^4 - 6\pi x(\pi x^2 + 1)^{-4}$.

**Problem 5.**

$f(x) = e^x (3x^2 + 1)^{-1} \Rightarrow f'(x) = e^x((3x^2 + 1)^{-1}) + (-(3x^2 + 1)^{-2} \left( \frac{d}{dx}(3x^2 + 1) \right)) e^x =$

$e^x((3x^2 + 1)^{-1} - 6x(3x^2 + 1)^{-2}) = \frac{e^x(3x^2 - 6x + 1)}{(3x^2 + 1)^2}$.

**Problem 7.**

$f(x) = \left( 1 - \frac{1}{x} \right) e^{-x} \Rightarrow f'(x) = x^{-2}e^{-x} + \left( e^{-x} \left( \frac{d}{dx}(-x) \right) \right) \left( 1 - \frac{1}{x} \right) =$

$x^{-2}e^{-x} + (-e^{-x}) \left( 1 - \frac{1}{x} \right) = e^{-x} \left( x^{-2} + \frac{1}{x} - 1 \right) = \frac{e^{-x}(x^2 - x - 1)}{x^2}$.

**Problem 9.**

$f(x) = 5\ln(2x^2 + 3x) \Rightarrow f'(x) = 5 \left( \frac{1}{2x^2 + 3x} \right) \left( \frac{d}{dx}(2x^2 + 3x) \right) =$

$5 \left( \frac{1}{2x^2 + 3x} \right) (4x + 3) = \frac{20x + 15}{4x + 3}$.

**Problem 10.**

$f(x) = (3x^3 + 2x)^{13} \Rightarrow f'(x) = 13(3x^3 + 2x)^{12} \left( \frac{d}{dx}(3x^3 + 2x) \right) = 13(3x^3 + 2x)^{12}(9x^2 + 2)$.

**Problem 12.**

$f(x) = \frac{\pi^2}{3(x^3 + 2)^6} = \frac{\pi^2}{3}(x^3 + 2)^{-6} \Rightarrow f'(x) = \frac{\pi^2}{3} \left( -6(x^3 + 2)^{-7} \left( \frac{d}{dx}(x^3 + 2) \right) \right) =$

$-2\pi^2(x^3 + 2)^{-7}(3x^2) = -6\pi^2 x^2 (x^3 + 2)^{-7}$.

**Problem 14.**

$f(x) = \frac{3^x}{2^{x+1}} = 3^x 2^{-(x+1)} \Rightarrow f'(x) = ((\ln 3)3^x)(2^{-(x+1)}) + \left( (\ln 2)2^{-(x+1)} \left( \frac{d}{dx}(-x-1) \right) \right) (3^x) =$

$(\ln(3)3^x)(2^{-(x+1)}) + (-\ln(2)2^{-(x+1)})(3^x) = 3^x 2^{-x-1}(\ln(3) - \ln(2)) = \frac{3^x \ln(3/2)}{2^{x+1}}$.

**Problem 16.**

$f(x) = \frac{4}{\sqrt{e^x + 1}} = 4(e^x + 1)^{-1/2} \Rightarrow f'(x) = 4 \left( -\frac{1}{2} \right) (e^x + 1)^{-3/2} \left( \frac{d}{dx}(e^x + 1) \right) =$

$-2e^x(e^x + 1)^{-3/2}$.

**Problem 17.**

$f(x) = \sqrt{e^x + \ln(x+1)^2} = (e^x + 2\ln(x+1))^{1/2} \Rightarrow f'(x) = \frac{1}{2}(e^x + 2\ln(x+1))^{-1/2} \left(\frac{d}{dx}(e^x + 2\ln(x+1))\right) =$
$\frac{1}{2}(e^x + 2\ln(x+1))^{-\frac{1}{2}} \left(e^x + \frac{2}{x+1}\left(\frac{d}{dx}(x+1)\right)\right) = \frac{1}{2}(e^x + 2\ln(x+1))^{-\frac{1}{2}} \left(e^x + \frac{2}{x+1}\right) =$
$\frac{xe^x + e^x + 2}{2x(x+1)\sqrt{e^x + \ln(x+1)^2}}$.

**Problem 19.**

$f(x) = \ln(e^x + x^2) \Rightarrow f'(x) = \frac{1}{e^x + x^2}\left(\frac{d}{dx}(e^x + x^2)\right) = \frac{e^x + 2x}{e^x + x^2}$.

**Problem 20.**

$f(x) = x\ln\left(\frac{x}{x^2+1}\right) = x(\ln(x) - \ln(x^2+1)) \Rightarrow$
$f'(x) = (1)(\ln(x) - \ln(x^2+1)) + (x)\left(\frac{1}{x} - \frac{1}{x^2+1}\left(\frac{d}{dx}(x^2+1)\right)\right) = \ln(x) - \ln(x^2+1) + 1 - \frac{x}{x^2+1}(2x) =$
$\ln(x) - \ln(x^2+1) + 1 - \frac{2x^2}{x^2+1}$.

**Problem 22.**

$f(x) = x(x+3)^2 = x^3 + 6x^2 + 9x \Rightarrow$

$f'(x) = (1)(x+3)^2 + x(2)(x+3) = (x+3)((x+3) + 2x) = 3(x+3)(x+1) = 3x^2 + 12x + 9.$

$f'(x) = 0$ when $x = -3$ or $x = -1$. Now $f''(x) = 6x + 12$, and hence $f''(-3) = 6(-3) + 12 = -6 < 0$ and $f''(-1) = 6(-1) + 12 = 18 > 0$. Therefore, $(-3, f(-3)) = (-3, 0)$ is a local maximum point and $(-1, f(-1)) = (1, -4)$ is a local minimum point.

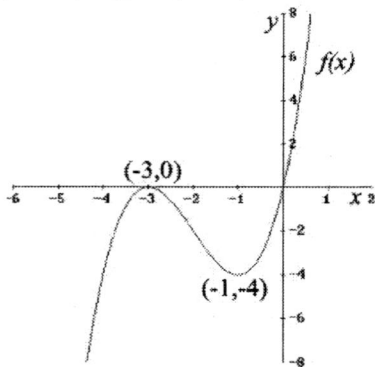

**Problem 24.**

$f(x) = (3-x)^2(x-1) = (x-3)^2(x-1) \Rightarrow f'(x) = 2(x-3)(x-1) + (x-3)^2(1) = (x-3)(2(x-1) + (x-3)) = (x-3)(3x-5) = 3x^2 - 14x + 15.$ $f'(x) = 0$ when $x = 3$ or $x = \frac{5}{3}$. Now $f''(x) = 6x - 14$, and hence $f''(3) = 4 > 0$ and $f''\left(\frac{5}{3}\right) = -4 < 0$. Therefore $(3, f(3)) = (3, 0)$ is a local minimum point and $\left(\frac{5}{3}, f\left(\frac{5}{3}\right)\right) = \left(\frac{5}{3}, \frac{32}{27}\right)$ is a local maximum point.

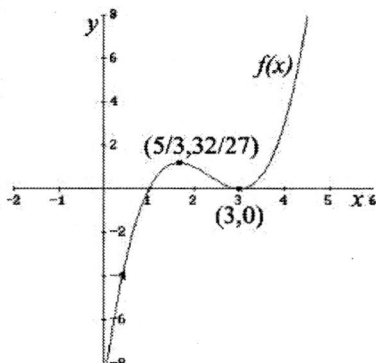

**Problem 26.**

If $f(x) = (x-a)^3(x-b)$, then $f'(x) = 3(x-a)^2(x-b) + (x-a)^3(1) = (x-a)^2(3x-3b+x-a) = (x-a)^2(4x-3b-a)$. Now $f''(x) = 2(x-a)(4x-3b-a) + (x-a)^2(4) = 2(x-a)(4x-3b-a+2x-2a) = 2(x-a)(6x-3b-3a) = 6(x-a)(2x-b-a)$. Now $f''(a) = 0$. Note that as $a \neq b$. $2a-b-a = a-b \neq 0$, and hence only one power of $(x-a)$ is a factor of $f''(x)$. Moreover, $f''(x) = 0$ also when $2x-a-b = 0$ when $x = \frac{a+b}{2}$. Furthermore, as the two zeros of the quadratic polynomial $f''(x)$ are distinct, we can conclude that $f''(x) < 0$ on the open interval between $a$ and $\frac{a+b}{2}$ and $f''(x) > 0$ outside the closed interval between these two points. Hence $f$ has points of inflection at both $x = a$ and $\frac{a+b}{2}$.

**Problem 28.**

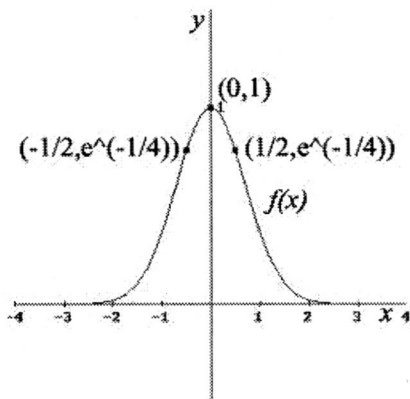

(a) Domain: $(-\infty, \infty)$; range: $(0, 1]$

(b) $f$ is an even function: $f(-x) = e^{-(-x)^2} = e^{-x^2} = f(x)$.

(c) $f$ is increasing on $(-\infty, 0)$ and is decreasing on $(0, \infty)$.

(d) $f'(x) = 0$ when $x = 0$. From part (b), we see that $(0, f(0)) = (0, 1)$ is a local maximum point. There are no local minima.

(e) The absolute maximum value is $f(0) = 0$. To see this, note that the local maximum point $x = 0$ is the only critical point, and that $f'(x) > 0$ for $x > 0$ and $f'(x) < 0$ for $x > 0$. $f$ does not have an absolute minimum because $\lim_{|x| \to \infty} f(x) = 0$, which is not attained as a function value of $f$.

(f) $f''(x) = -2(e^{-x^2}) - 2x(-2x)e^{-x^2} = (4x^2 - 2)e^{-x^2}$. Now $f''(x) = 0$ when $x = -\frac{\sqrt{2}}{2}$ or $x = \frac{\sqrt{2}}{2}$. Moreover, $f''(x) > 0$ when $|x| > \frac{\sqrt{2}}{2}$ and $f''(x) < 0$ when $|x| < \frac{\sqrt{2}}{2}$. Therefore, $x = -\frac{\sqrt{2}}{2}$ and $x = \frac{\sqrt{2}}{2}$ are the $x$-coordinates of the point of inflection.

## Section 16.2    The Derivative of $x^n$ where $n$ is any Real Number

**Problem 2.**
$$y' = \sqrt{3}(2x^2 + 1)^{\sqrt{3}-1}(4x) = 4\sqrt{3}x(2x^2 + 1)^{\sqrt{3}-1}.$$

**Problem 4.**

(a) $y' = \pi e^{3t^2 + \pi}(6t) = 6\pi t e^{3t^2 + \pi}$.

(b) $y' = \frac{1}{e^t + 1}(e^t) = \frac{e^t}{e^t + 1}$.

(c) $y = \frac{\pi^2}{\sqrt{x^2+4}} = \pi^2(x^2 + 4)^{-1/2} \Rightarrow$
$y' = -\pi^2 \left(\frac{1}{2}\right)(x^2 + 4)^{-3/2}(2x) = -\pi^2 x(x^2 + 4)^{-3/2} = -\frac{\pi^2 x}{(x^2+4)^{3/2}}$.

(d) $y = \frac{1}{(\ln x)^{2.6}} = (\ln x)^{-2.6} \Rightarrow y' = -2.6(\ln x)^{-3.6}\left(\frac{1}{x}\right) = -\frac{2.6}{x(\ln(x))^{3.6}}$.

(e) $y = \frac{1}{(\ln x^2)^{1.5}} = (2\ln x)^{-1.5} \Rightarrow y' = -1.5(2\ln x)^{-2.5}\left(\frac{2}{x}\right) = -\frac{3}{x(2\ln(x))^{2.5}}$.

(f) $y' = \frac{1}{3}(\ln(e^t + 1))^{-2/3}\left(\frac{1}{e^t+1}(e^t)\right) = \frac{e^t}{(3e^t+3)(\ln(e^t+1))^{2/3}}$.

**Problem 6.**
$f(x) = (x^2 - 4)x^{\pi+1} \Rightarrow f'(x) = (2x)(x^{\pi+1}) + (x^2 - 4)(\pi+1)x^\pi = x^\pi((3+\pi)(x^2) - 4\pi - 4)$. Now for $x > 0$, $f'(x) = 0$ when $x = \sqrt{\frac{4\pi+4}{\pi+3}}$, which is the only critical point of $f$ in its domain. Note that $f'(x) > 0$ for $x > \sqrt{\frac{4\pi+4}{\pi+3}}$ and $f'(x) < 0$ on $\left(0, \sqrt{\frac{4\pi+4}{\pi+3}}\right)$. Hence this critical point is a local minimum point.

## Exploratory Problems for Chapter 16

**Problem 1.**
From the figure on the left, the distance $h$ is 6 miles, and from the Pythagorean Theorem, we see that $L = \sqrt{10^2 - 6^2} = \sqrt{64} = 8$.

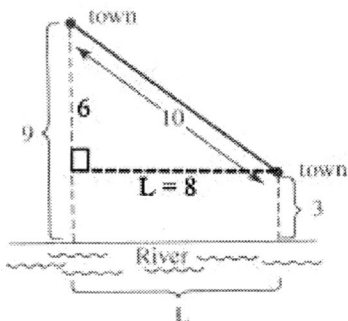

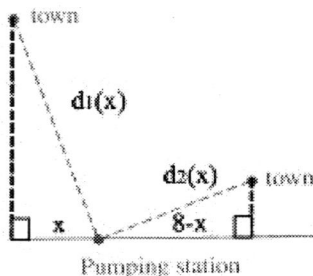

Now let $x$ be the distance in miles from the closest point on the shore to the town 9 miles from the river. We define $d_1(x)$ to be the distance from the town 9 miles from the shore to the pumping station and $d_2(x)$ to be the distance from the town 3 miles from the shore to the pumping station as labeled in the figure on the right. From the Pythagorean Theorem, we have

$$d_1(x) = \sqrt{81 + x^2}$$

and

$$d_2(x) = \sqrt{9 + (8-x)^2} = \sqrt{x^2 - 16x + 73}.$$

The total length of pipe needed is

$$d(x) = d_1(x) + d_2(x) = \sqrt{81 + x^2} + \sqrt{x^2 - 16x + 73}.$$

To determine the minimal length of pipe needed, we minimize $d(x)$ over the interval $[0, 8]$. Differentiating, we obtain

$$d'(x) = \frac{2x}{2\sqrt{81 + x^2}} + \frac{2x - 16}{2\sqrt{x^2 - 16x + 73}} = \frac{x\sqrt{x^2 - 16x + 73} + (x - 8)\sqrt{81 + x^2}}{\sqrt{81 + x^2}\sqrt{x^2 - 16x + 73}}.$$

The easiest way to identify the critical points is find where the graph of $d'$ crosses the $x-$axis. The figure below indicates that the only critical point on $(0, 8)$ is $x = 6$. This can be verified algebraically as well.

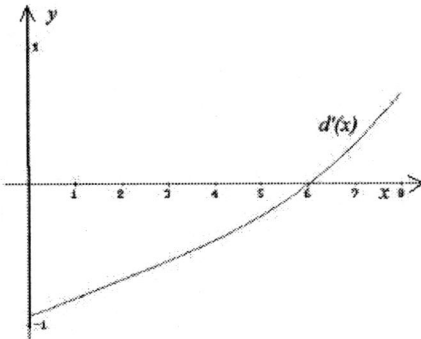

We next check our function values at $x = 6$ and the endpoints $x = 0$ and $x = 8$ to determine the minimum value of $d(x)$. We have

| $x$ | $d(x)$ |
|---|---|
| 0 | $9 + \sqrt{73} \approx 17.5$ |
| 6 | $4\sqrt{13} \approx 14.4$ |
| 8 | $3 + \sqrt{145} \approx 15.0$ |

Therefore, the pumping station should be placed on the point on the shore of the river that is 6 miles from the closet point on the shore to the towm that is 9 miles from the river.

## Section 16.3  Using the Chain Rule

**Problem 2.**

Let $x$ be the length of a vertical side of the rectangle as shown in the figure.

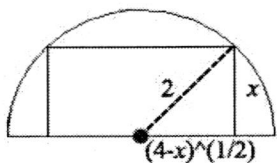

From the Pythagorean Theorem, we have that half the length of a horizontal side of the rectangle is $\sqrt{4-x^2}$. In terms of $x$, the area of the rectangle is $A(x) = 2x\sqrt{4-x^2}$. We now maximize $A(x)$ on the interval $[0,2]$. We have

$$A'(x) = 2\sqrt{4-x^2} + 2x\left(\frac{-2x}{2\sqrt{4-x^2}}\right) = \frac{2(4-x^2)-2x^2}{\sqrt{4-x^2}} = \frac{4(2-x^2)}{\sqrt{4-x^2}}.$$

On $(0,2)$, $A'(x) = 0$ when $x = \sqrt{2}$, and $A'(x) < 0$ on $(\sqrt{2},2)$ and $A'(x) > 0$ on $(0,\sqrt{2})$. Hence the area is maximized when $x = \sqrt{2}$. The dimensions of the rectangle of largest area are $2\sqrt{2} \times \sqrt{2}$, and the area of this rectangle is 4.

**Problem 4.**

(a) $y = \frac{1}{x\ln 2+1} = (x\ln 2+1)^{-1} \Rightarrow y' = -(\ln 2)(x\ln 2+1)^{-2} = -\frac{\ln 2}{(x\ln 2+1)^2}.$

(b) $y' = \left(\frac{1}{5x^3+8x}\right)(15x^2+8) = \frac{15x^2+8}{5x^3+8x}.$

(c) $y' = (\ln 2)2^x(x^2+x)^7 + 7(x^2+x)^6(2x+1))(2^x) =$
$2^x(x^2+x)^6(x^2\ln 2 + x\ln 2 + 14x + 7).$

(d) $y = \sqrt{\ln 5x + e^{6x}} = \sqrt{\ln 5 + \ln x + e^{6x}} \Rightarrow y' = \frac{1}{2}(\ln(5) + \ln(x) + e^{6x})^{-\frac{1}{2}}\left(\frac{1}{x} + 6e^{6x}\right).$

(e) $y = \frac{7}{\sqrt{\ln x}} = 7(\ln x)^{-1/2} \Rightarrow y' = 7\left(-\frac{1}{2}(\ln x)^{-3/2}\left(\frac{1}{x}\right)\right) = -\frac{7}{2x(\ln x)^{3/2}}.$

(f) $y' = 4^{x/3}(\ln 4)(\ln 3x)\left(\frac{1}{3}\right) + 4^{x/3}\frac{3}{3x} = 4^{x/3}\left(\frac{1}{3}\ln 4\ln 3x + \frac{1}{x}\right).$

**Problem 6.**

The denominator is always at least 1 since $x^2 + 1 \geq 1$ for all $x$, so the function is no larger than 3. It attains this value at $x = 0$.

Now $f(x) = \frac{3}{\sqrt{x^2+1}} = 3(x^2+1)^{-1/2} \Rightarrow f'(x) = -\frac{3}{2}(x^2+1)^{-3/2}(2x) = -3x(x^2+1)^{-3/2}$. Now $f'(x) = 0$ when $x = 0$, and $f'(x) > 0$ for $x < 0$ and $f'(x) < 0$ for $x > 0$. Hence the global maximum value, 3, of $f$ is attained at $x = 0$.

**Problem 8.**

(a) $\frac{dy}{dx}\Big|_{x=0} = f'(g(0))g'(0) = f'(2)(-1) = 2.5(-1) = 2.5.$

(b) $y(0) = g(f(0)) = g(3) = 1.$

(c) $\frac{d}{dx}(f(x)(g(x))^{-1})\Big|_{x=3} = (f'(3)(g(3))^{-1} - ((g(3))^{-2}g'(3)f(3)) = 5(1) - (1)^2(4)(3) = -7.$

(d) $\frac{dy}{dx} = 2f(x)f'(x)g(x^2) + [f(x)]^2 g'(x^2)(2x)$. $\frac{dy}{dx}\Big|_{x=2} = 2f(2)f'(2)g(4) + [f(2)]^2 g'(4)(4) = 2(0.3)(2.5)(3) + (0.3)^2(5)(4) = 6.3.$

(e) $y'(x) = \frac{1}{2}(f(x^2))^{-1/2}f'(x^2)(2x) = \frac{xf'(x^2)}{\sqrt{f(x^2)}}$. $y'(2) = \frac{4}{\sqrt{11}} = \frac{4}{\sqrt{11}} = \frac{4\sqrt{11}}{11}.$

**Problem 10.**

(a) $f'(x) = 4[g(x)]^3(g'(x))$.

(b) The absolute minimum value of $g$ is attained at $x = 3$. As the only zeros $g$ are $x = 1$ and $x = 4$ and the only local extrema of $g$ are minimum at $x = 3$ and a maximum at $x = 7$, we see that $g(3) < 0$ and $g(3)$ is the absolute minimum value on $[1, 4]$. Moreover, $g(x) > 0$ for $x < 1$ and for $x > 4$; otherwise, the continuous function $g$ would have zeros other than $x = 1$ and $x = 4$. As it is possible that $\lim\limits_{x \to -\infty} g(x) = \infty$, we cannot definititively determine whether $g$ has an absolute maximum value.

(c) $f'(x) = 0$ when $g(x) = 0$ or $g'(x) = 0$ or where $g'(x)$ is undefined. We know that $g(x) = 0$ for $x = 1$ and $x = 4$. Because $g$ has local extrema at $x = 3$ and $x = 7$, $g'$ is either zero or undefined for these values. It is possible that $g$ has other critical points which do not correspond to local extrema (these would be points where $g'$ is either zero or undefined, but no sign change occurs); these, too, would be critical points of $f$. Hence the critical points of $f$ are $x = 1, 3, 4, 7$, and all critical points of $g$ which do not correspond to local extrema.

(d) $f(x)$ is increasing on $(1, 3)$ and $(4, 7)$, and decreasing on $(-\infty, 1)$, $(3, 4)$, and $(7, \infty)$.

(e) The local maxima of $f$ occur at $x = 3$ $x = 7$; the local minima of $f$ occur at $x = 1$ and $x = 4$.

(f) Yes; $f(x) \geq 0$ for all $x$, and $f(1) = f(4) = 0$ is indeed the minimum value.

**Problem 12.**

(a)

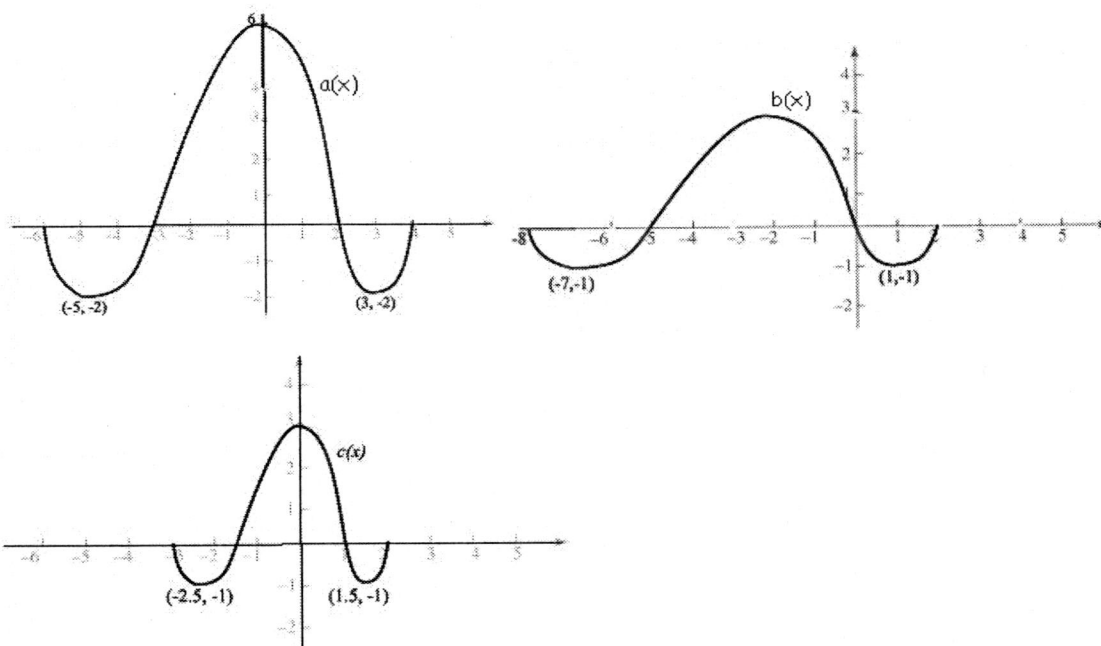

(b)  (i) $a'(-2) = 2f'(-2) = 2(3/2) = 3$

(ii) $b'(-2) = f'(-2 + 2)(1) = f'(0) = 0$.

(iii) $c'(-2) = 2f'(2(-2)) = 2f'(-4) = 2(1/2) = 1$.

**Problem 14.**

$$h'(x) = \frac{1}{2\sqrt{f(x)g(x)}}(f'(x)g(x) + f(x)g'(x)) = \frac{f'(x)g(x)+f(x)g'(x)}{2\sqrt{f(x)g(x)}}.$$

**Problem 16.**

$$h'(x) = f'(x^2)2xe^{g(x)} + f(x^2)g'(x)e^{g(x)} = e^{g(x)}(2xf'(x^2) + f(x^2)g'(x)).$$

**Problem 18.**

$$h'(x) = 3[f(x)]^2 f'(x)g(2x) + 2[f(x)]^3 g'(2x).$$

**Problem 19.**

$$y = 3\ln\left(\frac{x^2-1}{x+2}\right) = 3(\ln(x^2+1) - \ln(x+2)) \Rightarrow y' = 3\left(\frac{2x}{x^2+1} - \frac{1}{x+2}\right) = \frac{3x^2+12x-3}{(x^2+1)(x+2)}.$$

**Problem 21.**

$$y = \frac{5x^\pi - x^3 - 1}{x^2} = 5x^{\pi-2} - x - x^{-2} \Rightarrow y' = 5(\pi-2)x^{\pi-3} - 1 + 2x^{-3} = \frac{(5\pi-10)x^\pi - x^3 + 2}{x^3}.$$

**Problem 23.**

$$y' = \left(\frac{f(x^2)}{x}\right)\left(\frac{(1)f(x^2) - xf'(x^2)(2x)}{[f(x^2)]^2}\right) = \frac{f(x^2) - 2x^2 f'(x^2)}{xf(x^2)}.$$

**Problem 25.**

$$y = \ln\left(\frac{x \cdot f(x)}{\sqrt{3x^3+2}}\right) = \ln(x \cdot f(x)) - \ln(\sqrt{3x^3+2}) = \ln x + \ln f(x) - \tfrac{1}{2}\ln(3x^3+2) \Rightarrow$$

$$y' = \frac{1}{x} + \frac{f'(x)}{f(x)} - \frac{1}{2}\left(\frac{9x^2}{3x^3+2}\right) = \frac{1}{x} + \frac{f'(x)}{f(x)} - \frac{9x^2}{6x^3+4}.$$

**Problem 27.**

$$f'(x) = e^{(g(x))^2}(2g(x)g'(x)) = 2g(x)g'(x)e^{(g(x))^2}.$$

**Problem 29.**

(a) $\frac{f(2.0001)-f(2)}{0.0001} \approx 6.773$

(b) $2(2^{2-1}) = 4 \neq 6.773$; $f'(x) \neq x \cdot x^{x-1}$ because the exponent $x$ is not a constant.

(c) $\ln 2 \cdot 2^2 \approx 2.773 \neq 6.773$; $f'(x) \neq \ln x \cdot x^x$ because the base $x$ is not a constant.

(d) $f(x) = x^x = e^{\ln x^x} = e^{x\ln x} \Rightarrow f'(x) = e^{x\ln x}\left((1)(\ln x) + \left(\frac{1}{x}\right)(x)\right) = e^{x\ln x}(\ln x + 1).$

**Problem 31.**

Let $b$ be the base and $h$ the height of the inscribed triangle.

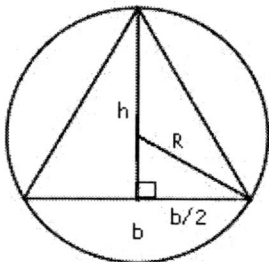

From the figure and Pythagorean Theorem, we have that $\frac{1}{2}b = \sqrt{R^2 - (h-R)^2} = \sqrt{2hR - h^2}$. The area of the triangle is $A(h) = \frac{1}{2}bh = h\sqrt{2hR - h^2}$. We now maximize $A(h)$ on the interval $[0, 2R]$. Differentiating, we obtain $A'(h) = (1)(\sqrt{2hR - h^2}) + (\frac{1}{2}(2hR - h^2)^{-1/2}(2R - 2h))(h) = \sqrt{2hR - h^2} + (2hR - h^2)^{-\frac{1}{2}}h(R - h)$

$A'(h) = 0$ when:

$\sqrt{2hR - h^2} = -(2hR - h^2)^{-\frac{1}{2}}h(R - h)$

$2hR - h^2 = -h(R - h)$

$3hR = 2h^2$

$h = \frac{3}{2}R$ or $h = 0$.

To maximize $A(h)$ we compute its value at its critical points $h = 0, \frac{3}{2}R$, and $2R$.

| $h$ | $A(h)$ |
|-----|--------|
| $0$ | $0$ |
| $\frac{3}{2}R$ | $\frac{3\sqrt{3}}{4}R^2$ |
| $2R$ | $0$ |

Hence $A(h)$ is maximized when $h = \frac{3}{2}R$. Now the length of the base is $b =$

$2\sqrt{3R^2 - \frac{9}{4}R^2} = \sqrt{3}R$. By the Pythagorean Theorem, we compute the length of one of the two equal-length sides to be

$$L = \sqrt{\frac{b^2}{4} + h^2} = \sqrt{\frac{3R^2}{4} + \frac{9R^2}{4}} = \sqrt{3}R.$$

As all sides of triangle have equal lengths, the triangle is equilateral.

**Problem 33.**

Let $b$ be the base radius and $h$ the height of the inscribed cylinder. From the Pythagorean Theorem and the figure of the cross-section of the ornament below, we see that $b^2 = 4R^2 - h^2$.

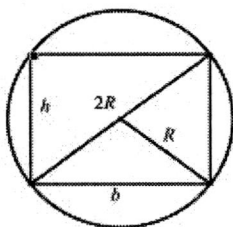

The surface area of the open hollow cylinder is $A(h) = \pi b^2 h = \pi(4R^2 - h^2)h = \pi(4R^2h - h^3)$. We now maximize $A(h)$ on $[0, 2R]$. Differentiating, we have

$$A'(h) = \pi(4R^2 - 3h^2).$$

Now $A'(h) = 0$ when $4R^2 - 3h^2 = 0$. Hence $h = \frac{2}{\sqrt{3}}R$ is the only critical point of $A(h)$.

To maximize $A(h)$ we compute the values of $A(h)$ at $h = \frac{2}{\sqrt{3}}R$ and the endpoints $h = 0$ and $h = 2R$.

| $h$ | $A(h)$ |
|-----|--------|
| $0$ | $0$ |
| $\frac{2}{\sqrt{3}}R$ | $\frac{16\pi\sqrt{3}}{9}R^3$ |
| $2R$ | $0$ |

Therefore, the maximum surface area of the cylinder is $\frac{16\pi\sqrt{3}}{9}R^3$ square inches.

**Problem 34.**

Let $c$ be the cost in dollars per mile of cable under ground. The cost of the installing the cable under water is $1.6c$ dollars per mile.

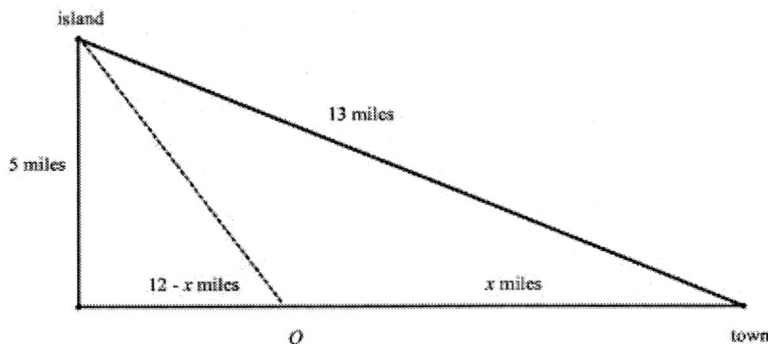

From the Pythagorean Theorem and the figure above, we see that the distance from the point $P$ closest to the island on the shore is $\sqrt{13^2 - 5^2} = 12$ miles. Suppose the cable is run from the island to a point $Q$ on shore $x$ miles from town. Also from the Pythagorean Theorem and the figure above, we see that the distance the cable runs underwater is $d_w(x) = \sqrt{5^2 + (12 - x)^2} = \sqrt{x^2 - 24x + 169}$. The total cost of installing the cable is $C(x) = cx + 1.6cd_w(x) = cx + 1.6c\sqrt{x^2 - 24x + 169}$. We now minimize $C(x)$ on $[0, 12]$. Differentiating, we obtain

$$C'(x) = c + 1.6c(\frac{1}{2}(x^2 - 24x + 169)^{-\frac{1}{2}}(2x - 24)) = c + 1.6c(x^2 - 24x + 169)^{-\frac{1}{2}}(x - 12) = \frac{c(5\sqrt{x^2 - 24x + 169} - 8(x - 12))}{5\sqrt{x^2 - 24x + 169}}.$$

Now $C'(x) = 0$ when $5\sqrt{x^2 - 24x + 169} - 8(x - 12) = 0$. We now solve this last equation for $x$ to determine the critical points of $C(x)$ on $(0, 12)$.

$$5\sqrt{x^2 - 24x + 169} = 8(x - 12)$$

$$\sqrt{x^2 - 24x + 169} = 1.6(x - 12)$$

$$x^2 - 24x + 169 = 1.6^2(144 - 24x + x^2)$$

$$1.56x^2 - 37.44x + 199.64 = 0$$

Using the quadratic formula, we obtain $x = \frac{37.44 \pm \sqrt{(-37.44)^2 - 4(1.56)(199.64)}}{2(1.56)} = 12 \pm \frac{25\sqrt{39}}{39}$. Decimal approximations of these values are $x = 16.0032$ and $x = 7.9968$. As the domain of $C(x)$ is $[0, 12]$, the only critical points in $(0, 12)$ is $x = 25 - \frac{25\sqrt{39}}{39} \approx 7.9968$.

To minimize $C(x)$, we compute $C(x)$ at the endpoints $x = 0$ and $x = 12$ and at the critical point $x = 25 - \frac{25\sqrt{39}}{39} \approx 7.9968$

| $x$ | $C(x)$ |
|---|---|
| 0 | 20.8c |
| $25 - \frac{25\sqrt{39}}{39}$ | $(12 + \sqrt{39})c \approx 18.245c$ |
| 12 | 20c |

Therefore, the cable should run $25 - \frac{25\sqrt{39}}{39} \approx 7.9968$ miles along the shore.

## Problem 36.

(a) $P = P(v(r))$. Therefore, $\frac{dP}{dr} = P'(v(r))v'(r)$. In Leibniz notation, this last equation can be expressed as $\frac{dP}{dr} = \frac{dP}{dv}\frac{dv}{dr}$.

(b) $P = P(v(r(t)))$. Therefore, $\frac{dP}{dt} = P'(v(r(t)))v'(r(t))r'(t)$. In Leibniz notation, this last equation can be expressed as $\frac{dP}{dr} = \frac{dP}{dv}\frac{dv}{dr}\frac{dr}{dt}$.

# CHAPTER 17

# Implicit Differentiation and its Applications

## Section 17.1    Introductory Example

**Problem 1.**

(a) $y = 3^x \Rightarrow y' = (\ln 3)3^x$ by formula or think of $y = 3^x = e^{x(\ln 3)}$

(b) $y = x^3 \Rightarrow y' = 3x^2$

(c) $y = x^x = e^{x(\ln x)} \Rightarrow y' = e^{x(\ln x)}[(1)\ln x + x\frac{1}{x}] = x^x[\ln x + 1]$

**Problem 3.**
$$y = (3x^2 + 2)^x = (e^{\ln(3x^2+2)})^x = e^{x\ln(3x^2+2)}$$

$$\Rightarrow y' = e^{x\ln(3x^2+2)}[(1)\ln(3x^2+2) + x\frac{6x}{3x^2+2}] = (3x^2+2)^x[\ln(3x^2+2) + \frac{6x^2}{3x^2+2}]$$

## Section 17.2    Logarithmic Differentiation

**Problem 1.**

(a) $f(x) = 2x^x \Rightarrow \ln(f(x)) = \ln(2x^x) = \ln 2 + x\ln x$

$$\Rightarrow \frac{1}{f(x)}f'(x) = (1)\ln x + x\frac{1}{x} = \ln x + 1 \Rightarrow f'(x) = [\ln x + 1]f(x) = [\ln x + 1]2x^x$$

(b) $\ln(f(x)) = \ln(5(x^2+1)^x) = \ln 5 + x\ln(x^2+1) \Rightarrow \frac{f'(x)}{f(x)} = 0 + (1)\ln(x^2+1) + x\frac{1}{(x^2+1)}(2x)$

$$= \ln(x^2+1) + \frac{2x^2}{x^2+1} \Rightarrow f'(x) = [\ln(x^2+1) + \frac{2x^2}{x^2+1}]f(x) = [\ln(x^2+1) + \frac{2x^2}{x^2+1}]5(x^2+1)^x$$

(c) $\ln(f(x)) = \ln((2x^4+5)^{3x+1}) = (3x+1)\ln(2x^4+5) \Rightarrow \frac{f'(x)}{f(x)} = 3\ln(2x^4+5) + (3x+1)\frac{8x^3}{2x^4+5}$

$$\Rightarrow f'(x) = (2x^4+5)^{3x+1}[3\ln(2x^4+5) + (3x+1)\frac{8x^3}{2x^4+5}]$$

**Problem 3.**

(a) $\ln(g(t)) = \ln(\frac{2^t}{t^{2t}}) = t\ln 2 - 2t\ln t \Rightarrow \frac{g'(t)}{g(t)} = \ln 2 - [2\ln t + 2t\frac{1}{t}] \Rightarrow g'(t) = \frac{2^t}{t^{2t}}[\ln 2 - 2\ln t - 2]$

(b) $g(t) = \ln(t+1)^{t^2+1} = (t^2+1)\ln(t+1) \Rightarrow g'(t) = 2t\ln(t+1) + (t^2+1)\frac{1}{t+1} = 2t\ln(t+1) + \frac{t^2+1}{t+1}$

**Problem 4.**

(a) $\ln y = \ln(x^{\ln \sqrt{x}}) = \ln x^{1/2} \ln x = \frac{1}{2} \ln x \ln x = \frac{1}{2}(\ln x)^2$

$$\Rightarrow \frac{y'}{y} = \frac{1}{2}(2 \ln x)\frac{1}{x} = \frac{\ln x}{x} \Rightarrow y' = \frac{\ln x}{x} x^{\ln \sqrt{x}} = (\ln x)x^{\ln \sqrt{x} - 1}$$

(b) $\ln y = \ln \left( \frac{xe^{5x}}{(x+1)^2 \sqrt{x-2}} \right) = \ln x + \ln e^{5x} - \ln(x+1)^2 - \ln \sqrt{x-2} = \ln x + 5x - 2\ln(x+1) - \frac{1}{2}\ln(x-2)$

$$\Rightarrow \frac{y'}{y} = \frac{1}{x} + 5 - 2\frac{1}{x+1} - \frac{1}{2}\frac{1}{x-2} \Rightarrow y' = (\frac{1}{x} + 5 - \frac{2}{x+1} - \frac{1}{2x-4})\left( \frac{xe^{5x}}{(x+1)^2 \sqrt{x-2}} \right)$$

(c) $\ln y = \ln(e^{2x}(x^2 + 3)^5(2x^2 + 1)^3) = \ln e^{2x} + \ln(x^2 + 3)^5 + \ln(2x^2 + 1)^3$

$$= 2x + 5\ln(x^2 + 3) + 3\ln(2x^2 + 1) \Rightarrow \frac{y'}{y} = 2 + 5\frac{2x}{x^2+3} + 3\frac{4x}{2x^2+1}$$

$$\Rightarrow y' = (2 + \frac{10x}{x^2+3} + \frac{12x}{2x^2+1})(e^{2x}(x^2 + 3)^5(2x^2 + 1)^3)$$

**Problem 7.**

$$y = f(x)^{g(x)} \Rightarrow \ln y = \ln \left( f(x)^{g(x)} \right) = g(x) \ln(f(x)) \Rightarrow \frac{y'}{y} = g'(x) \ln(f(x)) + g(x)\frac{f'(x)}{f(x)}$$

$$\Rightarrow y' = \left( g'(x) \ln(f(x)) + g(x)\frac{f'(x)}{f(x)} \right) f(x)^{g(x)}$$

## Section 17.3    Implicit Differentiation

**Problem 1.**

(a) $x^2 + y^2 = 169 \Rightarrow y^2 = 169 - x^2 \Rightarrow y = \pm \left( 169 - x^2 \right)^{1/2}$. This is not a function.

(b) $\frac{dy}{dx} = \begin{cases} \frac{1}{2}(169 - x^2)^{-1/2}(-2x) = -x(169 - x^2)^{-1/2} = \frac{-x}{(169-x^2)^{1/2}} & \textit{if } y = (169 - x^2)^{1/2} \\ -\frac{1}{2}(169 - x^2)^{-1/2}(-2x) = x(169 - x^2)^{-1/2} = \frac{x}{(169-x^2)^{1/2}} & \textit{if } y = -(169 - x^2)^{1/2} \end{cases}$

(c) $2x + 2y\frac{dy}{dx} = 0 \Rightarrow 2y\frac{dy}{dx} = -2x \Rightarrow \frac{dy}{dx} = \frac{-x}{y}$

(d) Method (c) is easier. To find $\frac{dy}{dx}$ in terms of $x$ substitute for $y$ in each of the two cases.

(e) $\frac{dy}{dx}\big|_{(5,12)} = \frac{-5}{12}$ and $\frac{dy}{dx}\big|_{(5,-12)} = \frac{-5}{-12} = \frac{5}{12}$

**Problem 2.**

$$x^2 + y^2 = 25 \Rightarrow 2x + 2y\frac{dy}{dx} = 0 \Rightarrow \frac{dy}{dx} = \frac{-x}{y} \Rightarrow \frac{dy}{dx}\big|_{(4,-3)} = \frac{-4}{-3} = \frac{4}{3}$$

$$x^2 + 4y^2 = 25 \Rightarrow 2x + 8y\frac{dy}{dx} = 0 \Rightarrow \frac{dy}{dx} = \frac{-x}{4y} \Rightarrow \frac{dy}{dx}\big|_{(4,-\frac{3}{2})} = \frac{-4}{4(-\frac{3}{2})} = \frac{2}{3}$$

The slope is larger at (4, -3). You can see the same result if you carefully sketch the graphs.

**Problem 4.**

$$(x - 2)^2 + (y - 3)^2 = 25. \text{ When } x = 6, \ 16 + (y - 3)^2 = 25 \Rightarrow (y - 3)^2 = 9$$

$\Rightarrow y - 3 = \pm 3 \Rightarrow y = 0, 6$. From implicit differentiation $2(x - 2) + 2(y - 3)\frac{dy}{dx} = 0$

At $(6, 0)$, $2(4) + 2(-3)\frac{dy}{dx}|_{(6,0)} = 0 \Rightarrow \frac{dy}{dx}|_{(6,0)} = \frac{4}{3}$

At $(6, 6)$, $2(4) + 2(3)\frac{dy}{dx}|_{(6,6)} = 0 \Rightarrow \frac{dy}{dx}|_{(6,6)} = \frac{-4}{3}$

## Problem 6.

$x^3 + 3y + y^2 = 6 \Rightarrow 3x^2 + 3\frac{dy}{dx} + 2y\frac{dy}{dx} = 0 \Rightarrow (3 + 2y)\frac{dy}{dx} = -3x^2 \Rightarrow \frac{dy}{dx} = \frac{-3x^2}{3+2y}$.

Hence $\frac{dy}{dx}|_{(2,-1)} = \frac{-3(4)}{3+2(-1)} = -12$.

$\frac{dy}{dx} = 0$ when $x = 0 \Rightarrow 0 + 3y + y^2 = 6 \Rightarrow y^2 + 3y - 6 = 0 \Rightarrow y = \frac{-3 \pm \sqrt{9 - 4(6)}}{2} = \frac{-3 \pm \sqrt{33}}{2}$.

Hence the slope $= 0$ at points $(0, \frac{-3+\sqrt{33}}{2})$ and $(0, \frac{-3-\sqrt{33}}{2})$.

## Problem 9.

(a) $x^2 + y^2 = 4 \Rightarrow 2x + 2y\frac{dy}{dx} = 0 \Rightarrow \frac{dy}{dx} = \frac{-x}{y}$

(b) $\frac{d^2y}{dx^2} = \frac{d}{dx}\left(-\frac{x}{y}\right) = \frac{(-1)y - (-x)\frac{dy}{dx}}{y^2} = \frac{-y + x\left(-\frac{x}{y}\right)}{y^2} = \frac{\frac{-y^2 - x^2}{y}}{y^2} = \frac{-(y^2 + x^2)}{y^3} = \frac{-4}{y^3}$.

We replaced $x^2 + y^2$ by 4 because of our original relationship.

## Problem 10.

$2(x^2 + y^2)^2 = 25(x^2 - y^2) \Rightarrow 4(x^2 + y^2)(2x + 2y\frac{dy}{dx}) = 25(2x - 2y\frac{dy}{dx})$. At point (-3, 1)

$$4(9 + 1)(-6 + 2\frac{dy}{dx}) = 25(-6 - 2\frac{dy}{dx}) \Rightarrow -240 + 80\frac{dy}{dx} = -150 - 50\frac{dy}{dx} \Rightarrow \frac{dy}{dx} = \frac{9}{13}.$$

## Problem 12.

(a) $6x + 12y\frac{dy}{dx} + 3y + 3x\frac{dy}{dx} = 0 \Rightarrow (12y + 3x)\frac{dy}{dx} = -6x - 3y \Rightarrow \frac{dy}{dx} = \frac{-(2x+y)}{(x+4y)}$

(b) $(x - 2)^3 = (y - 2)^{-3} \Rightarrow 3(x - 2)^2 = -3(y - 2)^{-4}\frac{dy}{dx} \Rightarrow \frac{dy}{dx} = -(x - 2)^2(y - 2)^4 = -\frac{(y-2)}{(x-2)}$

(c) $y^2 + x(2y\frac{dy}{dx}) + 2\frac{dy}{dx} = 2xy + x^2\frac{dy}{dx} \Rightarrow (2xy + 2 - x^2)\frac{dy}{dx} = 2xy - y^2 \Rightarrow \frac{dy}{dx} = \frac{2xy - y^2}{2xy + 2 - x^2}$

(d) $2(x^2y^3 + y)(2xy^3 + x^2(3y^2\frac{dy}{dx}) + \frac{dy}{dx}) = 3 \Rightarrow 2xy^3 + (3x^2y^2 + 1)\frac{dy}{dx} = \frac{3}{2(x^2y^3+y)}$

$$\Rightarrow \frac{dy}{dx} = \frac{\frac{3}{2(x^2y^3+y)} - 2xy^3}{3x^2y^2 + 1} = \frac{3 - 4x^3y^6 - 4xy^4}{2(x^2y^3 + y)(3x^2y^2 + 1)} = \frac{3 - 4x^3y^6 - 4xy^4}{6x^4y^5 + 8x^2y^3 + 2y}$$

(e) $e^{xy}(y + x\frac{dy}{dx}) = 2y\frac{dy}{dx} \Rightarrow xe^{xy}\frac{dy}{dx} - 2y\frac{dy}{dx} = -ye^{xy} \Rightarrow \frac{dy}{dx} = \frac{ye^{xy}}{2y - xe^{xy}}$

(f) $x(\ln x + 3\ln y) = y^2 \Rightarrow x\ln x + 3x\ln y = y^2 \Rightarrow$

$$\ln x + x\frac{1}{x} + 3\ln y + 3x\frac{1}{y}\frac{dy}{dx} = 2y\frac{dy}{dx} \Rightarrow \frac{dy}{dx} = \frac{\ln x + 1 + 3\ln y}{2y - \frac{3x}{y}} = \frac{y(1 + \ln x + 3\ln y)}{2y^2 - 3x}$$

(g) $\frac{1}{xy}(y + x\frac{dy}{dx}) = y^2 + 2xy\frac{dy}{dx} \Rightarrow \frac{1}{x} + \frac{1}{y}\frac{dy}{dx} = y^2 + 2xy\frac{dy}{dx} \Rightarrow (\frac{1}{y} - 2xy)\frac{dy}{dx} = y^2 - \frac{1}{x} \Rightarrow \frac{dy}{dx} = \frac{xy^3 - y}{x - 2x^2y^2}$

## Problem 14.

$x^{2/3} + y^{2/3} = 5 \Rightarrow \frac{2}{3}x^{-1/3} + \frac{2}{3}y^{-1/3}\frac{dy}{dx} = 0 \Rightarrow \frac{dy}{dx} = \frac{-y^{1/3}}{x^{1/3}} \Rightarrow \frac{dy}{dx}|_{(8,1)} = \frac{-1}{2}$. The line is $y = -\frac{1}{2}x + 5$.

## Section 17.4    Implicit Differentiation in Context: Related Rates of Change

**Problem 1.**

$A = \pi r^2 \;\Rightarrow\; \frac{dA}{dt} = \pi 2r\frac{dr}{dt}$. At $d = 4$, $r = 2$ so $\frac{dA}{dt} = \pi 2(2)(3) = 12\pi \; {}^{ft^2}/_{sec}$

**Problem 3.**

$x^2 + y^2 = 169 \;\Rightarrow\; 2x\frac{dx}{dt} + 2y\frac{dy}{dt} = 0 \;\Rightarrow\; \frac{dx}{dt} = -\frac{y}{x}\frac{dy}{dt}$.

(a) At point $(-5, 12)$, $\frac{dy}{dt} = -3$ so the bug is going counterclockwise.

(b) $\frac{dx}{dt}\big|_{(-5,12)} = -\frac{12}{-5}(-3) = -\frac{36}{5}\frac{units}{sec}$.

**Problem 5.**

(a) $V = \pi r^2 h = \pi(3)^2 h = 9\pi h$

(b) $\frac{dV}{dh} = 9\pi$ is constant as a change in h will cause a similar change in volume.

(c) i.$\frac{dV}{dh} = 9\pi\frac{dh}{dt}$; $\frac{dV}{dh} = 9\pi(-\frac{1}{2}) = -\frac{9}{2}\pi\frac{ft^3}{hr}$. ii. $-3 = 9\pi\frac{dh}{dt} \;\Rightarrow\; \frac{dh}{dt} = -\frac{1}{3\pi}\frac{ft}{hr}$

**Problem 6.**

$\frac{db}{dt} = -40\frac{m}{hr}$, $\frac{da}{dt} = 60\frac{m}{hr}$, $a^2 + b^2 = c^2$, when $t = \frac{1}{2}$,

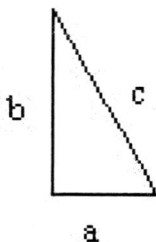

$$b = 100 - 40(\tfrac{1}{2}) = 80, \quad a = 60(\tfrac{1}{2}) = 30, \quad c^2 = 80^2 + 30^2 \;\Rightarrow\; c = 10\sqrt{73}$$

$$2c\frac{dc}{dt} = 2a\frac{da}{dt} + 2b\frac{db}{dt} \;\Rightarrow\; \frac{dc}{dt} = \frac{a\frac{da}{dt} + b\frac{db}{dt}}{c} = \frac{30(60)+80(-40)}{10\sqrt{73}} = \frac{-140}{\sqrt{73}}\frac{m}{hr} \approx -16.4\frac{m}{hr}$$

**Problem 8.**

$V = \frac{1}{3}\pi r^2 h$. By similar triangles $\frac{h}{r} = \frac{12}{6} \;\Rightarrow\; h = 2r$ so $r = \frac{1}{2}h$

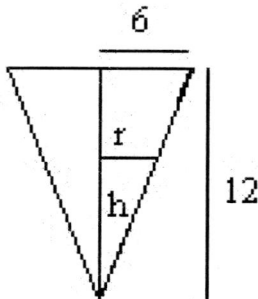

Substituting gives $V = \frac{1}{3}\pi(\frac{1}{2}h)^2 h = \frac{1}{12}\pi h^3$

$\Rightarrow \frac{dV}{dt} = \frac{1}{12}\pi 3h^2 \frac{dh}{dt} \Rightarrow -2 = \frac{1}{4}\pi(10)^2\frac{dh}{dt} \Rightarrow \frac{dh}{dt} = \frac{-2}{25\pi}\frac{cm}{\sec}$

**Problem 11.**

(a) $V = \frac{1}{3}\pi r^2 h \Rightarrow \frac{dV}{dt} = \frac{1}{3}\pi(2r\frac{dr}{dt}h + r^2\frac{dh}{dt})$.

(b) By similar triangles $\frac{r}{h} = \frac{12}{24} \Rightarrow r = \frac{1}{2}h$

  substituting gives $V = \frac{1}{3}\pi(\frac{1}{2}h)^2 h = \frac{1}{12}\pi h^3$

(c)  $\frac{dV}{dt} = \frac{1}{4}\pi h^2\frac{dh}{dt}$

  when $h = 10, \quad -5 = \frac{1}{4}\pi(10)^2\frac{dh}{dt} \Rightarrow \frac{dh}{dt} = -\frac{20}{100\pi} = -\frac{1}{5\pi}\frac{in}{hr}$

  when $h = 4, \quad -5 = \frac{1}{4}\pi(4)^2\frac{dh}{dt} \Rightarrow \frac{dh}{dt} = -\frac{5}{4\pi}\frac{in}{hr}$

  Hence the height is decreasing much faster when $h = 4$ than when $h = 10$. There is a much smaller volume of oil per vertical inch when $h = 4$.

**Problem 12.**

  $z^2 = 1^2 + x^2 \Rightarrow 2z\frac{dz}{dt} = 2x\frac{dx}{dt} \Rightarrow \frac{dz}{dt} = \frac{x}{z}\frac{dx}{dt}$ when $z = 10, \quad 10^2 = 1 + x^2 \Rightarrow 99 = x^2$

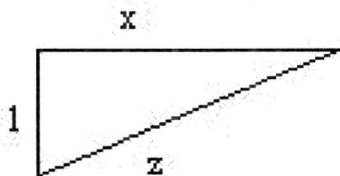

  hence $\frac{dz}{dt} = \frac{\sqrt{99}}{10}(600) = 180\sqrt{11} \approx 597\frac{m}{hr}$

**Problem 14.**

  By similar triangles $\frac{x+l}{14} = \frac{l}{5.5} \Rightarrow 5.5x + 5.5l = 14l \Rightarrow 8.5l = 5.5x$

$$\Rightarrow 8.5\frac{dl}{dt} = 5.5\frac{dx}{dt} \Rightarrow \frac{dl}{dt} = \frac{5.5(4)}{8.5} = \frac{44}{17} \approx 2.588\frac{ft}{\sec}$$

# CHAPTER 18
# Geometric Sums, Geometric Series

## Section 18.1    Geometric Sums

**Problem 1.**
   geometric a = 1 r = -10

**Problem 3.**
   geometric a = 0.3 r = 0.1

**Problem 4.**
   geometric a = 2/3 r = 1/3

**Problem 8.**
   geometric a = 2/3 r = 3/2

**Problem 10.**
   not geometric

**Problem 14.**
   geometric a = -2 r = -2

**Problem 16.**
   not geometric

**Problem 18.**
   $a = 1$, r = -10, $S - (-10)S = 1 - (-10)^{11} \Rightarrow S = 9090909091$

**Problem 20.**
   $a = 2/3$, r = 3, $S - 3S = {}^{2}\!/_{3} - {}^{2}\!/_{3}(3)^{101} \Rightarrow S \approx 5.154 \times 10^{47}$

**Problem 22.**
   $a = 2e$, r = 2e, $S - 2eS = 2e - (2e)^{n+1} \Rightarrow S = \frac{2e-(2e)^{n+1}}{1-2e}$

**Problem 25.**
   $a = 5$, r = 3, $S - 3S = 5 - 5(3^{11}) \Rightarrow S = 442865$

**Problem 29.**
   $a = -5/2$, r $= -2/v$, $S - (-\frac{2}{v})S = -\frac{5}{2} - \frac{5(-2)^{n+1}}{v^{n+2}} \Rightarrow S = \frac{-\frac{5}{2} - \frac{5(-2)^{n+1}}{v^{n+2}}}{1-(-\frac{2}{v})}$

**Problem 31.**
   $a = $ , r $= mq^3$, $S - mq^3 S = mq - m^{12}q^{34} \Rightarrow S = \frac{mq-m^{12}q^{34}}{1-mq^3}$

**Problem 33.**
   [10 ft.] down + [10 ft.](0.7) up +[10ft.](0.7) down + [10 ft.(0.7)](0.7) up + [10 ft.(0.7)](0.7) down = $10 + 20(0.7) + 20(0.7)^2 = 33.8$ ft. travelled when it hits the ground the $3^{rd}$ time.

   $10 + 20[(0.7) + (0.7)^2 + \cdots + (0.7)^{11}] = 10 + 20[\frac{0.7-(0.7)^{12}}{1-0.7}] \cong 55.74$ft. traveled by the $12^{th}$ hit.

## Section 18.2    Infinite Geometric Series

**Problem 1.**

$r = -10 \Rightarrow |r| > 1 \Rightarrow$ diverges

**Problem 3.**

$r = \frac{1}{3}$, $a = \frac{2}{3}$, $\Rightarrow$ converges to $\frac{a}{1-r} = \frac{\frac{2}{3}}{1-(\frac{1}{3})} = 1$

**Problem 6.**

$r = -\frac{1}{2}$, $a = \frac{1}{4}$, $\Rightarrow$ converges to $\frac{\frac{1}{4}}{1-(-\frac{1}{2})} = \frac{1}{6}$

**Problem 7.**

$r = \frac{3}{2} \Rightarrow |r| > 1 \Rightarrow$ diverges

**Problem 9.**

$r = \frac{1}{e}$, $a = e \Rightarrow$ converges to $\frac{e}{1-\frac{1}{e}} = \frac{e^2}{e-1}$

**Problem 11.**

$r = \frac{1}{2e}$, $a = \frac{1}{(2e)^2}$, $\Rightarrow$ converges to $\frac{\frac{1}{(2e)^2}}{1-\frac{1}{2e}} = \frac{1}{4e^2-2e}$

**Problem 12.**

(a) $r = 3$, $a = 3 \Rightarrow S = \frac{3-3^{21}}{1-3} = 5230176600$ (infinite sum diverges as $|r| > 1$)

(b) $r = \frac{2}{3}$, $a = \frac{2}{3}$, $\Rightarrow$ converges to $\frac{\frac{2}{3}}{1-(\frac{2}{3})} = 2$

(c) $r = 10 \Rightarrow |r| > 1 \Rightarrow$ diverges

(d) $r = 0.8$, $a = 3$, $\Rightarrow$ converges to $\frac{3}{1-(0.8)} = \frac{3}{0.2} = 15$

(e) $r = 1.3 \Rightarrow |r| > 1 \Rightarrow$ diverges

(f) $r = x^2$, $a = 1$, $\Rightarrow$ converges to $\frac{1}{1-(x^2)}$ for $-1 < x < 1$

**Problem 14.**

(a) $r = 0.1$, $a = 2$, $\Rightarrow 2.222\bar{2}\cdots = \frac{2}{1-0.1} = \frac{20}{9}$

(b) first we separate the 3 from the rest of the sum, then

$3.12\overline{12}\cdots = 3 + \frac{0.12}{1-0.01} = 3 + \frac{4}{33} = \frac{103}{33}$ (used $r = 0.01$, $a = 0.12$)

## Section 18.3     A More General Discussion of Infinite Series

**Problem 2.**

Looking at the sequence of terms, $\lim\limits_{n\to\infty} a_n = \lim\limits_{n\to\infty} \frac{n}{1000} \neq 0$, hence the series diverges.

**Problem 4.**

For each $n$, $S_n = \frac{1}{2\cdot2^2} + \frac{1}{3\cdot2^3} + \frac{1}{4\cdot2^4} + \frac{1}{5\cdot2^5} + \cdots \frac{1}{n\cdot2^n} < \frac{1}{2^2} + \frac{1}{2^3} + \frac{1}{2^4} + \frac{1}{2^5} + \cdots\frac{1}{2^{n+1}} + \cdots = \frac{1}{2}$ (geometric series).

As the sequence of partial sums is increasing and bounded the series converges.

**Problem 6.**

Looking at the sequence of terms, $\lim\limits_{n\to\infty} a_n \neq 0$, hence the series diverges.

**Problem 9.**

For each $n$, $a_n = (1+\frac{1}{n})^n \geq 1 \;\Rightarrow\; \lim\limits_{n\to\infty} a_n \neq 0$, hence the series diverges.

**Problem 12.**

There are many examples for each of these, such as:

(a)  Use any geometric series with $0 < r < 1$.

(b)  Use any geometric series with $-1 < r < 0$.

(c)  Use a series similar to a multiple of the harmonic series.

(d)  Any oscillating series like $a - a + a - a + a - \cdots$ will work.

## Section 18.4     Summation Notation

**Problem 1.**
$$\sum_{k=2}^{19} (k)^3$$

**Problem 3.**
$$\sum_{k=2}^{100} k^{k+1}$$

**Problem 5.**
$$\sum_{k=0}^{\infty} (-10)^k$$

**Problem 6.**
$$\sum_{k=1}^{\infty} 3\,(0.1)^k$$

**Problem 8.**

(a)  $\sum\limits_{k=0}^{\infty} \left(-\frac{1}{2}\right)^k$

(b)  $\sum\limits_{k=2}^{\infty} \left(-\frac{1}{2}\right)^k$

**Problem 9.**

(a) $\displaystyle\sum_{k=0}^{\infty} \frac{2}{3}\left(\frac{3}{2}\right)^k$

(b) $\displaystyle\sum_{k=0}^{\infty} \frac{3}{2}\left(-\frac{1}{2}\right)^k$

**Problem 10.**

(a) $\displaystyle\sum_{k=-1}^{\infty} e^{-k}$

(b) $\displaystyle\sum_{k=1}^{\infty} 2\,e^k$ (c) $\displaystyle\sum_{k=2}^{\infty} (2e)^{-k}$

**Problem 11.**

(a) $-\frac{1}{3} + \frac{1}{9}$, geometric with $r = -\frac{1}{3}$ hence converges to $\dfrac{-\frac{1}{3}}{1-\left(-\frac{1}{3}\right)} = -\frac{1}{4}$

(b) $\frac{1}{9} + \left(-\frac{1}{27}\right)$, geometric with $r = -\frac{1}{3}$ hence converges to $\dfrac{\frac{1}{9}}{1-\left(-\frac{1}{3}\right)} = \frac{1}{12}$

**Problem 13.**
$\frac{9}{4} + \frac{27}{16}$, geometric with $r = \frac{3}{4}$ hence converges to $\dfrac{\frac{9}{4}}{1-\frac{3}{4}} = 9$

**Problem 15.**
$\frac{10}{100} + \frac{10}{101}$, diverges as it is 10 times a tail of the harmonic series which diverges. The first 99 terms of a series are not a factor in deciding whether a series converges or not. However, they are important if we need to know to what the series converges to if it converges.

**Problem 17.**

(a) $a = q^5$, $r = -q^2$, hence sum $= \dfrac{q^5 - q^{43}}{1-(-q^2)} = \dfrac{q^5 - q^{43}}{1+q^2}$

(b) $\displaystyle\sum_{k=0}^{18} q^5(-q^2)^k = \sum_{k=0}^{18} (-1)^k q^{2k+5}$

(c) $\displaystyle\sum_{k=0}^{18} (-1)^{k+1} q^{2k+5}$

**Problem 20.**

(a) $\displaystyle\sum_{k=1}^{20} 500(e^{0.1})^k = \dfrac{500 - 500e^{2.1}}{1-e^{0.1}}$

(b) $\displaystyle\sum_{k=0}^{\infty} \frac{5}{3}\left(-\frac{1}{2}\right)^k = \dfrac{\frac{5}{3}}{1-\left(-\frac{1}{2}\right)} = \frac{10}{9}$

## Section 18.5    Applications of Geometric Sums and Series

**Problem 1.**

A1(a)  $\frac{10000}{(1.04)^3} + \frac{10000}{(1.04)^4} + \cdots + \frac{10000}{(1.04)^8} = 10000\left(\frac{\frac{1}{(1.04)^3} - \frac{1}{(1.04)^9}}{1 - \frac{1}{1.04}}\right) = \$48466.50$

A2(a)  $\frac{20000}{(1.04)^2} + \frac{20000}{(1.04)^4} + \frac{20000}{(1.04)^6} = 20000\left(\frac{\frac{1}{(1.04)^2} - \frac{1}{(1.04)^8}}{1 - \frac{1}{(1.04)^2}}\right) = \$51393.50$

A1(b)  $10000(1.04)^5 + 10000(1.04)^4 + \cdots + 10000 = 10000\left(\frac{1 - (1.04)^6}{1 - 1.04}\right) = \$66329.75$

A2(b)  $20000(1.04)^6 + 20000(1.04)^4 + 20000(1.04)^2 = 20000\left(\frac{(1.04)^2 - (1.04)^8}{1 - (1.04)^2}\right) = \$70335.55$

**Problem 3.**

(a)  $\frac{6000}{e^{(0.04)(5)}} = \$4912.38$

(b)  $\frac{2000}{e^{(0.04)(3)}} + \frac{2000}{e^{(0.04)(4)}} + \frac{2000}{e^{(0.04)(5)}} = 2000\left(\frac{\frac{1}{e^{0.12}} - \frac{1}{e^{0.24}}}{1 - \frac{1}{e^{0.04}}}\right) = \$5115.59$

(c)  b, as we get paid the same amount of money but some comes sooner in part b.

**Problem 6.**
$$10000 = \frac{P}{(1.06)} + \frac{P}{(1.06)^2} + \cdots + \frac{P}{(1.06)^8} = \frac{P\left(\frac{1}{1.06} - \left(\frac{1}{1.06}\right)^9\right)}{1 - \left(\frac{1}{1.06}\right)} = P(6.20979) \Rightarrow P \approx \$1610.36$$

**Problem 8.**
$$PV = \frac{1000}{(1.06)} + \frac{1000}{(1.06)^2} + \cdots + \frac{1000}{(1.06)^{10}} = \frac{1000\left(\frac{1}{1.06} - \left(\frac{1}{1.06}\right)^{11}\right)}{1 - \left(\frac{1}{1.06}\right)} = \$7360.09$$

**Problem 10.**
$$PV = \frac{1000}{(1 + \frac{0.06}{12})^{12}} + \frac{1000}{(1.005)^{24}} + \cdots + \frac{1000}{(1.005)^{360}} = \frac{1000\left(\left(\frac{1}{1.005}\right)^{12} - \left(\frac{1}{1.005}\right)^{372}\right)}{1 - \left(\frac{1}{1.005}\right)^{12}} = \$13521.20$$

**Problem 11.**
$$4000 = P\left(1 + \frac{.045}{12}\right)^{23} + P(1.00375)^{22} + \cdots P = \frac{P(1 - (1.00375)^{24})}{1 - (1.00375)} = 25.06P \Rightarrow P = \$159.59$$

**Problem 13.**

(a)  $\frac{1}{2}(3) = 3e^{k(7)} \Rightarrow \ln\left(\frac{1}{2}\right) = k(7) \Rightarrow k = \frac{\ln(0.5)}{7} \cong -0.099021 \Rightarrow T \approx 3e^{(-0.099)t}$

(b)  $A = 3e^{(-0.099)4} + 3e^{(-0.099)3} + \cdots + 3 = \frac{3(1 - e^{(-0.099)5})}{1 - e^{(-0.099)1}} \approx 12.426$

(c)  $Before = 3e^{(-0.099)30} + 3e^{(-0.099)29} + \cdots + 3e^{(-0.099)1} = \frac{3(e^{(-0.099)1} - e^{(-0.099)31})}{1 - e^{(-0.099)1}} \approx 27.344$

$After = Before + 3 \approx 30.344$

(d)  $Long\,Term = 3 + 3e^{(-0.099)1} + 3e^{(-0.099)2} + \cdots = \frac{3}{1 - e^{(-0.099)}} \approx 31.821$

**Problem 14.**

$$PV = \frac{10000}{(1+\frac{0.06}{4})^{36}} + \frac{10000}{(1.015)^{40}} + \cdots + \frac{10000}{(1.015)^{64}} = \frac{10000\left((\frac{1}{1.015})^{36} - (\frac{1}{1.015})^{68}\right)}{1-(\frac{1}{1.015})^4} = \$38355.13$$

**Problem 16.**

(a) $100000 = \frac{P}{(1.01)^{360}} + \frac{P}{(1.01)^{359}} + \cdots + \frac{P}{(1.01)^1} = \frac{P\left((\frac{1}{1.01})^1 - (\frac{1}{1.01})^{361}\right)}{1-(\frac{1}{1.01})} \Rightarrow P \approx \$1028.61$

(b) $100000 = \dfrac{Q\left(\dfrac{1}{1+\frac{0.0675}{12}} - \left(\dfrac{1}{1+\frac{0.0675}{12}}\right)^{361}\right)}{1-\left(\dfrac{1}{1+\frac{0.0675}{12}}\right)} \Rightarrow Q \approx 648.60$, hence she has monthly savings of $380.01

**Problem 19.**

Using the half-life, $1 = 2e^{3k} \Rightarrow \ln(\frac{1}{2}) = 3k \Rightarrow k = \frac{1}{3}\ln(\frac{1}{2})$.

(a) $SAR = 2 + 2e^{3k} + 2e^{6k} + \cdots = 2 + 1 + \frac{1}{2} + \frac{1}{4} + \cdots = \frac{2}{1-\frac{1}{2}} = 4$ is upper limit, never quite reached.

(b) $SAB = 2 + 2e^{2k} + 2e^{4k} + \cdots = \frac{2}{1-e^{2k}} \approx 5.4048$, so he eventually gets to a level above the equivalent of taking 5 all at once.

**Problem 21.**

$3000/30 = 100$. She would need to put aside $100/month if she gained no interest, hence she will need to put away less/month when gaining interest.

$$3000 = P(1 + \tfrac{.04}{12})^{29} + P(1 + \tfrac{.04}{12})^{28} + \cdots P = \frac{P(1-(1+\frac{.04}{12})^{30})}{1-(1+\frac{.04}{12})} \Rightarrow P = \$95.25$$

**Problem 22.**

$18000/6 = 3000$. If there was no interest you would pay 6 payments of $3000. Hence you will pay more. Note that the last payment is made at the beginning of the $10^{th}$ year or end of $9^{th}$ year. If you made one payment at the end it would be $18000(1.08)^9 = \$35982.08$ so divide by 6 we get an upper bound of $5997.02.

$$18000 = \frac{P}{(1.08)^4} + \frac{P}{(1.08)^5} + \cdots + \frac{P}{(1.08)^9} = \frac{P\left((\frac{1}{1.08})^4 - (\frac{1}{1.08})^{10}\right)}{1-(\frac{1}{1.08})} \Rightarrow P \approx \$4904.91$$

**Problem 25.**

$\frac{1}{2} = e^{10k} \Rightarrow k = \frac{1}{10}\ln(\frac{1}{2}) \approx -0.0693147$

$$L = P + Pe^k + Pe^{2k} + \cdots = \frac{P}{1-e^k} \approx 14.93P \Rightarrow P \approx \frac{L}{14.93}$$

**Problem 27.**

$$3000 = \frac{P}{(1.005)^{12}} + \frac{P}{(1.005)^{13}} + \cdots + \frac{P}{(1.005)^{35}} = \frac{P\left((\frac{1}{1.005})^{12} - (\frac{1}{1.005})^{36}\right)}{1-(\frac{1}{1.005})}$$

$$\Rightarrow P = 3000/\left(\frac{\left((\frac{1}{1.005})^{12} - (\frac{1}{1.005})^{36}\right)}{1-(\frac{1}{1.005})}\right) \text{ is the exact answer which is} \approx \$140.46$$

# CHAPTER 19
# Trigonometry: Introducing Periodic Functions

## Section 19.1    The Sine and Cosine Functions: Definitions and Basic Properties

**Problem 2.**

(a) When the horizontal displacement is 0.3, arc $\approx$1.27 or 5.01

(b) When the vertical displacement is 0.7, arc $\approx$ 0.78 or 2.36

(c) When the vertical displacement is -0.7, arc $\approx$ 3.90 or 5.50

**Problem 3.**

(a) $\frac{12}{13}$

(b) $\frac{5}{13}$

(c) $-\frac{12}{13}$

(d) $\frac{5}{13}$

(e) $\frac{12}{13}$

(f) $\frac{5}{13}$

(g) $2w$ is in quadrant II so $cos(2w) <0$

**Problem 5.**

The point associated with arc $--x$ on the unit circle has the same horizontal displacement as $x$ and the opposite vertical displacement. Therefore:

(a) not true

(b) true

(c) true

(d) not true

**Problem 7.**

(a) undefined, oscillates

(b) undefined

(c) $+\infty$

(d) undefined, oscillates

(e) $\cos \pi = -1$

## Section 19.2     Modifying the Graphs of Sine and Cosine

**Problem 3.**

(a) A vertical shift, up if C > 0 or down if C < 0.

(b) A horizontal shift, to the left if D > 0 or right if $D < 0$

**Problem 5.**

(a) Domain = R, Range = [-3,3]

(b) Domain = R, Range = [0,2]

(c) Domain = R, Range = [-1,1]

(d) Domain = R, Range = [-3,1]

(e) Domain = $\cdots \cup [-2\pi, -\pi] \cup [0, \pi] \cup [2\pi, 3\pi] \cup \cdots$, Range = [0,1]

**Problem 6.**

(a) $b =$ half period $\Rightarrow$ $B(b) = \pi$ $\Rightarrow B = \frac{\pi}{b}$

$2(\text{amplitude}) = |k|$

$y = \frac{|k|}{2} \cos\left(\frac{\pi}{b}x\right) + \frac{k}{2}$ (note k is neg)

(b) $2 =$ half period $\Rightarrow$ $B(2) = \pi$ $\Rightarrow B = \frac{\pi}{2}$

$2(\text{amplitude}) = \frac{6\pi}{4}$

$y = -\frac{3\pi}{4} \sin\left(\frac{\pi}{2}x\right) - \frac{\pi}{2}$

(c) $3 =$ half period $\Rightarrow$ $B(3) = \pi$ $\Rightarrow B = \frac{\pi}{3}$

$2(\text{amplitude}) = 3$

$y = -\frac{3}{2} \cos\left(\frac{\pi}{3}x\right) + \frac{3}{2}$

(d) $12 = \frac{3}{2}$ period $\Rightarrow$ $B(12) = 3\pi$ $\Rightarrow B = \frac{\pi}{4}$

$2(\text{amplitude}) = 6$

$y = -3\sin\left(\frac{\pi}{4}x\right) + 6$

**Problem 9.**

(a) p = $\frac{2\pi}{3}$, a = 0.5, balance value = 0

(b) p = $6\pi$, a = 4, balance value = 0

(c) p = $10\pi$, a = $\frac{1}{\pi}$, balance value = 1

(d) p = 2, a = 4, balance value = - 4

(e) p = 2, a = 4, balance value = 0

**Problem 11.**

(a) Domain is

$$\cdots \cup [-2\pi, -\pi] \cup [0, \pi] \cup [2\pi, 3\pi] \cup \cdots$$

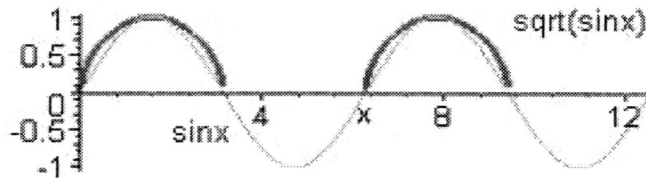

(b)

**Problem 13.**

(a) *ii*

(b) *iv*

(c) *v*

(d) *iii*

(e) *i*

(f) *vi*

**Problem 14.**

$$B(12) = 2\pi \;\Rightarrow\; B = \tfrac{\pi}{6}. \;\; A = \tfrac{450-130}{2} = 110$$
$$\text{baseline value} = \tfrac{450+230}{2} = 340 \text{ so}$$

(a) $R(t) = 110\cos(\tfrac{\pi}{6}t) + 340$

(b) average $= 340$ mm

(c) $340(12) = 4080$,

$\tfrac{4080-4370}{4080} = -7.1\%$. Prediction is too low.

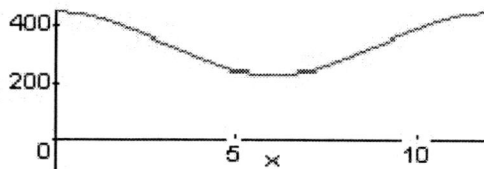

**Problem 16.**

The period $=1/$frequency, so the period $= \tfrac{1}{10} \;\Rightarrow\; B(\tfrac{1}{10}) = 2\pi \;\Rightarrow\; B = 20\pi \;\Rightarrow\; y = 3\sin(20\pi t)$
A natural alternative would be to use cos instead of sin and/or use phase shifts.

**Problem 18.**

(a) To make calculations easier we will assume $1990 \to t = 0$, $2000 \to t = 10$, $2020 \to t = 30$

Alex: $m = \frac{200}{10} = 20$, hence $L(t) = 20t + 800$

Jamey: $E(t) = 800e^{kt}$ so at $t = 0$, $100 = 800e^{k(10)}$. $\ln\frac{1000}{800} = k(10) \Rightarrow k = \frac{1}{10}\ln\frac{5}{4}$, hence

$E(t) = 800e^{\frac{1}{10}\ln\frac{5}{4}t}$

Mike: $B(20) = 2\pi \Rightarrow B = \frac{\pi}{10}$

$A = 100$, balance value $= 900$, hence $T(t) = -100\cos(\frac{\pi}{10}t) + 900$

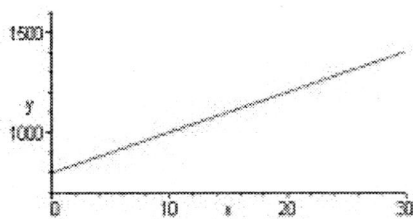

Alex

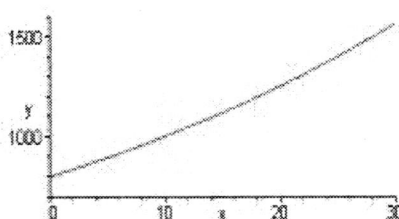

Jamey

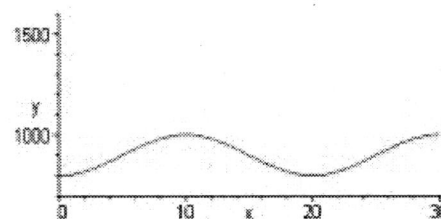

Mike

(b) $L(13) = 1060$, $E(13) \approx 1069$, $T(13) \approx 959$

(c) $L(30) = 1400$, $E(30) \approx 1563$, $T(30) = 1000$

(d) (*i*) exponential, (*ii*) trigonometric, (*iii*) linear

**Problem 19.**

(a) periodic, $p = 2\pi$

(b) periodic except at the origin

(c) periodic, $p = \pi$

(d) periodic, $p = \pi$

(e) not periodic

(f) periodic, $p = \pi$

**Problem 22.**

(a) amp$= \frac{120-80}{2} = 20$

(b) period $= 1/$frequency $= \frac{1}{70}$, so $B(\frac{1}{70}) = 2\pi \Rightarrow B = 140\pi$.

(c) $B(t) = 20\cos(140\pi t) + 100$ or $B(t) = 20\sin(140\pi t) + 100$ (balance value $= \frac{120+80}{2} = 100$)

# Section 19.3    The function f(x) = tan x

**Problem 1.**

(a) $\sin x$ has infinitely many zeros while a polynomial of degree n has at most n zeros.

All polynomials are unbounded as $x \to \pm\infty$ while $\sin x$ is bounded

(b) $\tan x$ has infinitely many zeros and infinitely many vertical asymptotes. The number of zeros of a rational function is $\leq$ the degree of the numerator and the number of asymptotes is $\leq$ the degree of the denominator.

(c) trigonometric functions are periodic.

**Problem 3.**

(a) $p = 2 = \frac{\pi}{\frac{\pi}{2}} \Rightarrow y = A\tan(\frac{\pi}{2}x)$

(b) $p = 4 = \frac{\pi}{\frac{\pi}{4}}$, with horizontal shift of 2 units, $\Rightarrow y = A\tan(\frac{\pi}{4}[x-2])$ or $y = A\tan(\frac{\pi}{4}[x+2])$

**Problem 4.**

The period of tan is $\pi$ and tan is an odd function

(a) $\tan(\alpha + \pi) = \tan(\alpha) = b$

(b) $\tan(-\alpha) = -\tan(\alpha) = -b$

(c) $\tan(\pi - \alpha) = \tan(-\alpha) = -b$

**Problem 7.**

(a) $\tan x = 1$ when terminal point on $y = x$ line, so $x = \frac{\pi}{4} + k\pi$, $k \in \mathbf{Z}$

(b) $\tan x = -1$ when terminal point is on $y = -x$, so $x = \frac{3\pi}{4} + k\pi$, $k \in \mathbf{Z}$.

**Problem 10.**

(a) undefined at $x = k\pi$, $k \in \mathbf{Z}$

(b) zeros at $x = \frac{\pi}{2} + k\pi$, $k \in \mathbf{Z}$

(c) period $= \pi$

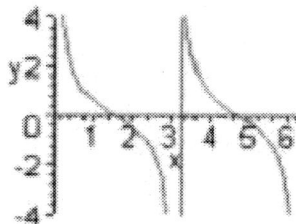

asymptotes at

$x = 0, \pi, 2\pi$

**Problem 12.**

(a) $x = \beta + k\pi$, $k \in \mathbf{Z}$

(b) $x = -\beta + k\pi, \quad k \in \mathbf{Z}$

**Problem 13.**

(a) even

(b) neither

(c) odd

(d) even

(e) neither

## Section 19.4    Angles and Arc Lengths

**Problem 1.**

(a) (i) $60° = 60° \frac{\pi}{180°} = \frac{\pi}{3}$ (ii) $\frac{\pi}{6}$ (iii) $\frac{\pi}{4}$ (iv) $-\frac{2\pi}{3}$

(b) $2^{rad} = 2\frac{180°}{\pi} = \frac{360°}{\pi} \approx 114.59°$

**Problem 2.**

(a) $-60° = -60° \frac{\pi}{180°} = -\frac{\pi}{3}$

(b) $\frac{\pi}{4}$

(c) $-\frac{3\pi}{2}$

(d) $\frac{2\pi}{9}$

(e) $-\frac{2\pi}{3}$

**Problem 4.**

(a) 20 seconds yields $\frac{2\pi}{3}$ radians, hence distance $= \frac{2\pi}{3}(6) = 4\pi in.$

(b) distance $= 70° \frac{\pi}{180°}(6) = \frac{7\pi}{3} in.$

(c) $120\pi$ for second hand,  $\frac{\pi}{6}$ for hour hand

**Problem 5.**
circumference $= 2\pi(4) in$, distance $= (8\pi \frac{in}{rev})(50\frac{rev}{\min})(2min) = 800\pi in.$

**Problem 7.**
distance $= \frac{150,000,000(2\pi)}{365} \approx 2,582,131 km$

## Problem 9.

(a) $\pi t = 2\pi \;\Rightarrow\; t = 2\,\text{sec}$

(b) circumference for one revolution $= 10\pi$

(c) once

(d) $\dfrac{chain\ length}{circumference} = \dfrac{10\pi}{4\pi} = \dfrac{5}{2}rev,\ \dfrac{chain\ length}{circumference} = \dfrac{5\pi}{4\pi} = \dfrac{5}{4}rev$

(e) $h(t) = 2\cos(\frac{5\pi}{2}t)$

## Problem 11.
$P(\theta) = (-\frac{12}{13}, \frac{5}{13}),\ \ P(-\theta) = (-\frac{12}{13}, -\frac{5}{13}),\ \ P(\pi - \theta) = (\frac{12}{13}, \frac{5}{13})$

(a) $-\frac{12}{13}$

(b) $-\frac{5}{12}$

(c) $-\frac{12}{13}$

(d) $-\frac{5}{13}$

(e) $\frac{5}{12}$

# CHAPTER 20

# Trigonometry – Circles and Triangles

## Section 20.1    Right-Triangle Trigonometry: The Definitions

**Problem 1.**

|            | (a)           | (b)                              | (c)                          |
|------------|---------------|----------------------------------|------------------------------|
| $\sin\theta$ | $\frac{4}{5}$ | $\frac{Q}{\sqrt{1+Q^2}}$         | $\frac{\sqrt{R^2-1}}{R}$     |
| $\cos\theta$ | $\frac{3}{5}$ | $\frac{1}{\sqrt{1+Q^2}}$         | $\frac{1}{R}$                |
| $\tan\theta$ | $\frac{4}{3}$ | $Q$                              | $\sqrt{R^2-1}$               |
| $\csc\theta$ | $\frac{5}{4}$ | $\frac{\sqrt{1+Q^2}}{Q}$         | $\frac{R}{\sqrt{R^2-1}}$     |
| $\sec\theta$ | $\frac{5}{3}$ | $\sqrt{1+Q^2}$                   | $R$                          |
| $\cot\theta$ | $\frac{3}{4}$ | $\frac{1}{Q}$                    | $\frac{1}{\sqrt{R^2-1}}$     |

**Problem 2.**

$$\tan(0.7) = \frac{h}{20} \;\Rightarrow\; h = 20\tan(0.7) \approx 16.8\,ft$$

**Problem 3.**

$$\cos\theta = \frac{7}{25} \Rightarrow \theta \approx 73.74^\circ.\ \ h^2 + 7^2 = 25^2 \Rightarrow h = 24\,ft$$

**Problem 5.**

$$\tan 15^\circ = \frac{35}{x} \Rightarrow x \approx 130.6\,ft$$
$$\tan 45^\circ = \frac{y}{x} \Rightarrow y = x \approx 130.6\,ft$$
$$35 + y \approx 165.6\,ft$$

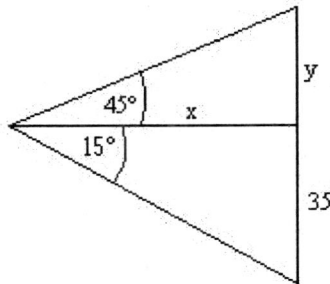

**Problem 6.**

$\tan(29°) = \frac{h}{x} \ \Rightarrow \ x\tan(29°) = h,$

$\tan(23°) = \frac{h}{500+x} \ \Rightarrow \ (500 + x)\tan(23°) = h$

equate $h$'s and solving for $x$ gives $x \approx 1635ft$ so $h \approx 906ft$, hence the building height is $\approx 911$

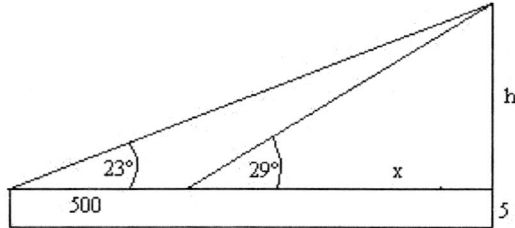

**Problem 8.**

If the astronomers knew the circumpherence of the earth and knew how long it took the moon to circle earth, they could calculate the distance $d$ using time. Then they could measure the angle between d and the hypotenuse. (They would need some fancy trigonometry to create a true right triangle. They would need to compensate for the curvature of the earth.) Then $h = d\tan(angle)$

## Section 20.2    Triangles We Know And Love, And The Information They Give Us

**Problem 1.**

(a) $\frac{1}{2}$

(b) $\frac{1}{\sqrt{2}} = \frac{\sqrt{2}}{2}$

(c) $\frac{\sqrt{3}}{2}$

(d) $-\frac{\sqrt{3}}{2}$

(e) $\frac{1}{2}$

(f) $\frac{1}{2}$

(g) $1$

(h) -1

(i) $-\sqrt{3}$

(j) $\frac{1}{2}$

**Problem 2.**

(a) $x = \frac{\pi}{3} + k\pi, \quad k \in \mathbf{Z}$

(b) $x = \frac{2\pi}{3} + k\pi, \quad k \in \mathbf{Z}$

**Problem 4.**

(a) $x = \frac{\pi}{6}$

(b) $x = \frac{\pi}{6}, \frac{5\pi}{6}$

(c) $x = \frac{\pi}{3}$

(d) $2x = \frac{\pi}{3}, \frac{5\pi}{3}, \frac{7\pi}{3}, \frac{11\pi}{3} \Rightarrow x = \frac{\pi}{6}, \frac{5\pi}{6}, \frac{7\pi}{6}, \frac{11\pi}{6}$

**Problem 6.**

| $\theta$ in degrees | $\theta$ in radians | $\sin\theta$ | $\cos\theta$ | $\tan\theta$ |
|---|---|---|---|---|
| 30° | $\frac{\pi}{6}$ | $\frac{1}{2}$ | $\frac{\sqrt{3}}{2}$ | $\frac{\sqrt{3}}{3}$ |
| 45° | $\frac{\pi}{4}$ | $\frac{\sqrt{2}}{2}$ | $\frac{\sqrt{2}}{2}$ | 1 |
| 60° | $\frac{\pi}{3}$ | $\frac{\sqrt{3}}{2}$ | $\frac{1}{2}$ | $\sqrt{3}$ |

**Problem 8.**

By symmetry the extra length on the end of top is 0.5 0.5 Then $\tan(30°) = \frac{0.5}{x} \Rightarrow x = \frac{\sqrt{3}}{2}$ sq ft.
Area = (average of bases)(height) = $\left(\frac{2.5+1.5}{2}\right) \frac{\sqrt{3}}{2} = \sqrt{3}$  $\overset{x}{30°}$

**Problem 10.**

(a) $\sin(-48°) = -\sin(48°)$

(b) $\cos(-48°) = \cos(48°)$

**Problem 12.**

(a) $\cos 130° = -\cos(50°)$

(b) $\cos(-130°) = \cos(130°) = -\cos 50°$

**Problem 13.**

(a) $\cos 200° = -\cos(200° - 180°) = -\cos 20°$

(b) $\sin 200° = -\sin(200° - 180°) = -\sin 20°$

## Section 20.3    Inverse Trigonometric Functions

**Problem 1.**

(a) $\frac{\pi}{2}$

(b) $\frac{\pi}{4}$

(c) $-\frac{\pi}{2}$

(d) $\pi$

(e) $-\frac{\pi}{6}$

(f) $\frac{2\pi}{3}$

(g) $\frac{\pi}{3}$

(h) $(\pi)^2 + (-\frac{\pi}{4})^2 = \frac{17\pi^2}{16}$

**Problem 2.**

(a) Find the arc length in quadrant I associated with a vertical displacement of 0.8 $\theta \approx 0.93$

(b) Find the arc length in quadrant I associated with a horizontal displacement of 0.8. $\theta \approx 0.93$

(c) Find the arc length in quadrant I associated with a $y$ to $x$ ratio of 2 to 1,$\theta \approx 1.1$

**Problem 4.**

(a) domain = [-1, 1], $\sin(\sin^{-1}(x)) = x$ for all $x$ in [-1,1]

(b) domain = all reals, $\sin^{-1}(\sin(x)) = x$ only for $x \in [-\frac{\pi}{2}, \frac{\pi}{2}]$, not = for any other $x$.

**Problem 7.**

(a) $\frac{3}{5}$

(b) $\sqrt{3}$

(c) $\sqrt{1-x^2}$,   $(-\frac{\pi}{2} \leq \theta \leq 0)$

(d) $\frac{w}{\sqrt{r^2-w^2}}$,   $(0 \leq \theta \leq \frac{\pi}{2})$

**Problem 8.**

(a) $\arctan(\cos(\pi)) = \arctan(-1) = -\frac{\pi}{4}$

(b) $\cos(\arctan(-\frac{a}{b})) = \frac{b}{\sqrt{a^2+b^2}}$ (positive in quadrant IV)

**Problem 10.**

(a) $\sin^{-1}(\sin(-x)) = \sin^{-1}(-\sin(x)) = -x$

(b) $\cos^{-1}(\cos(-x)) = \cos^{-1}(\cos(x)) = x$

**Problem 12.**

| Result depends on angle size | (a) $\arcsin(\sin(x))$ | (b) $\arccos(\cos(x))$ |
|---|---|---|
| $(\frac{\pi}{2}, \pi]$ | $\pi - x$ | $x$ |
| $[\pi, \frac{3\pi}{2}]$ | $\pi - x$ | $2\pi - x$ |
| $[\frac{3\pi}{2}, 2\pi)$ | $x - 2\pi$ | $2\pi - x$ |

**Problem 13.**

| Result depends on angle size | (a) $\arcsin(\sin(-x))$ | (b) $\arccos(\cos(-x))$ |
|---|---|---|
| $(\frac{\pi}{2}, \pi]$ | $x - \pi$ | $x$ |
| $[\pi, \frac{3\pi}{2}]$ | $x - \pi$ | $2\pi - x$ |
| $[\frac{3\pi}{2}, 2\pi)$ | $2\pi - x$ | $2\pi - x$ |

# Section 20.4 Solving Trigonometric Equations

**Problem 1.**

(a) $\cos x = \pm\frac{\sqrt{3}}{2} \Rightarrow x = \frac{\pi}{6},\ \frac{5\pi}{6},\ \frac{7\pi}{6},$ or $\frac{11\pi}{6}$

Solving Trigonometric Equations $2u^2 - u - 1 = 0 \Rightarrow u = -\frac{1}{2},\ 1 \Rightarrow \sin x = -\frac{1}{2},\ \sin x = 1 \Rightarrow x = \frac{7\pi}{6},\ \frac{11\pi}{6},$ or $\frac{\pi}{2}$

Solving Trigonometric Equations $x = \sin^{-1}\frac{2}{3} \approx .7297$ or $\pi - \sin^{-1}\frac{2}{3} \approx 2.4119$

**Problem 3.**

(a) $\cos^2 x = \frac{1}{2} \Rightarrow \cos x = \pm\frac{\sqrt{2}}{2} \Rightarrow x = \frac{\pi}{4} + k\frac{\pi}{2},\quad k \in \mathbf{Z},$

(b) $\cos x(\cos x - 0.2) = 0 \Rightarrow \cos x = 0$ or $0.2 \Rightarrow x = \frac{\pi}{2} + k\pi$ or

$$x = \cos^{-1}(0.2) + k2\pi \approx 1.369 + 2k\pi \text{ or } x = 2k\pi - \cos^{-1}(0.2) \approx -1.369 + 2k\pi.$$

(If we cancelled we would not find the solutions to $\cos x = 0$.)

(c) $(1 - \cos^2 x) = 3\cos x + 1 \Rightarrow \cos x(\cos x + 3) = 0 \Rightarrow$

$\cos x = 0$ or $\cos x = -3$ *(impossible)* $\Rightarrow x = \frac{\pi}{2} + k\pi,\quad k \in \mathbf{Z}$

**Problem 4.**

$\cos(x^2) = 0 \Rightarrow x^2 = \frac{\pi}{2}, \frac{3\pi}{2}, \frac{5\pi}{2}, \cdots \Rightarrow x = \pm\sqrt{\frac{\pi}{2}},\ \pm\sqrt{\frac{3\pi}{2}},\ \pm\sqrt{\frac{5\pi}{2}}$ (others outside interval)

**Problem 6.**

$\cos(2x) = 1 \Rightarrow 2x = 0,\ 2\pi,\ 4\pi \Rightarrow x = 0,\ \pi,\ 2\pi$

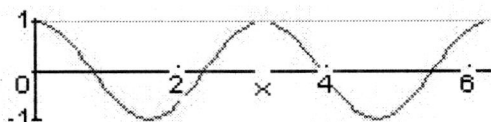

**Problem 8.**

$\tan^2 x = \frac{1}{3} \Rightarrow \tan x = \pm\frac{1}{\sqrt{3}}$

$\Rightarrow x = \frac{\pi}{6}, \frac{5\pi}{6}, \frac{7\pi}{6}, \frac{11\pi}{6}$

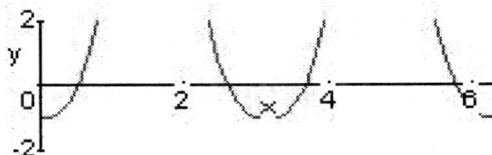

**Problem 10.**

$$\cos(3x) = 0.5 \ \Rightarrow \ 3x = \tfrac{\pi}{3} \ , \ \tfrac{5\pi}{3} + k2\pi$$
$$\Rightarrow x = \tfrac{\pi}{9}, \ \tfrac{5\pi}{9}, \ +k\tfrac{2}{3}\pi$$
$$\Rightarrow x = \tfrac{\pi}{9}, \ \tfrac{5\pi}{9}, \ \tfrac{7\pi}{9}, \ \tfrac{11\pi}{9}, \ \tfrac{13\pi}{9}, \ \tfrac{17\pi}{9}$$

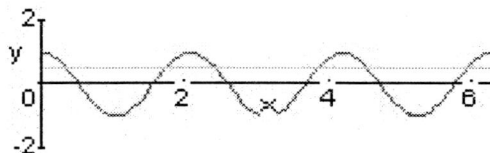

**Problem 12.**

$$u^2 - 4u - 5 = 0 \ \Rightarrow \ u = -1, \ 5$$
$$\Rightarrow \tan t = -1 \ \Rightarrow \ t = \tfrac{3\pi}{4}, \ \tfrac{7\pi}{4} \ . \text{ Also}$$
$$\tan t = 5 \ \Rightarrow \ t = \arctan 5, \ \arctan 5 + \pi$$

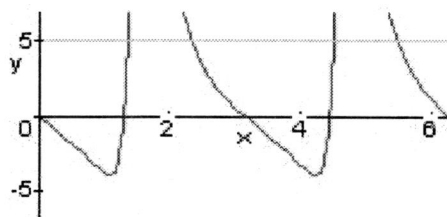

**Problem 13.**

$$1 - \sin^2 x = 1 + 2\sin x + \sin^2 x \Rightarrow$$
$$2\sin x(\sin x + 1) = 0 \Rightarrow \sin x = 0 \text{ or } \sin x = -1$$
$$x = 0, \ \pi, \ 2\pi \ or \ \tfrac{3\pi}{2} \ (\text{only } \pi \text{ doesn't work})$$
So solutions set is $\left\{0, \ 2\pi , \ \tfrac{3\pi}{2}\right\}$

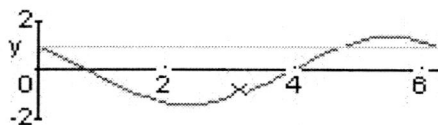

**Problem 15.**

$\cos^2(2t) = \tfrac{4}{3}$, which is impossible as $|\cos(2t)| \le 1$, hence no solution.

**Problem 16.**

$$2u^2 - 3u + 1 = 0 \ \Rightarrow \ u = \tfrac{1}{2}, \ 1 \ \Rightarrow \ \sin t = \tfrac{1}{2}, \ 1 \ \Rightarrow \ t = \tfrac{\pi}{6}, \ \tfrac{5\pi}{6} \text{ and } \tfrac{\pi}{2}$$

**Problem 18.**

$$\cos x(6\cos^2 x - \sin x - 5) = 0 \ \Rightarrow \ \cos x(6(1 - \sin^2 x) - \sin x - 5) = 0 \ \Rightarrow$$

$$\cos x(-6\sin^2 x - \sin x + 1) = 0 \ \Rightarrow \ \cos x(-3\sin x + 1)(2\sin x + 1) = 0 \ \Rightarrow \ \cos x = 0, \ \sin x = -\frac{1}{2}, \ \frac{1}{3}$$

$\Rightarrow x = \frac{\pi}{2}, \frac{3\pi}{2}$ or $x = \frac{7\pi}{6}, \frac{11\pi}{6}$ or $x = \arcsin\frac{1}{3} \approx 0.34$, $\pi - \arcsin\frac{1}{3} \approx 2.80$

**Problem 20.**

$0 \le x \le 2\pi \Rightarrow 0 \le 4x \le 8\pi$ hence we will look for all angles up to $8\pi$ which have output $\frac{1}{\sqrt{2}}$ from the sin function. $\sin(4x) = \frac{1}{\sqrt{2}} \Rightarrow 4x = \frac{\pi}{4}, \frac{3\pi}{4} + k2\pi \Rightarrow x = \frac{\pi}{16}, \frac{3\pi}{16} + \frac{k}{2}\pi$
$\Rightarrow x = \frac{\pi}{16}, \frac{9\pi}{16}, \frac{17\pi}{16}, \frac{25\pi}{16}$, and $x = \frac{3\pi}{16}, \frac{11\pi}{16}, \frac{19\pi}{16}, \frac{27\pi}{16}$

**Problem 21.**

$\tan\theta = \frac{-10}{50} \Rightarrow \theta = \tan^{-1}(-\frac{1}{5}) \approx 11.3°$

## Section 20.5    The Law of Cosines and the Law of Sines

**Problem 1.**

(a) $\frac{1}{2}(3)(4)\sin 120° = \frac{1}{2}(3)(4)\frac{\sqrt{3}}{2} = 3\sqrt{3} \approx 5.196$

(b) $\frac{1}{2}(10)(12)\sin 70° \approx \frac{1}{2}(10)(12)(.9497) = 56.38$

(c) $\frac{1}{2}(2)(2)\sin 60° = \frac{1}{2}(2)(2)\frac{\sqrt{3}}{2} = \sqrt{3} \approx 1.73$

**Problem 2.**

$\sin\theta = \sin(\pi - \theta) = \frac{4}{5}$, $\cos\theta = -\cos(\pi - \theta) = -\frac{3}{5}$, $\tan\theta = \frac{\sin\theta}{\cos\vartheta} = \frac{4}{5} \div (-\frac{3}{5}) = -\frac{4}{3}$

**Problem 4.**

(a) Let $A = 25°$, $C = 30°$ and $c = 4$.

Then $B = (180° - (25° + 30°)) = 125°$.

$$\frac{a}{\sin(25°)} = \frac{4}{\sin(30°)} \Rightarrow a = \frac{4\sin(25°)}{\sin(30°)} \approx 3.4$$

$$\frac{b}{\sin(125°)} = \frac{4}{\sin(30°)} \Rightarrow b = \frac{4\sin(125°)}{\sin(30°)} \approx 6.6$$

(b) Let $a = 2$, $b = 3$ and $c = 4$.

$$2^2 = 3^2 + 4^2 - 2(3)(4)\cos(A) \Rightarrow \cos(A) = \frac{-21}{-24} = \frac{7}{8} \Rightarrow A \approx 29.0°$$

$$3^2 = 2^2 + 4^2 - 2(2)(4)\cos(B) \Rightarrow \cos(B) = \frac{-11}{-16} = \frac{11}{16} \Rightarrow B \approx 46.6°$$

$$C = 180° - 29° - 46.6° = 104.4°$$

(c) Let $a = 6$, $b = 10$ and $B = 120°$.

$$\frac{\sin(120°)}{10} = \frac{\sin(A)}{6} \Rightarrow \sin(A) \approx .5196 \Rightarrow A \approx 31.3°$$

$$C = 180° - 120° - 31.3° = 28.7°$$

$$\frac{c}{\sin(28.7°)} = \frac{10}{\sin(120°)} \Rightarrow c \approx 5.5$$

(d) Let $a = 3$, $b = 4$ and $C = 25°$

$$c^2 = 3^2 + 4^2 - 2(3)(4)\cos(25°) \approx 3.249 \Rightarrow c \approx 1.8$$

$$\frac{\sin(25°)}{1.8} = \frac{\sin(A)}{3} \Rightarrow \sin(A) \approx .7034 \Rightarrow A \approx 44.7° \text{ (must be acute)}$$

$$B = 180° - 25° - 44.7° = 110.3°$$

**Problem 6.**
$$40^2 = 20^2 + 30^2 - 2(20)(30)\cos(D) \Rightarrow \cos(D) = \frac{300}{-1200} \Rightarrow D \approx 104.5°$$

**Problem 8.**
$$90^2 = 140^2 + 120^2 - 2(140)(120)\cos(D) \Rightarrow \cos(D) = \frac{-25900}{-33600} \approx .7708 \Rightarrow D \approx 39.6°$$
$$\text{Area} = \tfrac{1}{2}(140)(120)\sin 39.6° \approx \tfrac{1}{2}(140)(120)(.9497) = 5351\,ft^2 \Rightarrow \text{Cost} \approx \$133,775$$

**Problem 9.**
$$\sin(35°) = {}^{60}\!/_a \Rightarrow a = {}^{60}\!/_{\sin(35°)} = 104.6\,ft \quad 60$$
$$\sin(25°) = {}^{60}\!/_b \Rightarrow b = {}^{60}\!/_{\sin(25°)} = 142.0\,ft \quad a$$

$$C = 180° - 145° - 25° = 10°$$

$$c^2 = 104.6^2 + 142.0^2 - 2(104.6)(142.0)\cos(10°) \approx 1850.1 \quad b$$
$$\Rightarrow c \approx 43.0\,ft \Rightarrow speed = \tfrac{43}{15} \approx 2.9\tfrac{ft}{\min} \quad c$$

## Section 20.6    Trigonometric Identities

**Problem 1.**

(a) false, $\sin(\frac{\pi}{2} - \frac{\pi}{6}) = \frac{\sqrt{3}}{2} \neq 1 - \frac{1}{2} = \sin(\frac{\pi}{2}) - \sin(\frac{\pi}{6})$

(b) false, $\cos(\frac{\pi}{2} + \frac{\pi}{2}) = -1 \neq 0 = 0 + 0 = \cos(\frac{\pi}{2}) + \cos(\frac{\pi}{2})$

**Problem 3.**
$$\sin(x - 2y) = \sin x \cos 2y - \cos x \sin 2y = \sin x(\cos^2 y - \sin^2 y) - \cos x(2\sin y \cos y)$$

$$\cos(x - 2y) = \cos x \cos 2y + \sin x \sin 2y = \cos x(\cos^2 y - \sin^2 y) + \sin x(2\sin y \cos y)$$

**Problem 5.**
$$\tan(A + B) = \frac{\sin(A+B)}{\cos(A+B)} = \frac{(\sin A \cos B + \cos A \sin B)\frac{1}{\cos A \cos B}}{(\cos A \cos B - \sin A \sin B)\frac{1}{\cos A \cos B}} = \frac{\frac{\sin A \cos B}{\cos A \cos B} + \frac{\cos A \sin B}{\cos A \cos B}}{\frac{\cos A \cos B}{\cos A \cos B} - \frac{\sin A \sin B}{\cos A \cos B}}$$
$$= \frac{\frac{\sin A}{\cos A} + \frac{\sin B}{\cos B}}{1 - \frac{\sin A \sin B}{\cos A \cos B}} = \frac{\tan A + \tan B}{1 - \tan A \tan B}.$$ The second proof is similar with all second terms of opposite sign.

**Problem 7.**

(a) $\cos^2 x - \cos 2x = 0 \Rightarrow \cos^2 x - (\cos^2 x - \sin^2 x) = 0 \Rightarrow \sin^2 x = 0 \Rightarrow x = k\pi$

(b) $\sin x \cos x = \sqrt{3} \Rightarrow \sin(2x) = 2\sqrt{3} > 1$ which is impossible, hence no solution.

(c) $\sin^2 x - \cos 2x = 0 \Rightarrow \sin^2 x - (1 - 2\sin^2 x) = 0 \Rightarrow \sin^2 x = \frac{1}{3} \Rightarrow \sin x = \pm\sqrt{\frac{1}{3}}$

$\Rightarrow x = \sin^{-1}\sqrt{\frac{1}{3}} \approx 0.615$ or $\pi \pm \sin^{-1}\sqrt{\frac{1}{3}} \approx 2.526,\ 3.757$ or $2\pi - \sin^{-1}\sqrt{\frac{1}{3}} \approx 5.668$

(d) $-\cos^2 x + \frac{1}{2}\sin x + 1 = 0 \ \Rightarrow\ -(1 - \sin^2 x) + \frac{1}{2}\sin x + 1 = 0 \ \Rightarrow\ (\sin x)(\sin x + \frac{1}{2}) = 0$

$$\Rightarrow \sin x = 0 \ or \ \sin x = -\frac{1}{2} \ \Rightarrow\ x = 0, \ \pi, \ 2\pi \ \ or \ x = \frac{7\pi}{6}, \ \frac{11\pi}{6}$$

**Problem 8.**
$$\cos(3x) = \cos(2x + x) = \cos(2x)\cos x - \sin(2x)\sin x$$

$$= (2\cos^2 x - 1)\cos x - 2\sin x \cos x \sin x$$

$$= 2\cos^3 x - \cos x - 2\cos x \sin^2 x = 2\cos^3 x - \cos x - 2\cos x(1 - \cos^2 x)$$

$$= 2\cos^3 x - \cos x - 2\cos x + 2\cos^3 x = 4\cos^3 x - 3\cos x$$

## Section 20.7    A Brief Introduction to Vectors

**Problem 1.**
$$v_{hor} = 3\cos 60° = 3\left(\frac{1}{2}\right) = \frac{3}{2} \ v_{ver} = 3\sin 60° = 3\left(\frac{\sqrt{3}}{2}\right) = \frac{3\sqrt{3}}{2}$$

**Problem 3.**

(a) $v_{\vec{u}} = 7\cos\frac{\pi}{3} = 7\frac{1}{2} = \frac{7}{2}$

(b) $u_{\vec{v}} = 5\cos\frac{\pi}{3} = 5\frac{1}{2} = \frac{5}{2}$

**Problem 5.**

(a) $v_{\vec{u}} = 3\cos\frac{3\pi}{4} = 3\left(\frac{\sqrt{2}}{2}\right) = \frac{3\sqrt{2}}{2}$

(b) $u_{\vec{v}} = 2\cos\frac{3\pi}{4} = 2\left(\frac{\sqrt{2}}{2}\right) = \sqrt{2}$

**Problem 8.**
(What the velocity vector of the plane means is open to interpretation. Is it the resultant vector or the vector would have in the absence of wind. We will assume the first of these interpretations.)
(a) The component of the wind in the plane's direction is $50\cos 110° \approx 50(-3420) = -17.10 mph$

**Problem 10.**

(a) By right triangle the force of $A$ is 5 *pounds*. $\tan\alpha = \frac{4}{3} \Rightarrow \alpha \approx 53.13°$

(b) By right triangle the force of $B$ is 13 *pounds*. $\tan\alpha = \frac{12}{5} \Rightarrow \alpha \approx 67.38°$

(c) The angle between $A$ and $B$ is $\theta = 67.38° - 53.13° = 14.25°$

The component of force $A$ in the direction of force $B$ is $5\cos 14.25° \approx 5(.9692) = 4.846 pounds$

**Problem 11.**
$$10\cos\theta = 9 \Rightarrow \cos\theta = 0.9 \Rightarrow \theta \approx 25.84°$$

# CHAPTER 21

# Differentiation of Trigonometric Functions

## Section 21.1    Investigating the Derivative of sin x

**Problem 2.**

$f'(\pi) \cong \frac{\sin(\pi+0.0001)-\sin(\pi)}{0.0001} \cong -1.000$

**Problem 3.**

(a)

| $x$ | 1.998 | 1.999 | 2.000 | 2.001 | 2.002 |
|-----|-------|-------|-------|-------|-------|
| $f(x)$ | 0.910128 | 0.909713 | 0.909297 | 0.908881 | 0.908463 |

(b) $f'(1.999) \cong \frac{f(1.998)-f(1.999)}{1.998-1.999} \cong -0.415$, or $\frac{f(2)-f(1.999)}{2-1.999} \cong -0.416$, average is $-0.4155$

$f'(2) \cong \frac{f(1.999)-f(2)}{1.999-2} \cong -0.416$, or $\frac{f(2.001)-f(2)}{2.001-2} \cong -0.416$, they agree at $-0.416$

$f'(2.001) \cong \frac{f(2)-f(2.001)}{2-2.001} \cong -0.416$, or $\frac{f(2.002)-f(2.001)}{2.002-2.001} \cong -0.418$, average is $-0.417$

(c) $f''(2) \cong \frac{f'(1.999)-f'(2)}{1.999-2} \cong -0.5$, or $\frac{f'(2.001)-f'(2)}{2.001-2} \cong -1$, so averaging $f''(2) \approx -0.75$

Makes one curious what the results would be if used more digits of accuracy in the calculations.

**Problem 5.**

$\frac{d}{dx}\cos x|_{x=\pi} \approx \frac{\cos(\pi+.01)-\cos(\pi)}{.01} = 0.005$ or $\frac{\cos(\pi+.0001)-\cos(\pi)}{.0001} \approx 0.00005 \to 0$

## Section 21.2    Differentiating sin x and cos x

**Problem 1.**

$\frac{d}{dx}\tan x = \frac{d}{dx}\left(\frac{\sin x}{\cos x}\right) = \frac{\cos x \cos x - \sin x(-\sin x)}{(\cos x)^2} = \frac{1}{\cos^2 x} = \sec^2 x$

**Problem 3.**

(a) $y = 5\cos x$

$y' = -5\sin x$

$y'' = -5\cos x$

(b) $y = -3\sin(2x)$

$y' = -3\cos(2x)2 = -6\cos(2x)$

$y'' = 6\sin(2x)2 = 12\sin(2x)$

(c) $y = .5\tan x$

$y' = .5\sec^2 x$

$y'' = .5(2\sec x(\sec x \tan x)) = \sec^2 x \tan x$

(d) $y = 2\sin x \cos x$

$y' = 2(\cos^2 x - \sin^2 x)$

$y'' = 2(2\cos x(-\sin x) - 2\sin x \cos x) = -8\sin x \cos x$

OR equivalently $y = \sin(2x)$
$y' = 2\cos(2x)$
$y'' = -4\sin(2x)$

## Problem 4.

(a) $\frac{d}{dx}(\cos x)^2 = 2\cos x(-\sin x) = -2\cos x \sin x$

(b) $\frac{d}{dx}\cos(x^2) = -\sin(x^2)(2x) = -2x\sin(x^2)$

(c) $\frac{d}{dx}(x(\tan x)^2) = (1)(\tan x)^2 + x(2\tan x(\sec^2 x)) = \tan^2 x + 2x\tan x \sec^2 x$

(d) $\frac{d}{dx}(\sin(x^4))^3 = 3(\sin(x^4))^2\cos(x^4)4x^3 = 12x^3\sin^2(x^4)\cos(x^4)$

(e) $\frac{d}{dx}(7[\cos(5x)+3]^x) = \frac{d}{dx}7e^{x\ln(\cos(5x)+3)} = 7e^{x\ln(\cos(5x)+3)}[\ln(\cos(5x)+3) + x\frac{1}{\cos(5x)+3}(-\sin(5x)5)]$

$$= 7(\cos(5x)+3)^x[\ln(\cos(5x)+3) - \frac{5x\sin(5x)}{\cos(5x)+3}]$$

## Problem 6.

(a) $\frac{d}{dx}\sin(u(x)) = \cos(u(x))\frac{d}{dx}(u(x)) = \cos(u(x))u'(x)$

(b) $\frac{d}{dx}\cos(u(x)) = -\sin(u(x))\frac{d}{dx}(u(x)) = -\sin(u(x))u'(x)$

(c) $\frac{d}{dx}u(x)\sin x = u'(x)\sin x + u(x)\cos x$

## Problem 8.
$\frac{d}{dx}\sin(x^3 + \ln 3x) = \cos(x^3 + \ln 3x)[3x^2 + \frac{1}{3x}(3)] = \cos(x^3 + \ln 3x)(3x^2 + \frac{1}{x})$

## Problem 9.
$\frac{d}{dx}\cos^2(\sin x) = 2\cos(\sin x)(-\sin(\sin x)\cos x) = -2\cos(\sin x)\ \sin(\sin x)\ \cos x$

## Problem 11.
$\frac{d}{dx}\sqrt{\sin(2x^3)} = \frac{1}{2}(\sin(2x^3))^{-\frac{1}{2}}\cos(2x^3)6x^2 = \frac{3x^2\cos(2x^3)}{\sqrt{\sin(2x^3)}}$

## Problem 13.
$\frac{d}{dx}[e^{3x}\cos^2(7x)] = 3e^{3x}\cos^2(7x) + e^{3x}2\cos(7x)(-\sin(7x)7)$

$$= e^{3x}\cos(7x)[3\cos(7x) - 14\sin(7x)]$$

## Problem 15.

(a) $y' = \frac{1\sin x - x\cos x}{\sin^2 x}$

(b) $y' = 9\tan^2(x^2)\sec^2(x^2)2x = 18x\tan^2(x^2)\sec^2(x^2)$

(c) $y' = \sec^2\left(\frac{x}{3}\right)\left(\frac{1}{3}\right)\sec(3x) + \tan\left(\frac{x}{3}\right)\sec(3x)\tan(3x)3$

## Section 21.3    Applications

**Problem 1.**

(a)  $f'(x) = 1 + 2\cos x = 0 \Leftrightarrow \cos x = -\frac{1}{2} \Leftrightarrow x = \frac{2}{3}\pi + k(2\pi)$  *or*  $x = \frac{4}{3}\pi + k(2\pi)$

(b)  decreasing on $[\frac{2}{3}\pi + 2k\pi, \ \frac{4}{3}\pi + 2k\pi]$

increasing on $[-\frac{2}{3}\pi + 2k\pi, \ \frac{2}{3}\pi + 2k\pi]$

(c)  local maxima at $\frac{2}{3}\pi + 2k\pi$ and local minima at  $\frac{4}{3}\pi + 2k\pi$

(d)  no global max or min as keeps on climbing f)

(e)  $f''(x) = -2\sin x = 0$ for $x = k\pi$ $f$ is concave up on $[(2n-1)\pi, \ 2n\pi]$ and concave down on $[2n\pi, \ (2n+1)\pi]$

(f)

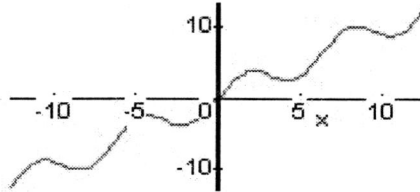

**Problem 3.**

$f'(x) = -\sin x + \sqrt{3}\cos x = 0 \ \Leftrightarrow \ \tan x = \sqrt{3}$

$\Rightarrow \ x = \frac{\pi}{3}, \ \frac{4\pi}{3}$

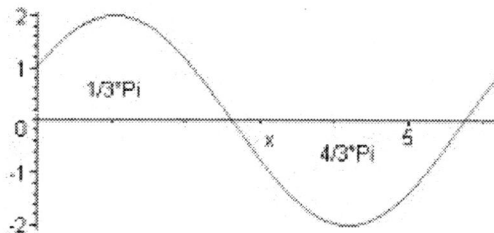

**Problem 5.**

$$f'(x) = -2\sin 2x + 2\sin x = -4\sin x \cos x + 2\sin x$$
$$= 2\sin x(-2\cos x + 1) = 0 \iff \sin x = 0 \; or \; \cos x = \tfrac{1}{2}$$
$$\Rightarrow \; x = 0, \pi, 2\pi, \; \tfrac{\pi}{3}, \tfrac{5\pi}{3}$$

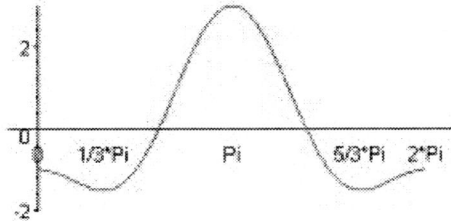

**Problem 6.**

$$f' = e^x \sin x + e^x \cos x = e^x(\sin x + \cos x) = 0$$
$$\iff \tan x = -1 \Rightarrow x = \tfrac{3\pi}{4}, \tfrac{7\pi}{4}$$

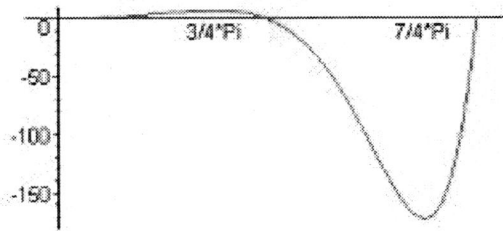

**Problem 8.**

$$\tan\theta = \tfrac{x}{30} \Rightarrow \sec^2\theta \tfrac{d\theta}{dt} = \tfrac{1}{30}\tfrac{dx}{dt} \Rightarrow \tfrac{d\theta}{dt} = \cos^2\theta \tfrac{1}{30}\tfrac{dx}{dt} \Rightarrow \tfrac{d\theta}{dt}\big|_{x=50} = \tfrac{900}{3400}\tfrac{1}{30ft}\left(-46\tfrac{ft}{\sec}\right) \approx -0.4059\tfrac{rad}{\sec}$$

(when $x = 50$, $z^2 = 900 + 2500 = 3400$ so $\cos^2\theta = \tfrac{900}{3400}$)

**Problem 9.**

$$f(x) = \sec x \Rightarrow f'(x) = \sec x \tan x = 0 \iff \sin(x) = 0 \iff x = k\pi.$$

$$f''(x) = \sec x \tan x \tan x + \sec x \sec^2 x = \sec x(\tan^2 x + \sec^2 x).$$

$f''(2\pi k) = 1(0+1) > 0$ so $f$ has a local minimum at $2\pi k$, while
$f''(\pi + 2\pi k) = -1(0+1) < 0$ so $f$ has a local maximum at $\pi + 2\pi k$.

**Problem 11.**

(a) $f(x) = 3\cos x + 2\sin x$ has period $2\pi$.

(b) $f'(x) = -3\sin x + 2\cos x = 0 \Rightarrow \tan x = \tfrac{2}{3} = \tfrac{-2}{-3}$ hence critical points are when

$\sin x = \tfrac{2}{\sqrt{13}}$ and $\cos x = \tfrac{3}{\sqrt{13}}$, or $\sin x = \tfrac{-2}{\sqrt{13}}$ and $\cos x = \tfrac{-3}{\sqrt{13}}$. Using these values we get

$f(crit \, pt) = 3(\tfrac{3}{\sqrt{13}}) + 2(\tfrac{2}{\sqrt{13}}) = \sqrt{13}$ or $3(\tfrac{-3}{\sqrt{13}}) + 2(\tfrac{-2}{\sqrt{13}}) = -\sqrt{13}$. Max $= \sqrt{13}$, min $= -\sqrt{13}$.

**Problem 13.**

$y = 80 \tan \theta \; \Rightarrow \; \frac{dy}{dt} = 80 \sec^2 \theta \frac{d\theta}{dt}$ . At the instant the runner is 100ft away

$\cos \theta = \frac{80}{100} = \frac{4}{5}$. Hence $9 = 80 \frac{25}{16} \frac{d\theta}{dt} \; \Rightarrow \; \frac{d\theta}{dt} = 0.072 \frac{rad}{\sec} \cong 0.01146 \frac{rev}{\sec}$

**Problem 15.**

$\frac{y}{0.5} = \tan \theta \; \Rightarrow \; y = 0.5 \tan \theta \; \Rightarrow \; \frac{dy}{dt} = 0.5 \sec^2 \theta \frac{d\theta}{dt}$

At the instant the window is one kilometer from the light $\cos \theta = \frac{1}{2} \; \Rightarrow \; \sec^2 \theta = 4 \; \Rightarrow \; \frac{dy}{dt} =$
$0.5(4)[6(2\pi)] = 24\pi \cong 75.4 \frac{km}{\min}$

**Problem 17.**

(a) $D(x) = \sin x - (-\cos x) \; \Rightarrow \; D'(x) = \cos x - \sin x = 0 \; \Leftrightarrow \; \tan x = 1 \; \Rightarrow \; x = \frac{\pi}{4}$

Hence the maximum distance is $D(\frac{\pi}{4}) = \frac{\sqrt{2}}{2} + \frac{\sqrt{2}}{2} = \sqrt{2}$

(b) Tangent lines to these two curves are parallel at $\frac{\pi}{4}$. This is not surprising as if they were not parallel, then movement in one of the directions would make the curves move further apart.

**Problem 19.**

(a) $y_1 = \sin kx \; \Rightarrow \; y_1' = k \cos kx \; \Rightarrow \; y_1'' = k(-k \sin kx) = -k^2 \sin kx$

Also $-k^2 y_1 = -k^2 \sin kx$. Therefore $y_1$ is a solution for $y'' = -k^2 y$

(b) $y_2 = C_1 \sin kx \; \Rightarrow \; y_2' = C_1(k \cos kx) \; \Rightarrow \; y_2'' = C_1(k(-k \sin kx)) = -k^2 C_1 \sin kx$

Also $-k^2 y_2 = -k^2 C_1 \sin kx$. Therefore $y_2$ is a solution for $y'' = -k^2 y$

(c) $y_3 = C_1 y_1 + C_2 y_2 \; \Rightarrow \; y_3'' = C_1 y_1'' + C_2 y_2'' = C_1(-k^2 y_1) + C_2(-k^2 y_2)) = -k^2(C_1 y_1 + C_2 y_2)$

Also $-k^2 y_3 = -k^2(C_1 y_1 + C_2 y_2)$ Therefore $y_3$ is a solution for $y'' = -k^2 y$

An argument like part (a) would show that $y = \cos kx$ is a solution for $y'' = -k^2 y$

Hence $y = C_1 \sin kx + C_2 \cos kx$ will be a solution by the result of this part.

**Problem 21.**

(a) satisfies (iii): $y_1 = 5 \sin 3t \; \Rightarrow \; y_1' = 5(3 \cos 3t) \; \Rightarrow \; y_1'' = 5(3(-3 \sin 3t)) = -45 \sin 3t$

while $-9y_1 = -9(5 \sin 3t) = -45 \sin 3t$. Therefore $y_1$ is a solution for $y'' = -9y$.

(b) satisfies (ii): $y_2 = e^{3t} \; \Rightarrow \; y_2' = 3e^{3t} \; \Rightarrow \; y_2'' = 9e^{3t}$,

while $9y_2 = 9e^{3t}$. Therefore $y_2$ is a solution for $y'' = 9y$.

(c) satisfies (iii): $y_3 = 2 \cos 3t \; \Rightarrow \; y_3' = 2(-3 \sin 3t) \; \Rightarrow \; y_3'' = 2(-3(3 \cos 3t)) = -18 \cos 3t$,

while $-9y_3 = -9(2 \cos 3t) = -18 \cos 3t$. Therefore $y_3$ is a solution for $y'' = -9y$.

(d) satisfies (i): $y_4 = 4.5t^2 + 3t + 8 \; \Rightarrow \; y_4' = 4.5(2t) + 3 \; \Rightarrow \; y_4'' = 9$. Thus $y_4$ is a solution for $y'' = 9$.

(e) satisfies (ii): $y_5 = 4e^{-3t} \; \Rightarrow \; y_5' = 4(-3e^{-3t}) \; \Rightarrow \; y_5'' = 4(3)^2 e^{-3t} = 36e^{-3t}$,

while $9y_5 = 9(4e^{-3t}) = 36e^{-3t}$. Therefore $y_5$ is a solution for $y'' = 9y$.

(f) satisfies (i): $y_6 = 4.5t^2 - t + 2 \; \Rightarrow \; y_6' = 4.5(2t) - 1 \; \Rightarrow \; y_6'' = 9$. Thus $y_6$ is a solution for $y'' = 9$.

**Problem 23.**

$z^2 = x^2 + y^2 \; \Rightarrow \; 2z\frac{dz}{dt} = 2x\frac{dx}{dt} + 2y\frac{dy}{dt}$ with $\frac{dx}{dt} = -5\frac{ft}{\sec}$ and $\frac{dy}{dt} = 10\frac{ft}{\sec}$

(a) When $x = 200$ and $y = 100$, $z = 100\sqrt{5}$ hence $100\sqrt{5}\frac{dz}{dt} = 200(-5) + 100(10) = 0$,

thus the distance between them is neither increasing or decreasing.

(b) $\tan\theta = \frac{y}{x} \Rightarrow \sec^2\theta \frac{d\theta}{dt} = \frac{x\frac{dy}{dt} - y\frac{dx}{dt}}{x^2} \Rightarrow \frac{d\theta}{dt} = \frac{x\frac{dy}{dt} - y\frac{dx}{dt}}{x^2}\cos^2\theta$.

At that instant $\cos\theta = \frac{200}{100\sqrt{5}}$, thus $\frac{d\theta}{dt} = \frac{200(10) - 100(-5)}{(200)^2}\left(\frac{2}{\sqrt{5}}\right)^2 = \frac{2500}{40000}\frac{4}{5} = \frac{5}{100}\frac{rad}{sec}$

## Section 21.4    Derivatives of Inverse Trigonometric Functions

**Problem 1.**

$\quad y' = 3\sec^2 x - 4\frac{1}{1+x^2}$

**Problem 3.**

$\quad y' = \cos x \arcsin x + \sin x \frac{1}{\sqrt{1-x^2}}$

**Problem 4.**

$\quad y = (\tan^{-1} x)^{1/2} \Rightarrow y' = \frac{1}{2}(\tan^{-1} x)^{-1/2}\frac{1}{1+x^2} = \frac{1}{2(1+x^2)\sqrt{\tan^{-1} x}}$

**Problem 6.**

$\quad y = \frac{1}{e}\arctan(e^x) \Rightarrow y' = \frac{1}{e}\frac{1}{1+(e^x)^2}e^x = \frac{e^{(x-1)}}{1+e^{2x}}$

**Problem 8.**

$\quad f'(x) = 3(-\sin\left(\frac{1}{1+x^2}\right)(-1(1+x^2)^{-2}(2x)) + (1)\arctan\left(\frac{1}{x}\right) + x\frac{1}{1+(\frac{1}{x})^2}(-1x^{-2})$

$$= \frac{6x}{(1+x^2)^2}\sin\left(\frac{1}{1+x^2}\right) + \arctan\left(\frac{1}{x}\right) - \frac{x}{x^2+1}$$

# CHAPTER 22

# Net Change in Amount and Area: Introducing the Definite Integral

## Section 22.1    Finding Net Change in Amount: Physical and Graphical Interplay

**Problem 1.**
These results are approximate, dependent upon eyeball interpretation of the graph in the text.
(a) 10 AM as after 10 AM the rate of arrival is greater than the service rate.

(b) $\approx$12 noon. After this the arrival rate decreases.

(c) $\approx$ 2 PM. After this the arrival rate is less than the service rate.

(d) $\approx$ 165 – 120 = 45 people. Arrivals(area under $r(t)$ 10 to 2) minus those served 10 to 2.

(e) Those arriving at 2 PM must wait through the longest line. $45/15 = 3$ hours to wait.

(f) $\approx$ 45 + 23 – 30 = 32. Those in line, plus arrivals minus those served.

(g) $\approx$ 225 people (area under curve.)

**Problem 2.**

(a) left sum is $10(2) + 9(2) + 7(2) + 4(2) = 60$ cm$^3$

(b) right sum is $9(2) + 7(2) + 4(2) + 2(2) = 44$ cm$^3$

**Problem 5.**

(a) $\frac{ds}{dt} = 3t \;\Rightarrow\; s(t) = \frac{3}{2}t^2 + s_0$
   (i) $s(0) = s_0$
   (ii) $s(k) = \frac{3}{2}k^2 + s_0$
   (iii) $s(k) - s(0) = \frac{3}{2}k^2$ is distance traveled

(b) The area under the curve is a triangle. So the area is
   $\frac{1}{2}(base)(height) = \frac{1}{2}(k)(3k) = \frac{3}{2}k^2$ which is the same as the above answer.

**Problem 7.**
Distance traveled = the area under the curve.
   (a) 4 (b) 30 (c) 8

# Section 22.2 The Definite Integral

**Problem 2.**

(a)

(i) $\sum_{k=3}^{300} (-1)^{k+1} k$

(ii) $\sum_{k=1}^{500} 2k$

(iii) $\sum_{k=1}^{500} (2k-1)$

(iv) $\sum_{k=1}^{15} (-1)^{k+1} \frac{2}{3^k}$

(v) $\sum_{k=1}^{40} x^k$

(vi) $\sum_{k=1}^{100} k^2$

(vii) $\sum_{k=0}^{n} a_k x^k$

(b)

(i) $2^2 + 3^2 + 4^2 + 5^2$

(ii) $2^0 + 2^1 + 2^2 + 2^3 + 2^4$

(iii) $a_0 + a_1 x + a_2 x^2 + a_3 x^3$

**Problem 4.**

(a) $L_4 = (0)^3 \frac{1}{2} + \left(\frac{1}{2}\right)^3 \frac{1}{2} + \left(\frac{2}{2}\right)^3 \frac{1}{2} + \left(\frac{3}{2}\right)^3 \frac{1}{2} = \frac{9}{4}$

$R_4 = \left(\frac{1}{2}\right)^3 \frac{1}{2} + \left(\frac{2}{2}\right)^3 \frac{1}{2} + \left(\frac{3}{2}\right)^3 \frac{1}{2} + \left(\frac{4}{2}\right)^3 \frac{1}{2} = \frac{25}{4}$

(b) $L_6 = \left(\frac{1}{3/3}\right) \frac{1}{3} + \left(\frac{1}{4/3}\right) \frac{1}{3} + \left(\frac{1}{5/3}\right) \frac{1}{3} + \left(\frac{1}{6/3}\right) \frac{1}{3} + \left(\frac{1}{7/3}\right) \frac{1}{3} + \left(\frac{1}{8/3}\right) \frac{1}{3} = \frac{341}{280} \approx 1.218$

$R_6 = \left(\frac{1}{4/3}\right) \frac{1}{3} + \left(\frac{1}{5/3}\right) \frac{1}{3} + \left(\frac{1}{6/3}\right) \frac{1}{3} + \left(\frac{1}{7/3}\right) \frac{1}{3} + \left(\frac{1}{8/3}\right) \frac{1}{3} + \left(\frac{1}{9/3}\right) \frac{1}{3} = \frac{2509}{2520} \approx 0.9956$

**Problem 6.**

(a) $L_4 < L_{20} < L_{100} < \int_0^2 x^3 dx < R_{100} < R_{20} < R_4$.

(b) $|R_4 - L_4| = \left|2^3 \frac{2}{4} - 0^3 \frac{2}{4}\right| = 4 = \frac{2^4}{4}$.

(c) $|R_{100} - L_{100}| = \left|2^3 \frac{2}{100} - 0^3 \frac{2}{100}\right| = \frac{2^4}{100} = 0.16$

(d) $|R_n - L_n| = \left|2^3 \frac{2}{n} - 0^3 \frac{2}{n}\right| = \frac{2^4}{n} < 0.05 \Leftrightarrow \frac{2^4}{0.05} < n \Leftrightarrow 320 < n$

(e) $R_4 = \sum_{k=1}^{4} \left(\frac{k}{2}\right)^3 \frac{1}{2} = \left(\frac{1}{2}\right)^3 \frac{1}{2} + \left(\frac{2}{2}\right)^3 \frac{1}{2} + \left(\frac{3}{2}\right)^3 \frac{1}{2} + \left(\frac{4}{2}\right)^3 \frac{1}{2} = \frac{25}{4}$

**Problem 8.**

(a) $R_4 < R_{20} < R_{100} < \int_0^2 \frac{1}{x+1} dx < L_{100} < L_{20} < L_4$.

(b) $|R_4 - L_4| = \left|\frac{1}{3} \frac{2}{4} - \frac{1}{1} \frac{2}{4}\right| = \frac{1}{3} = \frac{4}{3\cdot4}$.

(c) $|R_{100} - L_{100}| = \left|\frac{1}{3} \frac{2}{100} - \frac{1}{1} \frac{2}{100}\right| = \frac{4}{3\cdot100}$

(d) $|R_n - L_n| = \left|\frac{1}{3} \frac{2}{n} - \frac{1}{1} \frac{2}{n}\right| = \frac{4}{3\cdot n} < 0.05 \Leftrightarrow \frac{4}{0.15} < n \Rightarrow 27 \leq n$

(e) $R_4 = \sum_{k=1}^{4} (\frac{1}{\frac{K}{2}+1})^{\frac{1}{2}} = (\frac{1}{\frac{1}{2}+1})^{\frac{1}{2}} + (\frac{1}{\frac{2}{2}+1})^{\frac{1}{2}} + (\frac{1}{\frac{3}{2}+1})^{\frac{1}{2}} + (\frac{1}{\frac{41}{2}+1})^{\frac{1}{2}} = \frac{19}{20}$

## Section 22.3    The Definite Integral: Qualitative Analysis and Signed Area

**Problem 1.**

(a) Area under curve is triangle $\frac{1}{2}5(5) = \frac{25}{2}$

(b) Area above – area below $x$-axis $= 0$

(c) Area is $2[\frac{1}{2}(2)2] = 4$

(d) Area is $4(3)=12$

(e) Area above – area below $x$-axis $= 0$

(f) Area above – area below $x$-axis $= 0$

(g) Above – below $= \frac{1}{2}(3)(3) - \frac{1}{2}(1)(1) = 4$

(h) Area $= \frac{1}{2}(3)(3) + \frac{1}{2}(1)(1) = 5$

**Problem 3.**
$$b < c = e < d < a < f$$

**Problem 5.**
Signed area on [-8, -2] is 12, on [-2, 0] is 4, on [0, 3] is $-\frac{1}{4}\pi 3^2 = -\frac{9}{4}\pi$, and on [3, 6] is $-\frac{9}{4}\pi$

(a) 12

(b) 12+4=16

(c) $-\frac{9}{4}\pi + (-\frac{9}{4}\pi) = -\frac{9}{2}\pi$

(d) $-\frac{9}{4}\pi$

**Problem 7.**

(a) $x^2 + y^2 = 2^2$ or $x^2 + y^2 = 4$

(b) $y = \sqrt{4 - x^2} = f(x)$ on [-2, 2]

(c) $\int_{-2}^{2} f(x)dx = \int_{-2}^{0} f(x)dx + \int_{0}^{2} f(x)dx = \int_{-2}^{0} \sqrt{4 - x^2}dx + \int_{0}^{2} 2xdx =$ area of a quarter circle of radius 2
plus the area of a triangle with base 2 and height $4 = \frac{1}{4}\pi\, 2^2 + \frac{1}{2}(2)(4) = \pi + 4$

## Section 22.4    Properties of the Definite Integral

**Problem 2.**

(a) $\int_{0}^{1} \sqrt{1 - x^2}dx = \frac{1}{4}\pi(1)^2 = \frac{1}{4}\pi$

(b) $\int_{1}^{-1} \sqrt{1 - x^2}dx = -\int_{-1}^{1} \sqrt{1 - x^2}dx = -\frac{1}{2}\pi$

## Problem 3.

(a) $\int_{-a}^{a} \frac{1}{1+x^2} dx = 2\int_{0}^{a} \frac{1}{1+x^2} dx$ as $\frac{1}{1+x^2}$ is symmetric about y-axis.

(b) Integrand is odd function so total signed area is zero.

## Problem 5.

(a) $\int_{a}^{b} |f(t)| dt \geq \left| \int_{a}^{b} f(t) dt \right|$ because all areas for $\int_{a}^{b} |f(t)| dt$ are positive while $\left| \int_{a}^{b} f(t) dt \right|$ adds signed areas (negative area subtracts from positive area) then takes absolute value.

(b) They will be equal if $f(t) \geq 0$ on [a, b], OR $f(t) \leq 0$ on [a, b].

## Problem 7.

$\int_{\pi}^{0} \sin t \, dt = -\int_{0}^{\pi} \sin t \, dt.$

$\int_{\pi}^{0} \sin t \, dt < \int_{0}^{\pi} \sin(2t) \, dt < \int_{0}^{\pi} \sin t \, dt < \int_{0}^{\pi} 2\sin t \, dt$

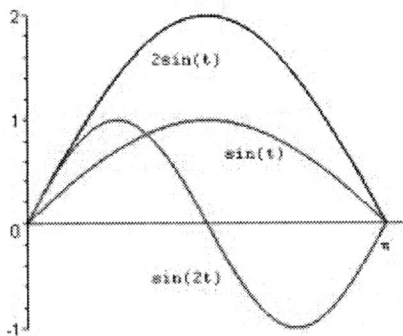

## Problem 8.

(a) $\int_{0}^{5} 7f(t) dt = 7\int_{0}^{5} f(t) dt = 7(10) = 70$

(b) $\int_{0}^{5} (f(t) + 7) dt = \int_{0}^{5} f(t) dt + \int_{0}^{5} 7 dt = 10 + 35 = 45$

(c) $\int_{0}^{5} f(t) dt + 7 = 10 + 7 = 17$

(d) Not enough information as we don't know $f$'s behavior on interval [7, 12].

(e) $\int_{-7}^{-2} 7f(t + 7) dt = \int_{0}^{5} 7f(t) dt = 7\int_{0}^{5} f(t) dt = 7(10) = 70$

# CHAPTER 23

# The Area Function and Its Characteristics

## Section 23.1    An Introduction to the Area Function $\int_a^x f(t)dt$

**Problem 1.**

(a) $_0A_f(x) = \int_0^x f(t)dt = \int_0^x 7dt = 7(x-0) = 7x,\ ;\ _0A_f(-3) = 7(-3) = -21$ works.

(b) $_2A_f(x) = \int_2^x f(t)dt = \int_2^x 7dt = 7(x-2) = 7x - 14,\ ;$ which works for $x < 2$ also.

(c) $_3A_f(x) = \int_3^x f(t)dt = \int_3^x 7dt = 7(x-3) = 7x - 21,\ ;$ which works for $x < 3$ also.

(d) The graphs are shown below in order left to right.

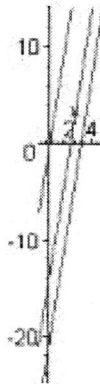

(e) The derivative of all three functions is the same constant function $f(x) = 7$.

**Problem 3.**
    $A_f(x)$ is the signed area on interval $[0,\ x]$.
(a) $A_f(0) = A_f(2\pi) = 0 < A_f(\frac{\pi}{2}) = A_f(\frac{3\pi}{2})$

(b) $A_f(x) = 0$ when signed area $= 0$. $x = k2\pi,\ k\varepsilon\, Z$.

(c) $A_f(x)$ is never negative.

(d) $A_f(x)$ is max. when $x = \pi + k2\pi = (2k+1)\pi,\ k\varepsilon\, Z$.

## Section 23.2    Characteristics of the Area Function

**Problem 1.**

$0,\ 8$ where $f(x) \geq 0$.

$8,\ 11$ where $f(x) \leq 0$.

(a) $A_f(0) = 0$, there is zero area on $[0, 0]$.

(b) No. The signed area is always positive on $[0, x]$ for $x$ in $[0, 11]$.

(c) Absolute minimum at $x = 0$, global maximum at $x = 8$, local minimum at $x = 11$.

**Problem 3.**

(a) C $F(x)$ is increasing on $[-1, 2]$, and $[5, 6]$; decreasing on $[-2, -1]$ and $[2, 5]$.

(b) C $G(x) = \int_{-2}^{0} f(t)dt + F(x)$ so is a vertical shift of $F(x)$ with $G(-2) = 0$.

## Section 23.3    The Fundamental Theorem of Calculus

**Problem 1.**

(a) Increasing on $[-4, 0]$ and $[4, 6]$. Decreasing on $[-6, -4]$ and $[0, 4]$.
   Concave up on $(-6, -2)$, $(1, 2)$ and $(3.5, 6)$. Concave down on $(-2, 1)$ and $(2, 3.5)$.

(b) Increasing on $[-4, 0]$ and $[4, 6]$. Decreasing on $[-6, -4]$ and $[0, 4]$.
   ($G(x)$ is a vertical shift of $F(x)$.)

(c) Local minimum at $x = -4$ and $x = 4$.

(d) $G(x)$ is a vertical shift of $F(x)$. The slopes of $F$ and $G$ are the same everywhere.

(e) $G(x)$ is a vertical shift of $F(x)$. $F(x) = G(x) + \left[\int_{-6}^{1} f(t)dt\right]$.

**Problem 2.**
   The graph must be increasing over the whole domain. It is concave down changing to concave up at the $q$ value where the original graph was lowest.

**Problem 4.**
   Approximate graphs are given below with time units in minutes.

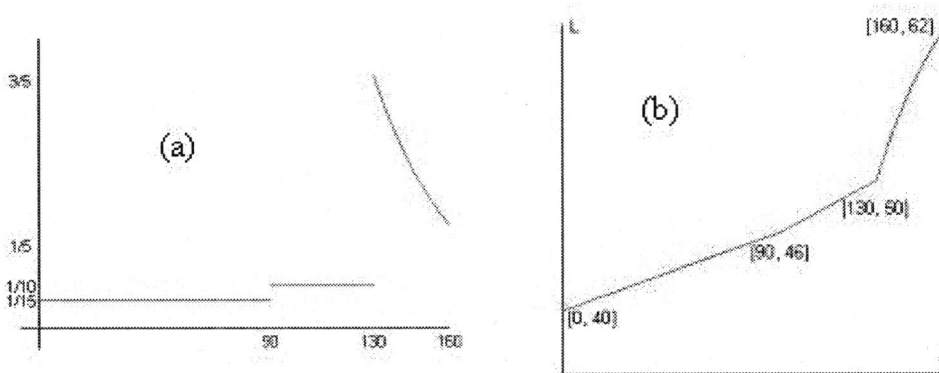

# CHAPTER 24

# The Fundamental Theorem of Calculus

## Section 24.1    Definite Integrals and the Fundamental Theorem

**Problem 2.**

The region above interval $[1, 9]$ under the graph is a trapezoid, hence the area is $\frac{1}{2}(13+93)(9-1) = 424$.

This supports: $\int_1^9 (10t + 3)dt = (5t^2 + 3t)|_1^9 = (405 + 27) - (5 + 3) = 424$

**Problem 3.**

$\int_1^2 t^3 dt = (\frac{1}{4}t^4)|_1^2 = 4 - \frac{1}{4} = \frac{15}{4}$.

**Problem 5.**

$\int_0^1 e^x dx = (e^x)|_0^1 = e^1 - e^0 = e - 1$.

**Problem 7.**

$\int_0^9 (1.5t + \sqrt{t})dt = \int_0^9 (\frac{3}{2}t + t^{\frac{1}{2}})dt = (\frac{3}{4}t^2 + \frac{2}{3}t^{\frac{3}{2}})|_0^9 = \frac{315}{4}$

**Problem 8.**

(a) Rate of change is negative so it is <u>cooling</u>.

(b) $\int_0^1 -2e^{-t}dt = 2e^{-t}|_0^1 = \frac{2}{e} - 2$ degrees.

(c) $\int_1^2 -2e^{-t}dt = 2e^{-t}|_1^2 = \frac{2}{e^2} - \frac{2}{e}$ degrees.

(d) 100 plus change $= 100 + \frac{2}{e} - 2 = 98 + \frac{2}{e}$ degrees.

**Problem 11.**

See graph for $r(t) = 2\sin(\frac{\pi}{4}t)$

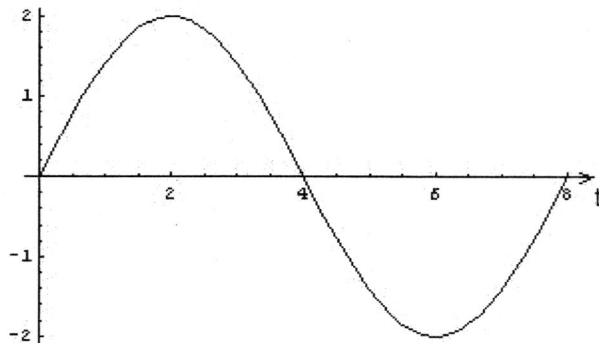

(a) Level is highest when $t = 4$ because the rate of change is positive before 4 and negative after 4.

(b) $\int_0^4 2\sin(\frac{\pi}{4}t)dt = -2\cos(\frac{\pi}{4}t)\frac{4}{\pi}|_0^4 = \frac{8}{\pi} - (-\frac{8}{\pi}) = \frac{16}{\pi}$

Hence the highest water level is $30 + \frac{16}{\pi}$ gallons.

(c) The minimum level will be 30 gallons.

(d) Looking at the graph we can see that the signed area from 4 to 5 will be the same size, just negative, as the area from 0 to 1. Hence the amount lost from 4 to 5 is the same as the amount gained from 0 to 1.

## Problem 13.

(a) See graph of $y = \sin(t^2)$.

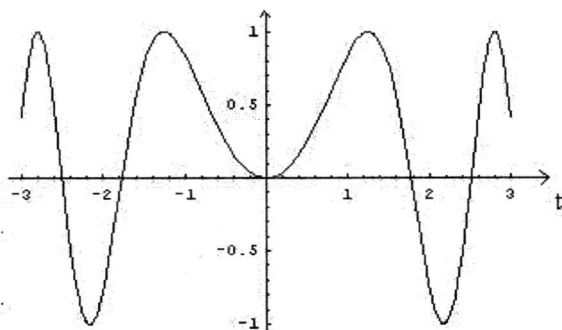

(b) $h(x) > 0$ on $(0, \infty)$ as positive signed area
always overpowers negative signed area.

(c) $h(-x) = \int_0^{-x} \sin(t^2)dt = -\int_{-x}^0 \sin(t^2)dt = $ - (area of the region reflected across the y-axis) $= -\int_0^x \sin(t^2)dt = -h(x)$, $h(x)$ is odd.

(d) $h(x)$ changes from increasing to decreasing to increasing where the graph of $\sin(t^2)$ crosses the $t$-axis. That is when $\sin(t^2) = 0$, when $t^2 = \pi$, $2\pi$, $3\pi \cdots$ increasing on $[0, \sqrt{\pi}]$, decreasing on $[\sqrt{\pi}, \sqrt{2\pi}]$ and increasing on $[\sqrt{2\pi}, 3]$.

(e) Maximum $= 1$ and minimum $= -1$.

(f) On $[0, 3]$ $h$ as a global min. at 0, a global max. at $\sqrt{\pi}$ and a local min at $\sqrt{2\pi}$. On $(0, \infty)$ 0 is no longer in the interval so we have no global minimum. We still have a global maximum at $\sqrt{\pi}$. There will be infinitely many more local maximum and minimum points. On $(-\infty, \infty)$ the global max is at $\sqrt{\pi}$ and the global min is at $-\sqrt{\pi}$.

(g) $h(\sqrt{\pi}) \cong 0.895$

**Problem 15.**

(a) Area of a rectangle (not shown) $(\pi + 1)(7 - (-3)) = 10(\pi + 1)$

(b) By area of two triangles (see graph) we get $\frac{1}{2}(1)(2) + \frac{1}{2}(9)(18) = 1 + 81 = 82$

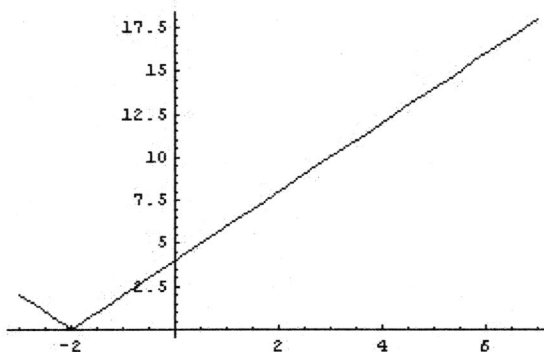

**Problem 17.**

(a) $\int_0^5 x\,dx = \lim_{n\to\infty} \sum_{i=1}^n f(x_i)\Delta x_i = \lim_{n\to\infty} \sum_{i=1}^n (i\frac{5}{n})\frac{5}{n} = \lim_{n\to\infty} \frac{25}{n^2} \sum_{i=1}^n i = \lim_{n\to\infty} \frac{25}{n^2}\frac{n(n+1)}{2} = \lim_{n\to\infty} \frac{25}{2}(1)(1 + \frac{1}{n})$
$= \frac{25}{2}$. Also $\int_0^5 x\,dx = \frac{1}{2}x^2\big|_0^5 = \frac{1}{2}5^2 - \frac{1}{2}0^2 = \frac{25}{2}$.

(b) $\int_0^5 x^2\,dx = \lim_{n\to\infty} \sum_{i=1}^n f(i\frac{5}{n})\frac{5}{n} = \lim_{n\to\infty} \sum_{i=1}^n (i\frac{5}{n})^2 \frac{5}{n} = \lim_{n\to\infty} \frac{125}{n^3} \sum_{i=1}^n i^2 = \lim_{n\to\infty} \frac{125}{n^3}\frac{n(n+1)(2n+1)}{6} = \lim_{n\to\infty} \frac{125}{6}(1)(1+$
$\frac{1}{n})(2 + \frac{1}{n}) = \frac{125}{3}$. Also $\int_0^5 x^2\,dx = \frac{1}{3}x^3\big|_0^5 = \frac{1}{3}5^3 - \frac{1}{3}0^3 = \frac{125}{3}$

(c) $\int_0^2 x^3\,dx = \lim_{n\to\infty} \sum_{i=1}^n f(i\frac{2}{n})\frac{2}{n} = \lim_{n\to\infty} \sum_{i=1}^n (i\frac{2}{n})^3 \frac{2}{n} = \lim_{n\to\infty} \frac{16}{n^4} \sum_{i=1}^n i^3 = \lim_{n\to\infty} \frac{16}{n^4}\left[\frac{n(n+1)}{2}\right]^2 = \lim_{n\to\infty} \frac{16}{4}\left[(1)(1 + \frac{1}{n})\right]^2 =$
4. Also $\int_0^2 x^3\,dx = \frac{1}{4}x^4\big|_0^2 = \frac{1}{4}2^4 - \frac{1}{4}0^4 = \frac{16}{4} = 4$

**Problem 18.**

(a) Let $P(n)$ be the statement: $1 + 2 + 3 + \cdots + n = \frac{n(n+1)}{2}$

$P(1)$ is true because $1 = \frac{1(1+1)}{2}$.

Assume that $P(k)$ is true. Then $1 + 2 + 3 + \cdots + k = \frac{k(k+1)}{2} \Rightarrow$

$$1 + 2 + 3 + \cdots + k + (k + 1) = \frac{k(k+1)}{2} + (k + 1) = (k + 1)\left(\frac{k}{2} + 1\right) = (k + 1)\left(\frac{k+2}{2}\right) = \frac{(k+1)((k+1)+1)}{2}$$

$\Rightarrow$   $P(k + 1)$ is true. Hence by induction, $P(n)$ is true for all positive integers $n$.

(b) Let $P(n)$ be the statement: $1^2 + 2^2 + 3^2 + \cdots + n^2 = \frac{n(n+1)(2n+1)}{6}$

$P(1)$ is true because $1 = \frac{1(1+1)(2(1)+1)}{6}$.

Assume that $P(k)$ is true. Then $1^2 + 2^2 + 3^2 + \cdots + k^2 = \frac{k(k+1)(2k+1)}{6} \Rightarrow$

$$1^2 + 2^2 + 3^2 + \cdots + k^2 + (k + 1)^2 = \frac{k(k+1)(2k+1)}{6} + (k + 1)^2 = (k + 1)\left(\frac{k(2k+1)}{6} + (k + 1)\right)$$

$= (k+1)\left(\frac{2k^2+k}{6} + \frac{6k+6}{6}\right) = (k+1)\left(\frac{2k^2+7k+6}{6}\right) = (k+1)\left(\frac{(k+2)(2k+3)}{6}\right) = \frac{(k+1)([k+1]+1)(2[k+1]+1)}{6}$

$\Rightarrow$   $P(k+1)$ is true. Hence by induction, $P(n)$ is true for all positive integers $n$.

(c) Let $P(n)$ be the statement that $1^3 + 2^3 + 3^3 + \cdots + n^3 = \left[\frac{n(n+1)}{2}\right]^2$

$P(1)$ is true because $1 = \left[\frac{1(1+1)}{2}\right]^2$.

Assume that $P(k)$ is true. Then $1^3 + 2^3 + 3^3 + \cdots + k^3 = \left[\frac{k(k+1)}{2}\right]^2 \Rightarrow$

$$1^3 + 2^3 + 3^3 + \cdots + k^3 + (k+1)^3 = \left[\frac{k(k+1)}{2}\right]^2 + (k+1)^3 = (k+1)^2\left(\frac{k^2}{4} + (k+1)\right)$$

$= (k+1)^2\left(\frac{k^2+4k+4}{4}\right) = (k+1)^2\left(\frac{(k+2)^2}{4}\right) = \left[\frac{(k+1)(k+2)}{2}\right]^2 \Rightarrow$   $P(k+1)$ is true.

Hence by induction, $P(n)$ is true for all positive integers $n$.

**Problem 21.**
  $\int_{0.5}^{2} \ln x\,dx < \int_{1}^{2} \ln x\,dx < \int_{1}^{2.5} \ln x\,dx$

**Problem 23.**
  *iii* $(x\ln x - x)' = (1)\ln x + x(\frac{1}{x}) - (1) = \ln x.$

$$\int_{1}^{6} \ln x\,dx = (x\ln x - x)\big|_{1}^{6} = (6\ln 6 - 6) - (1\ln 1 - 1) = 6\ln 6 - 5$$

## Section 24.2    The Average Value of a Function

**Problem 1.**
  By the graph we can see that as much area above the line $y = 5$ is
  gained on $[0, \pi]$ as is lost on $[\pi, 2\pi]$. Hence $y = 5$ is an average
  value. Also $v_{ave} = \frac{1}{2\pi}\int_{0}^{2\pi}(3\sin x + 5)dx = \frac{1}{2\pi}(-3\cos x + 5x)\big|_{0}^{2\pi}$
  $= \frac{1}{2\pi}[(-3(1) + 5(2\pi)) - (-3(1) + 5(0))] = 5$

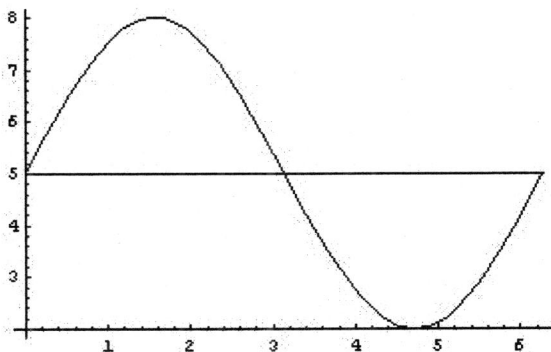

**Problem 3.**

$\frac{1}{\pi} \int_0^\pi \sin x dx = \frac{1}{\pi} \left( -\cos x \right)\big|_0^\pi = \frac{1}{\pi}(-(-1) - (-1)) = \frac{2}{\pi}$

**Problem 5.**

(a) Velocity of bicyclist at 5 PM.

(b) How many miles east from the starting point the bicyclist is at 5 PM.

(c) Total distance in miles the bicyclist has traveled from noon to 5 PM.

(d) Speed of bicyclist at 5 PM.

(e) Acceleration of bicyclist at 5 PM.

(f) Average velocity of bicyclist from noon to 5 PM.

**Problem 7.**

(a) $s(t) = |3\sin(\pi t)|$

(b) $\int_0^2 3\sin(\pi t)dt = -\frac{3}{\pi}\cos(\pi t)\big|_0^2 = -\frac{3}{\pi}\cos(2\pi) - (-\frac{3}{\pi}\cos(0\pi)) = 0$

(c) $\int_0^2 |3\sin(\pi t)|\, dt = \int_0^1 |3\sin(\pi t)|\, dt + \int_1^2 |3\sin(\pi t)|\, dt = \int_0^1 3\sin(\pi t)dt + \int_1^2 -3\sin(\pi t)dt$

$$= -\frac{3}{\pi}\cos(\pi t)\big|_0^1 + \frac{3}{\pi}\cos(\pi t)\big|_1^2 = \left[-\frac{3}{\pi}\cos(1\pi) - (-\frac{3}{\pi}\cos(0\pi))\right] + \left[\frac{3}{\pi}\cos(2\pi) - \frac{3}{\pi}\cos(1\pi)\right]$$

$$= \left[-\frac{3}{\pi}(-1) - (-\frac{3}{\pi}(1))\right] + \left[\frac{3}{\pi}(1) - \frac{3}{\pi}(-1)\right] = 4(\frac{3}{\pi}) = \frac{12}{\pi}$$

(d) $\frac{1}{2}\int_0^2 3\sin(\pi t)dt = 0$

(e) $\frac{1}{2}\int_0^2 |3\sin(\pi t)|\, dt = \frac{1}{2}\frac{12}{\pi} = \frac{6}{\pi}$

**Problem 8.**

III $<$ I $=$ IV $<$ II (same order as size of area under the curve.)

**Problem 10.**

The total distance is 2 miles. The total time is 10 minutes or $\frac{1}{6}$ hour.

$v_{ave} = 2/\frac{1}{6} = 12$ miles per hour.

**Problem 12.**

$\sin(3t)$ does 3 complete cycles on $[0, 2\pi]$, so average value will be the same as on $[0, \frac{2}{3}\pi]$. But $|\sin(3t)|$ is the same on $[0, \frac{1}{3}\pi]$ as on $[\frac{1}{3}\pi, \frac{2}{3}\pi]$, hence the average of $|\sin(3t)|$ on $[0, 2\pi]$ will be the same as the average of $\sin(3t)$ on $[0, \frac{1}{3}\pi]$ which is

$$\frac{1}{\frac{1}{3}\pi}\int_0^{\frac{1}{3}\pi}\sin(3t)dt = \frac{3}{\pi}\left(-\frac{1}{3}\cos(3t)\right)\big|_0^{\frac{1}{3}\pi} = \frac{3}{\pi}[-\frac{1}{3}\cos(\pi) - (-\frac{1}{3}\cos(0))] = \frac{3}{\pi}[\frac{1}{3} + \frac{1}{3}] = \frac{2}{\pi}.$$

**Problem 13.**

(a) $\int_0^{36} B(t)dt$

(b) $\frac{1}{10}\int_{20}^{30} D(t)dt$

(c) $4500 + \int_0^{36} B(t)dt - \int_0^{36} D(t)dt$

(d) 1972 ($t = 12$). Birth rate has been greater than death rate but switches at $t = 12$.

(e) Less, as had more deaths than births. $\int_0^{36} B(t)dt - \int_0^{36} D(t)dt < 0$

# CHAPTER 25

# Finding Antiderivatives: An Introduction to Indefinite Integration

## Section 25.1    A List of Basic Antiderivatives

**Problem 1.**
$$\frac{3}{4}x^4 + x^2 + \pi x + C$$

**Problem 3.**
$$3\ln|x| + C$$

**Problem 4.**
$$\frac{1}{2}\ln|x| + C$$

**Problem 7.**
$$\frac{2}{3}\sin w + C$$

**Problem 9.**
$$\frac{1}{2}e^p + C$$

**Problem 11.**
$$\sin t + \sec t + C$$

**Problem 12.**

(a)  $F(x) = \frac{1}{3}e^{3x}$

(b)  $f(x) = 3e^{-x} \Rightarrow F(x) = 3(-e^{-x}) = \frac{-3}{e^x}$

**Problem 13.**

(a)  $F(x) = -\frac{1}{2}\ln|x|$

(b)  $F(x) = 4\arctan x$

**Problem 14.**

(a)  $F(x) = -\frac{1}{2}\cos(2x)$

(b)  $F(x) = 3\sin(x/3)$

**Problem 16.**

(a)  $y' = -\frac{\pi}{3}(-\sin(3x)(3)) = \pi\sin(3x)$

(b)  $-\frac{A}{B}\cos(Bx) + C$

**Problem 17.**

(a) Net change $= \int_0^1 \frac{1}{1+t^2} dt = \arctan|_0^1 = \arctan(1) - \arctan(0) = \frac{\pi}{4} - 0 = \frac{\pi}{4}$ miles

As velocity is always positive the distance traveled is also the same.

(b) Net change $= \int_0^2 5t(t-1)dt = \int_0^2 (5t^2 - 5t)dt = (\frac{5}{3}t^3 - \frac{5}{2}t^2)|_0^2 = \frac{5}{3}8 - \frac{5}{2}4 = \frac{10}{3}$ meters.

As $5t(t-1)$ is negative on $[0,1]$ to find the total distance we need to separate the integral

Total distance $= -\int_0^1 (5t^2 - 5t)dt + \int_1^2 (5t^2 - 5t)dt = -(\frac{5}{3}t^3 - \frac{5}{2}t^2)|_0^1 + (\frac{5}{3}t^3 - \frac{5}{2}t^2)|_1^2 = 5$

**Problem 19.**

(a) $\int (x+\pi)x^2 dx = \int (x^3 + \pi x^2)\, dx = \frac{1}{4}x^4 + \frac{\pi}{3}x^3 + C$

(b) $\int \frac{kx}{\sqrt{x}} dx = \int kx^{1/2} dx = k\frac{2}{3}x^{\frac{3}{2}} + C$

(c) $\int \frac{3t^2 + t}{6t^3} dt = \int (\frac{1}{2}\frac{1}{t} + \frac{1}{6}t^{-2})dt = \frac{1}{2}\ln|t| + \frac{1}{6}(-1)t^{-1} + C = \frac{1}{2}\ln|t| - \frac{1}{6t} + C$

(d) $\int (2 - \frac{1}{x})\sqrt{x}\, dx = \int (2x^{\frac{1}{2}} - x^{-\frac{1}{2}})\, dx = 2(\frac{2}{3})x^{\frac{3}{2}} - (2)x^{\frac{1}{2}} + C = \frac{4}{3}x^{\frac{3}{2}} - 2x^{\frac{1}{2}} + C$

(e) $\int (x+1)\sqrt{5x}\, dx = \int (\sqrt{5}x^{\frac{3}{2}} + \sqrt{5}x^{\frac{1}{2}})\, dx = \frac{2\sqrt{5}}{5}x^{\frac{5}{2}} + \frac{2\sqrt{5}}{3}x^{\frac{3}{2}} + C$

## Section 25.2   Substitution: The Chain Rule in Reverse

**Problem 2.**

(a) $-\frac{3}{5}\cos(5t) + C$

(b) $\sin(\pi t) + C$

(c) Use $u = 3x + 5$, $du = 3\, dx$: $\int \frac{1}{3}u^{1/2} du = \frac{1}{3} \cdot \frac{2}{3} \cdot u^{3/2} + C = \frac{2}{9} \cdot (3x + 5)^{3/2} + C$

(d) $\int \pi \cdot e^{-x} dx = -\pi \cdot e^{-x} + C = -\frac{\pi}{e^x} + C$

(e) $-\frac{1}{3}e^{-3t} + C$

(f) $\int (e^t)^{1/2} dt = \int e^{\frac{1}{2}t} dt = 2e^{\frac{1}{2}t} + C = 2\sqrt{e^t} + C$

(g) $\int 6t^{-3/2} dt = -2 \cdot 6t^{-1/2} + C = -\frac{12}{\sqrt{t}} + C$

(h) Use $u = 3t + 8$, $du = 3dt$: $\int \frac{1}{3t+8} dt = \int \frac{1}{u} \cdot \frac{1}{3} du = \frac{1}{3}\ln|u| + C = \frac{1}{3}\ln|3t + 8| + C$

**Problem 4.**

(a) Use $u = 2x^2 + 1$, $du = 4x\, dx$: $\int \sqrt{u}\frac{1}{4} du = \frac{1}{4} \cdot \frac{2}{3}u^{\frac{3}{2}} + C = \frac{1}{6}(2x^2 + 1)^{\frac{3}{2}} + C$

(b) Use $u = 2x^2 + 1$, $du = 4x\, dx$: $\int u^{-\frac{1}{2}} \cdot \frac{1}{4} du = \frac{1}{4} \cdot 2u^{\frac{1}{2}} + C = \frac{1}{2}\sqrt{2x^2 + 1} + C$

(c) Use $u = \sqrt{x}$, $2du = \frac{1}{\sqrt{x}}dx$: $\int \cos(u) \cdot 2du = 2\sin(u) + C = 2\sin(\sqrt{x}) + C$

(d) Use $u = \cos x$, $-du = \sin x\, dx$: $\int -\sqrt{u}\, du = -\frac{2}{3}u^{\frac{3}{2}} + C = -\frac{2}{3}(\cos x)^{\frac{3}{2}} + C$

**Problem 6.**

(a) Use $u = t^3$, $du = 3t^2 dt$: $\int \sin u \cdot \frac{1}{3}du = -\frac{1}{3}\cos u + C = -\frac{1}{3}\cos(t^3) + C$

(b) Use $u = -x^2$, $du = -2x\, dx$: $\int e^u \cdot \frac{-1}{2}du = -\frac{1}{2}e^u + C = -\frac{1}{2}e^{-x^2} + C$

(c) $\ln|x+5| + C$

(d) Use $u = 2t^2 + 7$, $du = 4t\, dt$: $\int \frac{1}{u} \cdot \frac{1}{4}du = \frac{1}{4}\ln|u| + C = \frac{1}{4}\ln\left|2t^2 + 7\right| + C$

**Problem 7.**

(a) Use $u = t^2 + 1$, $du = 2t\, dt$: $\int u^{-2}\frac{1}{2}du = -\frac{1}{2}u^{-1} + C = \frac{-1}{2(t^2+1)} + C$

(b) Use $u = \cos(2t)$, $du = -2\sin(2t)\, dt$: $\int \frac{\sin(2t)}{\cos(2t)}dt = \int \frac{1}{u} \cdot \frac{-1}{2}du = -\frac{1}{2}\ln|u| + C = -\frac{1}{2}\ln|\cos(2t)| + C$

(c) Use $u = e^w + 1$, $du = e^w dw$: $\int \frac{1}{u}du = \ln|u| + C = \ln|e^w + 1| + C$

(d) Use $u = e^w$, $du = e^w dw$: $\int \frac{1}{u^2+1}du = \arctan(u) + C = \arctan(e^w) + C$

**Problem 9.**

(a) $3\int_0^1 \frac{1}{1+(2w)^2}dw = \frac{3}{2}\arctan(2w)\Big|_0^1 = \frac{3}{2}\arctan(2) - \frac{3}{2}\arctan(0) = \frac{3}{2}\arctan(2)$

(b) $\int_0^1 \frac{1}{3} + \frac{4}{3}w^2\, dw = \left(\frac{1}{3}w + \frac{4}{9}w^3\right)\Big|_0^1 = \frac{1}{3} + \frac{4}{9} = \frac{7}{9}$

(c) $\int_{\pi/2}^{3\pi} \cos\left(\frac{t}{2}\right) dt = 2\sin\left(\frac{t}{2}\right)\Big|_{\frac{\pi}{2}}^{3\pi} = 2\sin\left(\frac{3\pi}{2}\right) - 2\sin\left(\frac{\pi}{4}\right) = 2\cdot(-1) - 2\cdot\frac{\sqrt{2}}{2} = -2 - \sqrt{2}$

(d) $\frac{4}{3}\ln|3x+2|\Big|_1^3 = \frac{4}{3}(\ln 11 - \ln 5) = \frac{4}{3}\ln\left(\frac{11}{5}\right)$

(e) Use $u = 2x + 1$, $du = 2dx$: $\int_3^9 u^{-2} \cdot \frac{1}{2}du = -\frac{1}{2}u^{-1}\Big|_3^9 = -\frac{1}{2}\left(\frac{1}{9} - \frac{1}{3}\right) = -\frac{1}{2}\cdot\frac{-2}{9} = \frac{1}{9}$

**Problem 11.**
Use $u = e^{-x}$, $du = -e^{-x}dx$: $-\int \cos(u)du = -\sin(u) + C = -\sin(e^{-x}) + C$

**Problem 13.**
Use $u = \ln x$, $du = \frac{1}{x}dx$: $\int \sec^2 u\, du = \tan(u) + C = \tan(\ln|x|) + C$

**Problem 14.**
Use $u = \cos(x)$, $du = -\sin(x)\, dx$: $-\int u^{-\frac{1}{2}}du = -2u^{\frac{1}{2}} + C = -2\sqrt{\cos x} + C$

**Problem 16.**
Rewrite, then use: $u = e^{2x}$, $du = 2\cdot e^{2x}\, dx$:

$$\int e^{2x} \cdot e^{\frac{1}{2}x}dx = \int e^{2x} \cdot e^{\frac{1}{4}(2x)}dx = \frac{1}{2}\int u^{\frac{1}{4}}du = \frac{1}{2}\cdot\frac{4}{5}u^{\frac{5}{4}} + C = \frac{2}{5}\left(e^{2x}\right)^{\frac{5}{4}} + C = \frac{2}{5}e^{\frac{5}{2}} + C$$

**Problem 18.**

Use $u = e^x + 4$, $du = e^x dx$: $\int_5^9 3u^{-\frac{1}{2}} du = 6u^{\frac{1}{2}} \Big|_5^9 = 6\sqrt{9} - 6\sqrt{5} = 18 - 6\sqrt{5}$

**Problem 19.**

Substitute $u = e^{2x} + 1$, $du = 2e^{2x} dx$:

$$\int_{x=0}^{x=\ln 3} \frac{1}{u} \cdot \frac{1}{2} du = \frac{1}{2} \ln |u| \Big|_{x=0}^{x=\ln 3} = \frac{1}{2} \ln \left| e^{2x} + 1 \right| \Big|_0^{\ln 3} = \frac{1}{2} \ln(3^2 + 1) - \frac{1}{2} \ln(2) = \frac{1}{2}(\ln 10 - \ln 2) = \frac{1}{2} \ln 5$$

**Problem 21.**

Substitute $u = 3^x + 1$, $du = \ln 3 \cdot 3^x dx$: $\int \frac{1}{u} \cdot \frac{1}{\ln 3} du = \frac{1}{\ln 3} \cdot \ln |3^x + 1| + C$

**Problem 23.**

Substitute $u = \sqrt{x}$, $2du = \frac{1}{\sqrt{x}} dx$: $2 \int \sec^2 u \cdot \tan^2 u\, du = \frac{2}{3} \tan^3 u + C = \frac{2}{3} \tan^3 \sqrt{x} + C$

**Problem 24.**

$\int_0^x 100 \cdot e^{-.2t} dt = \frac{100}{-.2} \cdot e^{-.2t} \Big|_0^x = -500 \left( e^{-.2x} - 1 \right) = 500 - 500e^{-.2x}$, for $x = 3$ this is about 226 ants.

**Problem 26.**

$x$ hours from now: $4 + \int_0^x -e^{.2t} dt = 4 - \left( \frac{1}{.2} e^{.2t} \right) \Big|_0^x = 4 - 5(e^{.2x} - 1) = 9 - 5e^{.2x}$ gallons.

$9 - 5e^{.2x} = 0$, $e^{.2x} = \frac{9}{5}$, $x = \frac{\ln \frac{9}{5}}{.2} \approx 2.94$, so it takes almost 3 hours to dry out.

## Section 25.3   Substitution to alter the Form of an Integral

**Problem 1.**

Use $u = x + 1$, $du = dx$, $x = u - 1$:
$\int \frac{u-1}{u} du = \int (1 - \frac{1}{u}) du = u - \ln |u| + C = x + 1 - \ln |x + 1| + C$ or:$x - \ln |x + 1| + D$

**Problem 3.**

Use $u = 3x + 5$, $\frac{1}{3} du = dx$, $x = \frac{u-5}{3}$:

$$\int \frac{1}{3} \frac{u-5}{3} \sqrt{u}\, du = \int (\frac{1}{9} u^{\frac{3}{2}} - \frac{5}{9} u^{\frac{1}{2}}) du = \frac{1}{9} \cdot \frac{2}{5} u^{\frac{5}{2}} - \frac{5}{9} \cdot \frac{2}{3} u^{\frac{3}{2}} + C = \frac{2}{45} (3x+5)^{\frac{5}{2}} - \frac{10}{27} (3x+5)^{\frac{3}{2}} + C$$

**Problem 4.**

Use $u = 3 + x$, $du = dx$, $x = u - 3$:
$\int \frac{2(u-3)}{u} du = \int (2 - \frac{6}{u}) du = 2u - 6 \ln |u| + C = 2(3 + x) - 6 \ln |3 + x| + C$ or: $2x - 6 \ln |3 + x| + D$

**Problem 6.**

Use $u = 2t + 5$, $du = 2\, dt$, $t = \frac{u-5}{2}$:

$$\int \frac{u-5}{2} \sqrt{u} \frac{1}{2} du = \int (\frac{1}{4} u^{\frac{3}{2}} - \frac{5}{4} u^{\frac{1}{2}}) du = \frac{1}{4} \cdot \frac{2}{5} u^{\frac{5}{2}} - \frac{5}{4} \cdot \frac{2}{3} u^{\frac{3}{2}} + C = \frac{1}{10} (2t+5)^{\frac{5}{2}} - \frac{5}{6} (2t+5)^{\frac{3}{2}} + C$$

**Problem 8.**

Use $u = \sin(3x)$, $\frac{du}{dx} = 3\cos(3x)$, $\frac{1}{3}du = \cos(3x)\,dx$:

$$\int \frac{\cos(3x)}{\sin(3x)}dx = \int \frac{1}{u} \cdot \frac{1}{3}\,du = \frac{1}{3}\ln|u| + C = \frac{1}{3}\ln|\sin(3x)| + C$$

**Problem 10.**

Use $u = \sin t$, $du = \cos t$: $\int u^4 du = \frac{1}{5}u^5 + C = \frac{1}{5}(\sin t)^5 + C$

**Problem 11.**

Use $u = 2t + 6$, $du = 2dt$, $2t = u - 6$:

$$\int \frac{u-6}{\sqrt{u}} \cdot \frac{1}{2}du = \int (\frac{1}{2}u^{\frac{1}{2}} - 3u^{-\frac{1}{2}})du = \frac{1}{2} \cdot \frac{2}{3}u^{\frac{3}{2}} - 3 \cdot 2u^{\frac{1}{2}} + C = \frac{1}{3}(2t+6)^{\frac{3}{2}} - 6\sqrt{2t+6} + C$$

**Problem 13.**

For $\frac{A}{x} + \frac{B}{x+2} = \frac{A(x+2)+Bx}{x(x+2)} = \frac{(A+B)x+2A}{x(x+2)}$ to equal $\frac{2}{x(x+2)}$ we need $A + B = 0$ and $A = 1$, so $B = -1$.

$$\int \frac{2}{x(x+2)}dx = \int (\frac{1}{x} + \frac{-1}{x+2})dx = \ln|x| - \ln|x+2| + C = \ln\left|\frac{x}{x+2}\right| + C$$

# CHAPTER 26

# Numerical Methods of Approximating Definite Integrals

## Section 26.1   Approximating Sums: $L_n$, $R_n$, $T_n$ and $M_n$

**Problem 1.**

(a) recall $R_n = \sum_{i=1}^{n} f(t_i) \cdot \Delta t = f(t_1)\Delta t + f(t_2)\Delta t + \cdots + f(t_{n-1})\Delta t + f(t_n)\Delta t$

and $L_n = \sum_{i=0}^{n-1} f(t_i) \cdot \Delta t = f(t_0)\Delta t + f(t_1)\Delta t + \cdots + f(t_{n-2})\Delta t + f(t_{n-1})\Delta t$.

Then $R_n = L_n + f(t_n)\Delta t - f(t_0)\Delta t = L_n + f(b)\Delta t - f(a)\Delta t$

(b) Subtracting gives $R_n - L_n = [f(b) - f(a)]\,\Delta t = [f(b) - f(a)] \cdot \frac{|b-a|}{n}$

**Problem 2.**

(a) $\int_1^x \frac{1}{t}dt = \ln t \big|_1^x = \ln x - \ln 1 = \ln x$

(b) $L_4 = \left(\frac{1}{1} + \frac{1}{5/4} + \frac{1}{6/42} + \frac{1}{7/4}\right) \cdot \frac{1}{4} = (1 + \frac{4}{5} + \frac{4}{6} + \frac{4}{7}) \cdot \frac{1}{4}$

$= \frac{319}{420} \approx 0.7595$

$R_4 = \left(\frac{1}{5/4} + \frac{1}{6/42} + \frac{1}{7/4} + \frac{1}{2}\right) \cdot \frac{1}{4} = (\frac{4}{5} + \frac{4}{6} + \frac{4}{7} + \frac{1}{2}) \cdot \frac{1}{4}$

$= \frac{533}{840} \approx 0.6345$

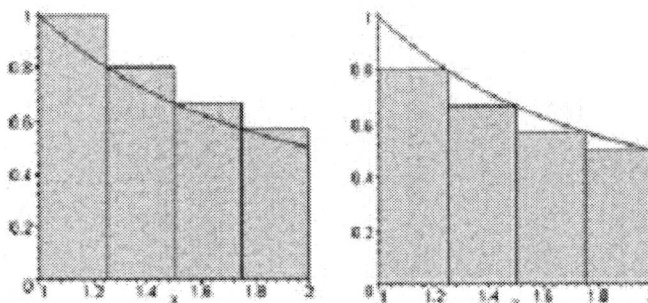

(c) Using the results of problem 1 above we get

$$0.01 \geq |R_p - L_p| = \left|\frac{1}{2} - \frac{1}{1}\right| \cdot \frac{2-1}{p} \Rightarrow \frac{1}{2} \cdot \frac{1}{.01} \leq p \Rightarrow p \geq 50.$$

(d) use a programmable calculator or a computer for this part!

$$T_{50} = \frac{R_{50} + L_{50}}{2} \approx 0.6931721794 \quad M_{50} \approx 0.6931346816$$

(e) $0.001 \geq |R_n - L_n| = \left|\frac{1}{3} - \frac{1}{1}\right| \cdot \frac{3-1}{n} \Rightarrow \frac{2}{3} \cdot \frac{2}{.001} \leq n \Rightarrow n \geq 1334.$

$$\left|\frac{1}{3} - \frac{1}{1}\right| \cdot \frac{3-1}{n} \leq D \Leftrightarrow n \geq \frac{4}{3D}.$$

**Problem 4.**

(a) $L_n < A < R_n$ since $f$ is increasing, and $T_n < A < M_n$ since $f$ is concave down.

   $T_n$ is the average of $R_n$ and $L_n$. Therefore: $L_n \leq T_n \leq A \leq R_n$.

(b) $R_n < A < L_n$ since $f$ is decreasing, and $M_n < A < T_n$ since $f$ is concave up: $R_n \leq A \leq L_n \leq T_n$.

**Problem 6.**

$0.01 \geq |R_n - L_n| = \left|\sqrt{2} - \sqrt{1}\right| \cdot \frac{2-1}{n} \Rightarrow n \geq \frac{\sqrt{2}-1}{.01} \approx 41.42$. Hence any number between $R_{42} \approx 1.094427336$ and $L_{42} \approx 1.084565108$ will approximate the integral with error less than .01.

**Problem 8.**

$f(x) = \frac{1}{\ln x}$ is decreasing, so $R_n < I < L_n$.

$0.05 \geq |R_n - L_n| = |f(b) - f(a)| \cdot \frac{b-a}{n} = \left|\frac{1}{\ln 3} - \frac{1}{\ln 2}\right| \cdot \frac{3-2}{n} \Rightarrow n \geq \left|\frac{1}{\ln 3} - \frac{1}{\ln 2}\right| \cdot \frac{1}{.05} \cong 10.65$. Use $n \geq 11$.

Lower bound: $R_{11} \approx 1.0947$ upper bound: $L_{11} \approx 1.1432$

**Problem 10.**

$R_n < L_n$ since $f$ is decreasing, and $T_n < M_n$ since $f$ is concave down.

   $T_n$ and $M_n$ are always between $R_n$ and $L_n$. Therefore: $R_n \leq T_n \leq M_n \leq L_n$.

$$R_n = 0.274760, \quad T_n = 0.3211885, \quad M_n = 0.341189, \quad L_n = 0.367617$$

**Problem 12.**

Strategy: $\int_0^1 + \int_1^2 = \int_0^2$, and from symmetry (see a graph) $2\int_0^2 = \int_{-2}^2$.

We will compute the first two with error less than $\frac{1}{400}$ and use to give the third with error less than $\frac{1}{200}$ and the fourth with error less than $\frac{1}{100}$.

(i) $|R_n - L_n| = \left|e^{-.5} - 1\right|\frac{1}{n} \leq \frac{1}{400} \Rightarrow n \geq 400\left|e^{-0.5} - 1\right| \cong 157.4$. Use $n \geq 158$. Then any number between

$R_{158} \approx 0.854377$ or $L_{158} \approx 0.856868$ approximates $\int_0^1 f(x)dx$ to within $\frac{1}{400}$.

(ii) $|R_n - L_n| = \left|e^{-2} - e^{-.5}\right|\frac{1}{n} \leq \frac{1}{400}$ gives $\Rightarrow n \geq 400 \cdot \left|e^{-2} - e^{-.5}\right| \approx 188.48$. Use $n \geq 189$. Then any

number between $R_{189} \approx 0.339417$ and $L_{189} \approx 0.341911$ approximates $\int_1^2 f(x)dx$   $(\leq \frac{1}{400})$.

(iii) Any number between the sum of the above right and left sums: $R \approx 1.193794$ and $L \approx 1.198779$

approximates $\int_0^2 f(x)dx$ to within $\frac{1}{400} + \frac{1}{400} = \frac{1}{200}$

(iv) Any number between $2R \approx 2.38758$ and $2L \approx 2.39755$ approximates $\int_{-2}^2 f(x)dx$ to within $\frac{1}{100}$.

**Problem 14.**

From symmetry $\int_{-1}^0 \arctan(x)dx = -\int_0^1 \arctan(x)dx$. Hence $\int_{-1}^2 \arctan(x)dx = \int_1^2 \arctan(x)dx$.

$f(x) = \arctan x$ is increasing on the whole interval which means $L_n < I < R_n$.

$0.05 \geq |R_n - L_n| = |f(2) - f(1)| \cdot \frac{2-1}{n} \Rightarrow n \geq \frac{\arctan(2) - \arctan(1)}{0.05} \cong 6.4$. Use $n \geq 7$.

Upper bound $R_7 \approx 0.9933$ lower bound $L_7 \approx 0.9473$.

**Problem 15.**

Since $f$ is decreasing, $R_n < L_n$ and since $f$ is concave down $T_n < M_n$.

   $T_n$ and $M_n$ are always between $R_n$ and $L_n$. Therefore: $R_n \leq T_n \leq I \leq M_n \leq L_n$.

   (a) true (b) true (c) true (d) false (e) false (f) true

**Problem 17.**
   (a) must be true, (b) must be true, (c) must be true, (d) not always for all functions.

## Section 26.2    Simpson's Rule and Error Estimates

**Problem 2.**
   (Be aware of the double use of $M_4$. Determine which one is needed from context.)

(a) For $f(x) = x^2$, $f''(x) = 2$ and $f^{(4)}(x) = 0$ so $M_2 = 2$ and $M_4 = 0$ in the error bounds.

$$T_4 = \frac{1}{2}\left(0^2 + 2\cdot\left(\frac{1}{2}\right)^2 + 2\cdot(1)^2 + 2\cdot\left(\frac{3}{2}\right)^2 + 2^2\right)\cdot\frac{1}{2} = \frac{11}{4} \text{ with } |T_4 - I| \le \frac{M_2(b-a)^3}{12n^2} = \frac{2\cdot 2^3}{12\cdot 4^2} = \frac{1}{12}$$

$$M_4 = \left(\left(\frac{1}{4}\right)^2 + \left(\frac{3}{4}\right)^2 + \left(\frac{5}{4}\right)^2 + \left(\frac{7}{4}\right)^2\right)\cdot\frac{1}{2} = \frac{21}{8} \text{ with } |M_4 - I| \le \frac{M_2(b-a)^3}{24n^2} = \frac{2\cdot 2^3}{24\cdot 4^2} = \frac{1}{24}$$

$$S_{2(4)} = \frac{2M_4 + T_4}{3} = \frac{8}{3} \text{ with } |S_8 - I| \le \frac{M_4(b-a)^5}{180\cdot n^4} = 0 \text{ since } f^{(4)}(x) = 0, \text{ no error!}$$

Exact answer is $\int_0^2 x^2 dx = \frac{1}{3}x^3\big|_0^2 = \frac{8}{3}$

(b) For $f(x) = \frac{1}{1+x}$, $f''(x) = \frac{2}{(x+1)^3}$ and $f^{(4)}(x) = \frac{24}{(x+1)^5}$ so $M_2 = |f''(0)| = 2$ and $M_4 = \left|f^{(4)}(0)\right| = 24$

$$T_4 = \frac{1171}{1680} \approx 0.6970238095 \text{ with } |T_4 - I| \le \frac{M_2(b-a)^3}{12n^2} = \frac{2}{(12)4^2} = \frac{1}{96}$$

$$M_4 = \frac{4448}{6435} \approx 0.691219891 \text{ with } |M_4 - I| \le \frac{M_2(b-a)^3}{24n^2} = \frac{2}{(24)4^2} = \frac{1}{192}$$

$$S_{2(4)} = \frac{2M_4 + T_4}{3} = \frac{1498711}{2162160} \approx 0.6931471806 \text{ with } |S_8 - I| \le \frac{M_4(b-a)^5}{(180)8^4} = \frac{24}{(180)8^4} = \frac{1}{30720} \approx 0.000032552$$

Exact answer is $\int_0^1 \frac{1}{x+1} dx = \ln(x+1)\big|_0^1 = \ln 2 - \ln 1 = \ln 2 \approx 0.6931471806$

**Problem 4.**
   For $f(x) = \cos(x^2)$, $f''(x) = -4x^2\cos(x^2) - 2\sin(x^2)$ so $M_2 = |f''(1)| \approx 3.8442$.
   $f^{(4)}(x) = 16x^4\cos(x^2) + 48x^2\sin(x^2) - 12\cos(x^2)$ so

(a) $M_{10} \approx 0.9052254887$ with error bound $|M_{10} - I| \le \frac{M_2}{(24)10^2} \approx 0.0016017$

(b) $T_{10} \approx 0.9031217571$ so $S_{20} = \frac{2M_4 + T_4}{3} \approx 0.9045242448$, with $|S_{20} - I| \le \frac{M_4}{180(20)^4} \approx 0.0000014775$

**Problem 5.**

(a) $T_6 = \frac{1}{2}\left(y_0 + 2y_1 + 2y_2 + 2y_3 + 2y_4 + 2y_5 + y_6\right)\Delta x$

$$= \frac{1}{2}(0.5 + 4(1.5) + 2(3) + 4(5) + 2(4) + 4(3) + 1)(10) \approx 86.25$$

(b) $S_{2(3)} = \frac{1}{6}\left(y_0 + 4y_1 + 2y_2 + 4y_3 + 2y_4 + 4y_5 + y_6\right)\Delta x$ (be careful with $\Delta x$.)

$$= \frac{1}{6}(0.5 + 4(1.5) + 2(3) + 4(5) + 2(4) + 4(3) + 1)(10) \approx 89.1666667$$

# CHAPTER 27

# Applying the Definite Integral: Slice and Conquer

## Section 27.1    Finding "Mass" When Density Varies

**Problem 2.**

(a)

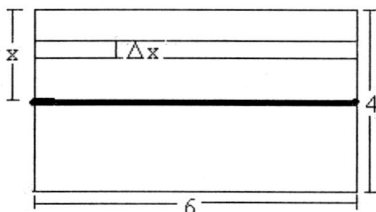

(b)  $A_i = 6 \Delta x$

(c)  $Pop_i \approx \rho(x_i)(6\Delta x) = (10000 - 800x_i)(6\Delta x)$

(d)  $2\sum\limits_{i=1}^{n}(10000 - 800x_i)(6\Delta x)$ (by symmetry just do twice one side)

(e)  $2\int_0^2 (10000 - 800x)(6dx) = 12\left(10000x - 400x^2\right)\big|_0^2 = 220800$ people

**Problem 4.**

(a)  $Y_i = f(x_i)1000\Delta x$, hence the limit of the sum becomes

$$\int_0^{800} f(x)1000dx = 1000\int_0^{800}(50 - 0.3\sqrt{x})dx = 1000(50x - 0.3\tfrac{2}{3}x^{\frac{3}{2}})\Big|_0^{800}$$

$$= 1000(40000 - 0.2(800)^{3/2} \approx 35474517 \text{ ears}$$

(b)  $Y_i = f(x_i)2\pi x_i\Delta x$, hence the limit of the sum becomes

$$\int_0^{80} f(x)2\pi\, x dx = \int_0^{80}(50 - 0.3\sqrt{x})2\pi\, x dx = 2\pi\int_0^{80}(50x - 0.3x^{3/2})dx$$

$$= 2\pi\left(25x^2 - 0.3\tfrac{2}{5}x^{5/2}\right)\Big|_0^{80} \approx 962149 \text{ ears}$$

**Problem 6.**

   $H_i = \rho(r_i)2\pi r_i\Delta r$, hence the limit of the sum becomes

$$\int_0^{10}\rho(r)2\pi\, rdr = \int_0^{10}\frac{1010}{\pi(r^2+1)^2}2\pi\, rdr = 1010\int_0^{10}\left(\frac{2r}{(r^2+1)^2}\right)dr$$

$1010 \int_0^{10} (r^2+1)^{-2} 2r dr = -1010 \, (r^2+1)^{-1} \big|_0^{10} = -1010 \left( \frac{1}{101} - \frac{1}{1} \right) = 1000$ holes

The above could be done using $u = (r^2 + 1)$

## Problem 8.

(a) $\sum\limits_{i=1}^{n} \delta(x_i)(7) \, \Delta x = \sum\limits_{i=1}^{n} (4000 - 2000x_i)(7) \, \Delta x$

(b) $\int_0^2 7(4000 - 2000x) dx = 7 \left( 4000x - 1000x^2 \right) \big|_0^2 = 28000$ people

## Problem 10.

(a) $W_i = g(x_i)(400)\Delta x$, hence the limit of the sum becomes

$2 \int_0^{50} g(x)(400) dx$ (by symmetry same below as above pipe)

(b) $2 \int_0^{50} \frac{50}{1+x^2}(400) dx = 40000 \int_0^{50} \frac{1}{1+x^2} \, dx = $

$40000 \, (\arctan x) \big|_0^{50} = 40000(\arctan 50 - \arctan 0) \approx 62032$ flowers

## Problem 12.

$M_i = \rho(r_i) 4\pi r_i^2 \Delta r$, hence the limit of the sum becomes (slice is like the peeling of a grapefruit)

$$\int_0^{8000} \frac{40000}{1+0.0001r^3} 4\pi r^2 dr = \frac{160000}{0.0003} \pi \int_0^{8000} \frac{0.0003r^2}{1+0.0001r^3} \, dr = \frac{160000}{0.0003} \pi \left( \ln(1 + 0.0001r^3) \right) \big|_0^{8000}$$

$\frac{160000}{0.0003} \pi (\ln(51200001) - \ln(1)) \approx 29742505033$ kilograms

## Problem 14.

For a horizontal slice, let $h$ be the distance from the top of the sphere to the slice, and let r be the radius of the disk. When working with a sphere it often helps to simplify if we use a variable that measures distance from the center. Hence we introduce $x$ to be the distance the slice is up, or down, from the center. Then $x + h = 2$ and $x^2 + r^2 = 4$. x is positive in the top half and negative in the bottom half.

$M_i = \rho(h_i)\pi(r_i)^2 \Delta h$, so limit of sum becomes

$$\int_{-2}^{2} \rho(x)\pi r^2 dx = \int_{-2}^{2} 5(2-x)\pi(\sqrt{4-x^2})^2 dx$$

$$= 5\pi \int_{-2}^{2} (2-x)(4-x^2) dx = 5\pi \int_{-2}^{2} (8 - 4x - 2x^2 + x^3) dx$$

$$= 5\pi \left( 8x - 2x^2 - \tfrac{2}{3}x^3 + \tfrac{1}{4}x^4 \right)\big|_{-2}^{2} = \tfrac{320}{3}\pi$$

## Problem 16.

$\frac{dW}{dt} = 30\cos(\frac{\pi}{2}t)$ is positive on intervals $[0, 1)$ and $(3, 5)$ and negative on interval $(1, 3)$.

$\frac{dW}{dt} = 30\cos(\frac{\pi}{2}t) = -15$ when $\cos(\frac{\pi}{2}t) = -\frac{1}{2}$ or when $\frac{\pi}{2}t = \frac{2}{3}\pi$ or $\frac{4}{3}\pi$, hence when $t = \frac{4}{3}$ or $\frac{8}{3}$.

Gallons flowing out $= \int_{4/3}^{8/3} 30\cos(\frac{\pi}{2}t) dt = \int_{2\pi/3}^{4\pi/3} 30\cos(u)\frac{2}{\pi} du = \frac{60}{\pi} (\sin u)\big|_{2\pi/3}^{4\pi/3} = -\frac{60}{\pi}\sqrt{3} \approx -33$

gallons

## Problem 17.

Take horizontal slices of the hemisphere. From a top view the slice will look like a disk. The volume of the slice will be $\pi r_i^2 \Delta h$

(a) $I_i = \rho(h_i)\pi\, r_i^2 \Delta h = 6(10^{-5})(200 - h_i)\pi(100^2 - h_i^2)\Delta h.$

Hence the limit of the sum becomes $\int_0^{100} 6(10^{-5})(200 - h)\pi(100^2 - h^2)dh$

(b) $6(10^{-5})\pi \int_0^{100}(2000000 - 10000h - 200h^2 + h^3)dh$

$$= 6(10^{-5})\pi \left(2000000h - 5000h^2 - \tfrac{200}{3}h^3 + \tfrac{1}{4}h^4\right)\Big|_0^{100} = 6500\pi$$

**Problem 19.**

Use a cylindrical shell. $M_i = \rho(x_i)4(2\pi\, x_i)\Delta x.$ Hence the limit of the sum becomes $\int_0^3 \rho(x)8\pi\, x\, dx$

**Problem 21.**

The volume of a cylindrical shell is (height)(circumference)(thickness).

$V_i = h(x_i)(2\pi x_i)\Delta x.$ Hence the limit of the sum becomes

$V = \int_0^4 \frac{8}{1+\frac{x^2}{16}}(2\pi\, x)dx = 128\pi \int_0^4 \frac{2x}{16+x^2}dx =$ (use $u = 16 + x^2$)

$128\pi \int_{16}^{32} \frac{1}{u}du = 128\pi\,(\ln u)\big|_{16}^{32} = 128\pi \ln 2 \approx 278.73$ cubic feet.

**Problem 22.**

For the whole pizza $G_i = g(x_i)(2\pi x_i)\Delta x.$ The limit of the sum becomes

$\int_0^7 g(x)(2\pi\, x)dx = \int_0^7 \frac{x}{(x^3+2)^2}(2\pi\, x)dx = \tfrac{2}{3}\pi \int_0^7 \frac{3x^2}{(x^3+2)^2}dx =$ (use $u = (x^3 + 2)$)

$\tfrac{2}{3}\pi \int_2^{345} \frac{1}{u^2}du = \tfrac{2}{3}\pi \int_2^{345} u^{-2}du = \tfrac{2}{3}\pi\left(-u^{-1}\right)\big|_2^{345} = \tfrac{2}{3}\pi(-\tfrac{1}{345} + \tfrac{1}{2}) = \tfrac{343}{1035}\pi \approx 1.0411$

Hence for one slice it will be $(1.0411)/6 = 0.1735$ ounces of garlic.

## Section 27.2    Slicing to Find the Area Between Two Curves

**Problem 1.**

Area $= \int_0^1 (e^x - (1 - x))dx = \left(e^x - x + \tfrac{1}{2}x^2\right)\Big|_0^1 = e - \tfrac{3}{2}$

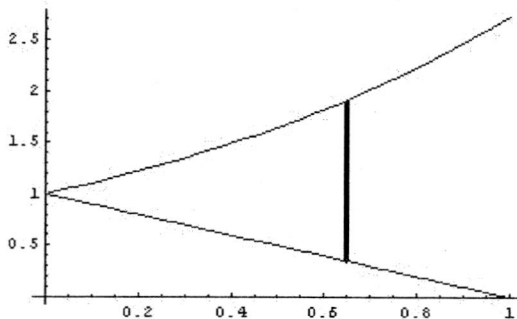

**Problem 3.**

$$-x + 6 = x^2 + 1 \;\Rightarrow\; x^2 + x - 5 = 0 \;\Rightarrow\; x = \frac{-1 \pm \sqrt{1-4(-5)}}{2} = \frac{-1 \pm \sqrt{21}}{2}$$

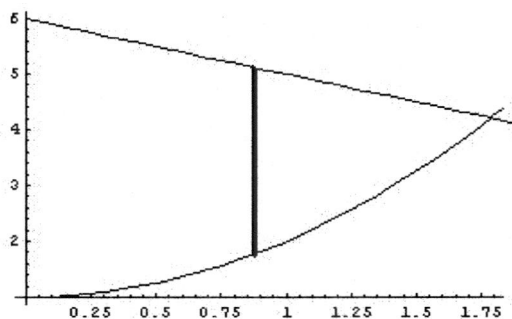

$$\text{Area} = \int_0^{\frac{-1+\sqrt{21}}{2}} ((-x+6) - (x^2+1))dx = \left. \left(-\tfrac{1}{3}x^3 - \tfrac{1}{2}x^2 + 5x\right) \right|_0^{\frac{-1+\sqrt{21}}{2}} \approx 5.436$$

**Problem 5.**

$$y = \arcsin x \;\Leftrightarrow\; x = \sin y$$

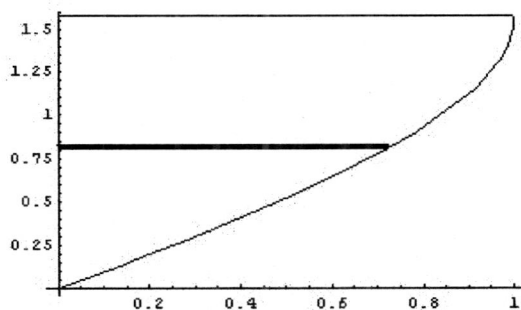

$$\text{Area} = \int_0^{\pi/2} \sin y \; dy = \left. (-\cos y) \right|_0^{\pi/2} = -\cos(\pi/2) + \cos(0) = 1$$

**Problem 6.**

(a)    $\text{Area} = \int_{-\pi/4}^{\pi/4} (\cos x - \sin x)dx$

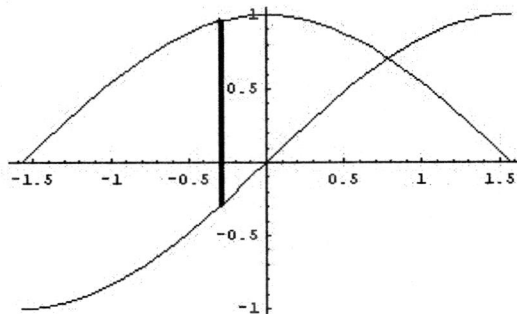

(b)    We need to split the

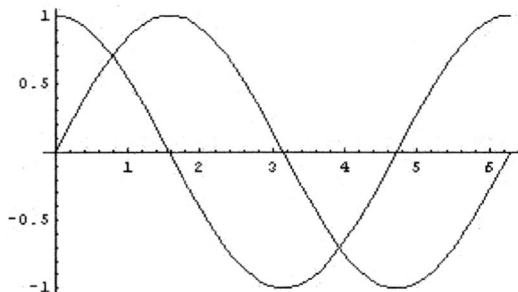

interval at $\pi/4$ and at $5\pi/4$ as the functions trade
which is higher. The combined area will be

$$\int_0^{\pi/4}(\cos x - \sin x)dx + \int_{\pi/4}^{5\pi/4}(\sin x - \cos x)dx + \int_{5\pi/4}^{2\pi}(\cos x - \sin x)dx$$

**Problem 8.**

See graph in text. First we find the intersection points.
$(-x^2+2) = x \Leftrightarrow x^2+x-2 = 0 \Rightarrow x = -2, 1.$ Also $y = -x^2+2 \Rightarrow x^2 = -y+2 \Rightarrow x = \pm\sqrt{-y+2}.$
$\int_{-2}^1 ((-x^2+2)-x)dx$ or $\int_{-2}^1 (y-(-\sqrt{-y+2}))dy + \int_1^2 (\sqrt{-y+2}-(-\sqrt{-y+2}))dy$

**Problem 10.**

$y = \ln x \Leftrightarrow x = e^y.$ Using $y$-slices the area is
$\int_0^{\ln 10}(10-e^y)dy = (10y-e^y)|_0^{\ln 10} = (10\ln 10 - 10) - (0-1) = 10\ln(10) - 9$

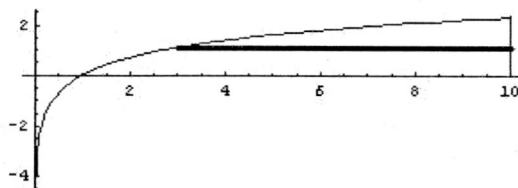

**Problem 12.**

Line is $(y-1) = \frac{(1-0)}{(e-2e)}(x-e) \Leftrightarrow x = -e(y-1)+e \Rightarrow x = -ey+2e$

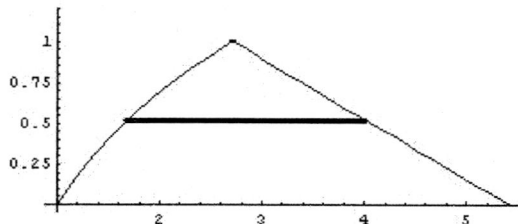

$\int_0^1 ((-ey+2e)-e^y)dy = (-\frac{e}{2}y^2+2ey-e^y)|_0^1 = -\frac{e}{2}+2e-e-(-1) = \frac{e}{2}+1$
OR $\int_1^e \ln x\,dx + \int_e^{2e}(-\frac{1}{e}x+2)dx = \frac{e}{2}+1$

**Problem 13.**

$y = \arctan x \iff \tan y = x$

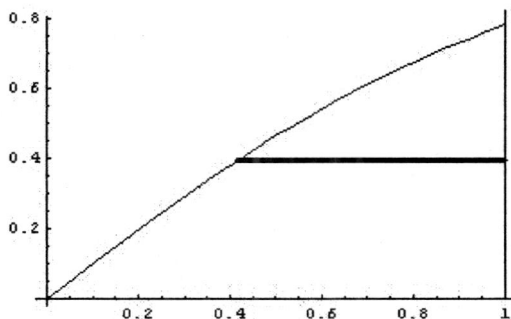

$\int_0^{\pi/4} (1 - \tan y) dy = \int_0^{\pi/4} (1) dy + \int_0^{\pi/4} (\frac{-\sin y}{\cos y}) dy = y\big|_0^{\pi/4} + (\ln(\cos y))\big|_0^{\pi/4}$

$= \pi/4 + \ln(\cos(\pi/4)) - (\ln(\cos(0)) = \pi/4 + \ln(\frac{\sqrt{2}}{2})$

**Problem 15.**

$\int_0^2 (2x - (-\frac{1}{9}(x^2 - 6x))) dx + \int_2^3 ((-3x + 10) - (-\frac{1}{9}(x^2 - 6x))) dx =$

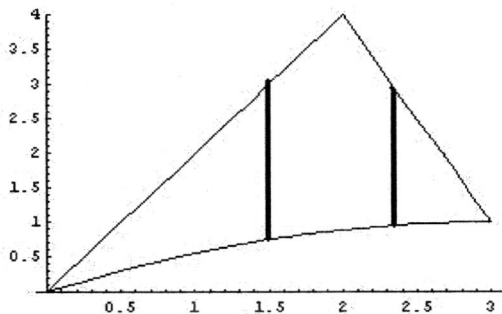

$\int_0^2 (\frac{4}{3}x + \frac{1}{9}x^2) dx + \int_2^3 (10 - \frac{11}{3}x + \frac{1}{9}x^2) dx =$

$(\frac{2}{3}x^2 + \frac{1}{27}x^3)\big|_0^2 + (10x - \frac{11}{6}x^2 + \frac{1}{27}x^3)\big|_2^3 = \frac{243}{54}$

# CHAPTER 28

# More Applications of Integration

## Section 28.1    Computing Volumes

**Problem 2.**

The base is part of a circle hence $x^2 + y^2 = 2^2 \Rightarrow y = \sqrt{4 - x^2}$

Cross sections of the volume above the base are right isosceles triangles so their heights are also $y$. Using symmetry we double one half to get

$$\text{Volume} = 2 \int_0^2 \tfrac{1}{2}(y)(y)dx = \int_0^2 (4 - x^2)dx = \left. \left( 4x - \tfrac{1}{3}x^3 \right) \right|_0^2 = \tfrac{16}{3}$$

**Problem 4.**

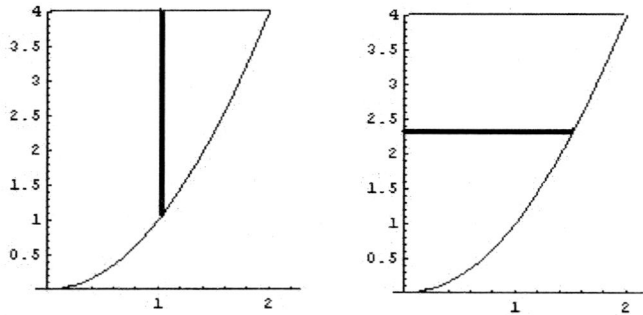

(a) If the $x$-slice is rotated about the $x$-axis it will form a washer. The volume will be $\int_0^2 \pi((4)^2 - (x^2)^2)dx = \pi \int_0^2 (16 - x^4)dx$

$$= \pi \left. \left( 16x - \tfrac{1}{5}x^5 \right) \right|_0^2 = \tfrac{128}{5}\pi$$

(b) If the $y$-slice is rotated about the $x$-axis it will form a shell. The volume will be

$$\int_0^4 2\pi \, y(\sqrt{y})dy = 2\pi \int_0^4 (y^{3/2})dy = \tfrac{128}{5}\pi$$

(c) The volume generated by rotating the region under $y = 4$ about the $x$-axis will be the volume of a cylinder of radius 4 and height 2, hence $\pi (4)^2(2) = 32\pi$. The volume not used is the region under $y = x^2$ about the $x$-axis $\int_0^2 \pi(x^2)^2 dx = \pi \int_0^2 x^4 dx = \pi \left. \left( \tfrac{1}{5}x^5 \right) \right|_0^2 = \tfrac{32}{5}\pi$. Subtracting these gives $\tfrac{128}{5}\pi$.

**Problem 6.**

$$\int_{1/2}^5 \pi(\tfrac{1}{x})^2 dx = \pi \int_{1/2}^5 x^{-2}dx = \pi \left. \left( -x^{-1} \right) \right|_{1/2}^5 = \pi(-(\tfrac{1}{5}) - (-2)) = \tfrac{9}{5}\pi$$

**Problem 7.**

(a) About the $y$-axis we get shells. The volume will be formed by rotating the

left side of region or the right side of the region. We will only use the right

side $\int_0^{\sqrt{2}} 2\pi \, x((4 - x^2) - x^2)dx = 2\pi \int_0^{\sqrt{2}} (4x - 2x^3)dx = 4\pi \left. \left( x^2 - \tfrac{1}{4}x^4 \right) \right|_0^{\sqrt{2}} = 4\pi$

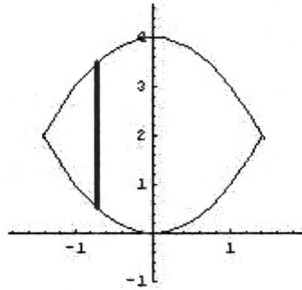

(b) About the $x$-axis we get washers. The volume is double the right half

$$2\int_0^{\sqrt{2}} \pi((4-x^2)^2 - (x^2)^2)dx = 2\pi \int_0^{\sqrt{2}} (16 - 8x^2)dx = 2\pi \left(16x - \tfrac{8}{3}x^3\right)\Big|_0^{\sqrt{2}} = \tfrac{64}{3}\sqrt{2}\pi$$

### Problem 8.

Think of rotating a piece of the circle of radius 5 about the $x$-axis. If we do $y$-slices, they will be disks. The volume will be $\int_3^5 \pi(\sqrt{25 - y^2})^2 dy = \pi \int_3^5 (25 - y^2)dy = \pi \left(25y - \tfrac{1}{3}y^3\right)\Big|_3^5 = \tfrac{52}{3}\pi$

### Problem 10.

Using the disk method, the volume for the whole cup is

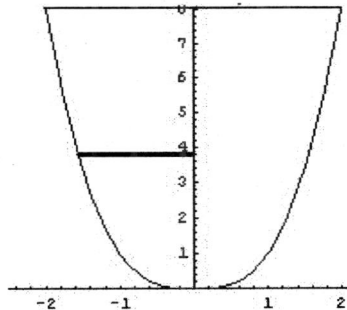

$\int_0^8 \pi (y^{1/3})^2 dy = \pi \int_0^8 y^{2/3} dy = \pi \tfrac{3}{5}y^{5/3}\Big|_0^8 = \tfrac{96}{5}\pi$. We will find $h$ so that

$$\tfrac{48}{5}\pi = \int_0^h \pi (y^{1/3})^2 dy = \pi \int_0^h y^{2/3} dy = \pi \tfrac{3}{5}y^{5/3}\Big|_0^h = \tfrac{3}{5}\pi h^{5/3} \Rightarrow$$

$$h^{5/3} = \tfrac{48}{3} = 16 \Rightarrow h = (16)^{3/5} \approx 5.278$$

### Problem 12.

(a) $\int_0^9 \pi(2\sqrt{y})^2 dy = \pi \int_0^9 4y\,dy = 2\pi\, y^2\big|_0^9 = 162\pi$

(b) $\int_0^4 \pi(2\sqrt{y})^2 dy = \pi \int_0^4 4y\,dy = 2\pi\, y^2\big|_0^4 = 32\pi$

Then $162\pi - 32\pi = 130\pi$ cubic inches was ladled out in the hour.

## Section 28.2    Arc Length, Work, and Fluid Pressure

**Problem 1.**

$$f(x) = mx + b \; \Rightarrow \; f'(x) = m. \;\; AL = \int_0^3 \sqrt{1 + (f'(x))^2}\,dx = \int_0^3 \sqrt{1 + m^2}\,dx = \sqrt{1 + m^2}\,x \big|_0^3 = 3\sqrt{1 + m^2}$$

**Problem 3.**

$$f(x) = x^2 \; \Rightarrow \; f'(x) = 2x. \;\; AL = \int_{-1}^1 \sqrt{1 + (2x)^2}\,dx = \int_{-1}^1 \sqrt{1 + 4x^2}\,dx \approx 2.96$$

**Problem 5.**

$$W = (200ft.)(12lb.) = 2400ft - lb.$$

**Problem 6.**

(a) $F = kx \; \Rightarrow \; 5 = k(3) \; \Rightarrow \; k = \frac{5}{3}\frac{lb.}{in.}$

(b) $W = \int_0^3 kx\,dx = \int_0^3 \frac{5}{3}x\,dx = \frac{5}{6}x^2 \big|_0^3 = \frac{15}{2}\,in - lb = \frac{5}{8}ft - lb.$

(c) $W = \int_3^6 kx\,dx = \int_3^6 \frac{5}{3}x\,dx = \frac{5}{6}x^2 \big|_3^6 = \frac{45}{2}\,in - lb = \frac{15}{8}ft - lb.$

(d) The distance is the same, but the force is greater on [3, 6] than on [0, 3].

**Problem 8.**

Think of $x$ as being the distance from the roof to the washer so $F(x) = 160 + 0.6x.$

(a) $W = \int_{25}^{50} F(x)dx = \int_{25}^{50} (160 + 0.6x)dx = \left(160x + 0.3x^2\right)\big|_{25}^{50} = \frac{9152}{2} = 4562.5\,ft - lb.$

(b) $W = \int_0^{50} F(x)dx = \int_0^{50} (160 + 0.6x)dx = \left(160x + 0.3x^2\right)\big|_0^{50} = 8750\,ft - lb.$

**Problem 10.**

(a) $HF = (62.4\frac{lb}{ft^3})(4ft)(75 \times 25ft^2) = 468000\,lb.$

(b) $HFS = \int_0^4 62.4(x)(25dx) = \int_0^4 1560x\,dx = 12480\,lb.$

(c) $HFL = \int_0^4 62.4(x)(75dx) = \int_0^4 4680x\,dx = 37440\,lb.$

(d) $W = \int_0^4 x(62.4)(1875)dx = \int_0^4 117000x\,dx = 936000\,ft - lb.$

**Problem 11.**

Let $h$ be the distance that a thin slice of gasoline is below the top of the tank. If $r$ is the radius of the slice then $r^2 + h^2 = 7^2$, so the area of the slice will be $\pi r^2 = \pi(49 - h^2)$. The slice will be lifted $h$ ft. $W = \int_1^7 h42\pi(49 - h^2)dh = 42\pi \int_1^7 (49h - h^3)dh = 42\pi \left(\frac{49}{2}h^2 - \frac{1}{4}h^4\right)\big|_1^7 = 24192\pi$

**Problem 13.**

Let h be the distance that a thin horizontal slice of milk pushing on the end is below the top of the tank. The length of this slice will be $2r$ where $h^2 + r^2 = 3^2$ (semicircular end).

$$F = \int_0^3 64.5(h)(2r)dh = 64.5 \int_0^3 (9 - h^2)^{1/2}(2h)dh = 64.5 \left((-\tfrac{2}{3})(9 - h^2)^{3/2}\right)\Big|_0^3 = 1161\,lb.$$

**Problem 15.**

Using a rectangle and similar triangles, a thin horizontal slice of water $h$ ft. above the bottom of the dam will have length $(\frac{3}{4}h + 30)\, ft$ and be $(40 - h)\, ft$ below the surface of the water.

$$F = \int_0^{40} (40 - h)62.4(\tfrac{3}{4}h + 30)dh = 62.4 \int_0^{40} (1200 - \tfrac{3}{4}h^2)dh = 62.4 \left(1200h - \tfrac{1}{4}h^3\right)\Big|_0^{40} = 1996800\, lb.$$

(This problem could be done with $h$ being the distance the slice is below the surface of the water. We chose the above way because it makes the similar triangle ratios easier.)

# CHAPTER 29
# Computing Integrals

## Section 29.1    Integration by Parts – The Product Rule in Reverse

**Problem 2.**

$u = x, \quad dv = \cos x \, dx$
$du = dx, \quad v = \sin x$

$$\int x \sin x \, dx = x(\sin x) - \int \sin x \, dx = x \sin x + \cos x + C$$

**Problem 3.**

$u = 3x, \quad dv = e^{-2x} \, dx$
$du = 3dx, \quad v = -\frac{1}{2}e^{-2x},$

$$\int 3xe^{-2x} \, dx = 3x(-\frac{1}{2}e^{-2x}) - \int 3(-\frac{1}{2}e^{-2x}) \, dx = -\frac{3}{2}xe^{-2x} - \frac{3}{4}e^{-2x} + C$$

**Problem 5.**

$u = \cos^{-1} x, \quad dv = 1 \, dx$
$du = \frac{-1}{\sqrt{1-x^2}}dx, \quad v = x$

$$\int_0^1 \cos^{-1} x \, dx = x \cos^{-1} x \Big|_0^1 - \int_0^1 x(\frac{-1}{\sqrt{1-x^2}}) \, dx = 0 - \int_0^1 -x(1-x^2)^{-1/2} \, dx = -(1-x^2)^{1/2}\Big|_0^1 = 1$$

(You can use regular u-substitution to do the last integral.)

**Problem 6.**

$u = \sin^{-1}\left(\frac{x}{2}\right), \quad dv = 1 \, dx$
$du \frac{\frac{1}{2}}{\sqrt{1-\left(\frac{x}{2}\right)^2}}dx, \quad v = x$

$$\int \sin^{-1}\left(\frac{x}{2}\right) \, dx = x \sin^{-1}\left(\frac{x}{2}\right) - \int x \left(\frac{\frac{1}{2}}{\sqrt{1-\left(\frac{x}{2}\right)^2}}\right) dx = x \sin^{-1}\left(\frac{x}{2}\right) - \int \frac{1}{2}x(1-\left(\frac{x}{2}\right)^2)^{-1/2}dx$$

(u-sub on this integral)

$$= x \sin^{-1}\left(\frac{x}{2}\right) + 2(1-\left(\frac{x}{2}\right)^2)^{1/2} + C = x \sin^{-1}\left(\frac{x}{2}\right) + 2\sqrt{1-\left(\frac{x}{2}\right)^2} + C$$

**Problem 8.**

$u = \cos x, \quad dv = e^{-x}dx$
$du = -\sin x \, dx, \quad v = -e^{-x}$ (Use $u = \sin x, \quad dv = e^{-x}dx$ on the second integral.)

$\int e^{-x}\cos x\,dx = -e^{-x}\cos x - \int e^{-x}\sin x\,dx$ $(du = \cos x\,dx, \quad v = -e^{-x}.)$

$= -e^{-x}\cos x - [-e^{-x}\sin x - \int -e^{-x}\cos x\,dx] = -e^{-x}\cos x + e^{-x}\sin x - \int e^{-x}\cos x\,dx$

$$\Rightarrow \quad 2\int e^{-x}\cos x\,dx = -e^{-x}\cos x + e^{-x}\sin x$$

$\Rightarrow \quad \int e^{-x}\cos x\,dx = -\frac{1}{2}e^{-x}\cos x + \frac{1}{2}e^{-x}\sin x + C$ (The C appears after we do the last integral.)

## Problem 10.
By logarithm property $\ln(\frac{1}{x}) = -\ln x$. Use $u = -\ln x, \ dv = x\,dx$.
$du = -\frac{1}{x}dx, \quad v = \frac{1}{2}x^2$

$\int x\ln(\frac{1}{x})dx = \int -x\ln x\,dx = (-\ln x)(\frac{x^2}{2}) - \int(-\frac{1}{x})(\frac{x^2}{2})dx = -\frac{x^2}{2}\ln x + \int \frac{1}{2}x\,dx = \frac{x^2}{2}\ln(\frac{1}{x}) + \frac{1}{4}x^2 + C$

## Problem 12.
$u = t^3, \quad dv = e^{-t}\,dt$
$du = 3t^2\,dt, \quad v = -e^{-t}$
$\int_0^1 t^3 e^{-t}\,dt = -t^3 e^{-t} - \int_0^1 -3t^2 e^{-t}\,dt = -t^3 e^{-t}\big|_0^1 + \int_0^1 3t^2 e^{-t}\,dt$ (Use $u = 3t^2, \quad dv = e^{-t}\,dt$)
$= -t^3 e^{-t}\big|_0^1 + [-3t^2 e^{-t}\big|_0^1 - \int_0^1 -6te^{-t}\,dt] = -t^3 e^{-t}\big|_0^1 -3t^2 e^{-t}\big|_0^1 + \int_0^1 6te^{-t}\,dt$ (Use $u = 6t, \quad dv = e^{-t}\,dt$)

$= -t^3 e^{-t}\big|_0^1 -3t^2 e^{-t}\big|_0^1 + [-6te^{-t}\big|_0^1 - \int_0^1 -6e^{-t}\,dt] = -t^3 e^{-t}\big|_0^1 -3t^2 e^{-t}\big|_0^1 -6te^{-t}\big|_0^1 - 6e^{-t}\big|_0^1 = 6 - 16\frac{1}{e}$

## Problem 13.
$u = \ln x, \quad dv = x^{1/2}dx$
$$du = \frac{1}{x}dx, \quad v = \frac{2}{3}x^{3/2}$$
$$\int x^{1/2}\ln x\,dx = \frac{2}{3}x^{3/2}\ln x - \int \frac{2}{3}x^{1/2}dx = \frac{2}{3}x^{3/2}\ln x - \frac{4}{9}x^{3/2} + C$$

## Problem 14.
$u = \cos(\ln x), \quad dv = 1dx$
$du = -\sin(\ln x)\frac{1}{x}dx, \quad v = x$
$\int \cos(\ln x)dx = x\cos(\ln x) - \int -x\sin(\ln x)\frac{1}{x}dx = x\cos(\ln x) + \int \sin(\ln x)dx$ (Use $u = \sin(\ln x), \ dv = 1dx$)

$= x\cos(\ln x) + x\sin(\ln x) - \int x\cos(\ln x)\frac{1}{x}dx = x\cos(\ln x) + x\sin(\ln x) - \int \cos(\ln x)dx$

$$\Rightarrow \int \cos(\ln x)dx = \frac{1}{2}x\cos(\ln x) + \frac{1}{2}x\sin(\ln x) + C$$

**Problem 16.**

(a) $\int \cos^2 x\, dx = \int \frac{1}{2}(1 + \cos(2x))dx = \frac{1}{2}x + \frac{1}{4}\sin(2x) + C$

(b) $u = \cos x, \quad dv = \cos x\, dx$
$\quad du = -\sin x\, dx, \quad v = \sin x$

$$\int \cos^2 x\, dx = \sin x \cos x - \int -\sin^2 x\, dx = \sin x \cos x + \int (1 - \cos^2 x)\, dx = \sin x \cos x + x - \int \cos^2 x\, dx$$

$$\Rightarrow\ 2\int \cos^2 x\, dx = \sin x \cos x + x$$

$$\Rightarrow\ \int \cos^2 x\, dx = \frac{1}{2}\sin x \cos x + \frac{1}{2}x + C$$

(c) $\left(\frac{1}{2}x + \frac{1}{4}\sin(2x)\right)' = \frac{1}{2} + \frac{1}{4}\cos(2x)2 = \frac{1}{2} + \frac{1}{2}\cos(2x)$

$\quad \left(\frac{1}{2}\sin x \cos x + \frac{1}{2}x\right)' = \frac{1}{2}[\cos x \cos x + \sin x(-\sin x)] + \frac{1}{2} = \frac{1}{2}\cos(2x) + \frac{1}{2}$ The same derivatives.

(d) $\frac{1}{2}x + \frac{1}{4}\sin(2x) = \frac{1}{2}x + \frac{1}{4}[2\sin x \cos x] = \frac{1}{2}x + \frac{1}{2}\sin x \cos x$. Differ by a constant of zero.

**Problem 18.**
$\quad u = (\ln x)^3, \quad dv = 1dx$
$\quad du = 3(\ln x)^2 \frac{1}{x}dx, \quad v = x$
$\quad \int (\ln x)^3 dx = x(\ln x)^3 - \int x3(\ln x)^2 \frac{1}{x}dx = x(\ln x)^3 - \int 3(\ln x)^2 dx$ (Use $u = (\ln x)^2, \quad dv = 1dx$)
$\quad = x(\ln x)^3 - [3x(\ln x)^2 - \int 3x \cdot 2\ln x \frac{1}{x}dx] = x(\ln x)^3 - 3x(\ln x)^2 + \int 6\ln x\, dx$ (Use $u = \ln x, \quad dv = 6dx$)

$$= x(\ln x)^3 - 3x(\ln x)^2 + 6x\ln x - \int 6x\frac{1}{x}dx = x(\ln x)^3 - 3x(\ln x)^2 + 6x\ln x - 6x + C$$

**Problem 19.**
$\quad$ First use $w$-substitution $w = \sqrt{x} \ \Rightarrow\ dw = \frac{1}{2}\frac{1}{\sqrt{x}}dx \ \Rightarrow\ 2w\,dw = dx$
$\quad \int \sqrt{x}e^{\sqrt{x}}dx = \int we^w 2w\, dw = \int 2w^2 e^w dw$ (Use $u = 2w^2, \quad dv = e^w dw$)
$\quad = 2w^2 e^w - \int 4we^w dw$ (Use $u = 4w, \quad dv = e^w dw$)
$\quad = 2w^2 e^w - [4we^w - \int 4e^w dw] = 2w^2 e^w - 4we^w + 4e^w + C$ (Substituting back)

$$= 2xe^{\sqrt{x}} - 4\sqrt{x}e^{\sqrt{x}} + 4e^{\sqrt{x}} + C$$

21. Use $u = \ln x, \quad dv = x^{-1/2}dx$

$$du = \frac{1}{x}dx, \quad v = 2x^{1/2}$$

$$\int \frac{\ln x}{\sqrt{x}}dx = 2x^{1/2}\ln x - \int 2x^{1/2}\frac{1}{x}dx = 2x^{1/2}\ln x - \int 2x^{-1/2}dx = 2x^{1/2}\ln x - 4x^{1/2} + C$$

**Problem 23.**
$\quad u = \ln \sqrt{w} = \frac{1}{2}\ln w, \quad dv = 1dw$
$\quad du = \frac{1}{2}\frac{1}{w}dw, \quad v = w$

$$\int_1^e \ln \sqrt{w}\, dw = w\ln \sqrt{w} - \int_1^e w\frac{1}{2}\frac{1}{w}dw = w\ln \sqrt{w} - \int_1^e \frac{1}{2}dw = w\ln \sqrt{w}\Big|_1^e - \frac{1}{2}w\Big|_1^e = \frac{1}{2}$$

**Problem 24.**

$$u = \ln x, \quad dv = \sqrt{x}\,dx$$
$$du = \tfrac{1}{x}dx, \quad v = \tfrac{2}{3}x^{3/2}$$

$$\int \sqrt{x}\ln x\,dx = \tfrac{2}{3}x^{3/2}\ln x - \int \tfrac{2}{3}x^{3/2}\tfrac{1}{x}dx = \tfrac{2}{3}x^{3/2}\ln x - \int \tfrac{2}{3}x^{1/2}dx = \tfrac{2}{3}x^{3/2}\ln x - \tfrac{4}{9}x^{3/2} + C$$

**Problem 26.**

$$w = \sqrt{x} \;\Rightarrow\; dw = \tfrac{1}{2}x^{-1/2}dx \;\Rightarrow\; 2w\,dw = dx$$
$$\int_0^1 \cos\sqrt{x}\,dx = \int_0^1 \cos w\,2w\,dw \;(\text{Use } u = 2w, \; dv = \cos w\,dw)$$

$$= 2w\sin w\big|_0^1 - \int_0^1 2\sin w\,dw = 2w\sin w\big|_0^1 + 2\cos w\big|_0^1 = 2\sin(1) + 2\cos(1) - 2$$

**Problem 26.**

$$\int_0^{\pi/2} x\sin x\cos x\,dx = \int_0^{\pi/2} x\tfrac{1}{2}\sin(2x)\,dx \;\text{Use } u = \tfrac{1}{2}x, \; dv = \sin(2x)dx$$

$$du = \tfrac{1}{2}dx, \quad v = -\tfrac{1}{2}\cos(2x)$$

$$\int_0^{\pi/2} \tfrac{1}{2}x\sin(2x)\,dx = \left(\tfrac{1}{2}x(-\tfrac{1}{2}\cos(2x))\right)\big|_0^{\pi/2} - \int_0^{\pi/2} -\tfrac{1}{4}\cos(2x)\,dx = \left(-\tfrac{1}{4}x\cos(2x)\right)\big|_0^{\pi/2} + \left(\tfrac{1}{8}\sin(2x)\right)\big|_0^{\pi/2} = \tfrac{\pi}{8}$$

$$\Rightarrow \int_0^{\pi/2} x\sin x\cos x\,dx = \tfrac{\pi}{8}$$

**Problem 30.**

(a) Use $u = \cos^{n-1}x, \quad dv = \cos x\,dx$

$$du = (n-1)\cos^{n-2}x(-\sin x)dx, \quad v = \sin x$$

$$\int \cos^n x\,dx = \sin x\cos^{n-1}x - \int \sin x(n-1)\cos^{n-2}x(-\sin x)dx$$

$$= \sin x\cos^{n-1}x + (n-1)\int \sin^2 x\cos^{n-2}x\,dx$$

$$= \sin x\cos^{n-1}x + (n-1)\int (1-\cos^2 x)\cos^{n-2}x\,dx$$

$$= \sin x\cos^{n-1}x + (n-1)\int \cos^{n-2}x\,dx - (n-1)\int \cos^n x\,dx$$

$$\Rightarrow n\int \cos^n x\,dx = \sin x\cos^{n-1}x + (n-1)\int \cos^{n-2}x\,dx$$

$$\Rightarrow \int \cos^n x\,dx = \tfrac{1}{n}\sin x\cos^{n-1}x + \tfrac{(n-1)}{n}\int \cos^{n-2}x\,dx$$

(b) $\int \cos^2 x\,dx = \tfrac{1}{2}\sin x\cos^1 x + \tfrac{1}{2}\int \cos^0 x\,dx = \tfrac{1}{2}\sin x\cos x + \tfrac{1}{2}\int 1\,dx = \tfrac{1}{2}\sin x\cos x + \tfrac{1}{2}x + C$

**Problem 31.**

(a) $\int_0^{\pi/2} \sin^3 x \, dx = -\frac{1}{3} \cos x \sin^2 x \big|_0^{\pi/2} + \frac{2}{3} \int_0^{\pi/2} \sin^1 x \, dx$

$$= -\frac{1}{3} \cos x \sin^2 x \big|_0^{\pi/2} + \frac{2}{3} (-\cos x)\big|_0^{\pi/2} = 0 + \frac{2}{3} = \frac{2}{3}$$

(b) $\int_0^{\pi/2} \sin^5 x \, dx = -\frac{1}{3} \cos x \sin^4 x \big|_0^{\pi/2} + \frac{4}{5} \int_0^{\pi/2} \sin^3 x \, dx = 0 + \frac{4}{5}(\frac{2}{3}) = \frac{2 \cdot 4}{3 \cdot 5}$

(c) By induction: Part (a) shows the statement is true for $n = 3$. Assume it is true for $n = k$ an odd integer. That means assume $\int_0^{\pi/2} \sin^k x \, dx = \frac{2 \cdot 4 \cdots (k-1)}{3 5 \cdots k}$ is true. Now for $n = k+2$, the next odd integer,

$\int_0^{\pi/2} \sin^{k+2} x \, dx = -\frac{1}{k+2} \cos x \sin^{k+1} x \big|_0^{\pi/2} + \frac{k+1}{k+2} \int_0^{\pi/2} \sin^k x \, dx = 0 + \frac{k+1}{k+2} \left( \frac{2 \cdot 4 \cdots (k-1)}{3 5 \cdots k} \right) = \frac{2 \cdot 4 \cdots (k-1) \cdot (k+1)}{3 \cdot 5 \cdots k \cdot (k+2)}.$
This shows the statement is true for $n = k+2$ whenever it is true for odd integer $n = k$. As is true for $n=3$, by induction it must be true for all positive odd integers $n¿1$. (It is trivial true for $n = 1$.)

**Problem 33.**
$u = (\ln x)^n, \quad dv = 1 dx$
$du = n(\ln x)^{n-1} \frac{1}{x} dx, \quad v = x$

$$\int (\ln x)^n dx = x(\ln x)^n - \int n(\ln x)^{n-1} \frac{1}{x} x dx = x(\ln x)^n - n \int (\ln x)^{n-1} dx$$

## Section 29.2    Trigonometric Integrals and Trigonometric Substitution

*There are many equivalent answers for most of these problems because of identity equivalences.*

**Problem 1.**

    1. $\int \cos^2 x \, dx = \int \frac{1}{2}(1 + \cos 2x) dx = \frac{1}{2} x + \frac{1}{4} \sin 2x + C$

**Problem 2.**

    $\int_0^\pi \sin^3 x \, dx = \int_0^\pi \sin x (1 - \cos^2 x) \, dx = \int_0^\pi \sin x \, dx + \int_0^\pi - \sin x \cos^2 x \, dx$ (Think $u = \cos x$)

$$= (-\cos x)\big|_0^\pi + \frac{1}{3} (\cos x)^3 \big|_0^\pi = 1 - (-1) + \frac{1}{3}(-1 - 1) = \frac{4}{3}$$

**Problem 4.**

    $\int \cos^3 x \sin^2 x \, dx = \int \cos x (1 - \sin^2 x) \sin^2 x \, dx = \int \cos x \sin^2 x \, dx + \int - \cos x \sin^4 x \, dx$
    $= \frac{1}{3} \sin^3 x - \frac{1}{5} \sin^5 x$ (Think $u = \sin x$)

**Problem 6.**

    $\int \cos^2 x \sin^2 x \, dx = \int \frac{1}{4} \sin^2 2x \, dx = \frac{1}{4} \int \frac{1}{2}(1 - \cos 4x) dx = \frac{1}{8} x - \frac{1}{32} \sin(4x) + C$ (or equivalent)

**Problem 8.**

    $\int \cos^3 x \sin^{11} x \, dx = \int \cos x (1 - \sin^2 x) \sin^{11} x \, dx$ (Think $u = \sin x$)

$$= \int \cos x \sin^{11} x \, dx - \int \cos x \sin^{13} x \, dx = \frac{1}{12} \sin^{12} x - \frac{1}{14} \sin^{14} x + C$$

**Problem 9.**

$\int \cos^3(3x)dx = \int \cos(3x)(1 - \sin^2(3x))dx$ (Think $u = \sin x$)

$$= \int \cos(3x)dx - \int \cos(3x)\sin^2(3x)dx = \frac{1}{3}\sin(3x) - \frac{1}{9}\sin^3(3x) + C$$

**Problem 11.**

$\int \frac{\sin x}{\sqrt{\cos^3 x}}dx = \int (\cos x)^{-3/2} \sin x\, dx = 2(\cos x)^{-1/2} + C = \frac{2}{\sqrt{\cos x}} + C$ (Think $u = \cos x$)

**Problem 13.**

$\int \tan 2x \sec 2x\, dx = \frac{1}{2}\sec 2x + C$

**Problem 15.**

$\int \tan x \sec^4 x\, dx = \int \sec^3 x(\tan x \sec x)\, dx = \frac{1}{4}(\sec x)^4 + C$ (Think $u = \sec x$)

**Problem 16.**

$\int \tan^3 x \sec^4 x\, dx = \int \tan^3 x(1 + \tan^2 x) \sec^2 x\, dx$ (Think $u = \tan x$)
$= \int \tan^3 x \sec^2 x\, dx + \int \tan^5 x \sec^2 x\, dx = \frac{1}{4}\tan^4 x + \frac{1}{6}\tan^6 x + C$ (or equivalent)

**Problem 18.**

$\int \tan^3 x \sec x\, dx = \int \tan x(\sec^2 x - 1) \sec x\, dx$ (Think $u = \sec x$)

$$= \int \sec^2 x(\sec x \tan x)\, dx - \int \sec x \tan x\, dx = \frac{1}{3}\sec^3 x - \sec x + C$$

**Problem 20.**

$\int \tan^8 x \sec^4 x\, dx = \int \tan^8 x(1 + \tan^2 x) \sec^2 x\, dx$ (Think $u = \tan x$)

$$= \int \tan^8 x \sec^2 x\, dx + \int \tan^{10} x \sec^2 x\, dx = \frac{1}{9}\tan^9 x + \frac{1}{11}\tan^{11} x + C$$

**Problem 21.**

$\int \frac{\sin x}{\cos^2 x}dx = \int (\cos x)^{-2} \sin x\, dx = (\cos x)^{-1} + C = \frac{1}{\cos x} + C$ (Think $u = \cos x$)

**Problem 23.**

$\int_{-\pi}^{\pi} \sin mx \sin nx\, dx = \int_{-\pi}^{\pi} \frac{1}{2}[\cos(m - n)x - \cos(m + n)x]\, dx$

$= \frac{1}{2}\left[\frac{1}{m-n}\sin(m - n)x - \frac{1}{m+n}\sin(m + n)x\right]\Big|_{-\pi}^{\pi} = 0$ if $m \neq n$ as the sin of any multiple of $\pi$ is 0

If $m = n$ then $\int_{-\pi}^{\pi} \frac{1}{2}[\cos(m - n)x - \cos(m + n)x]\, dx = \int_{-\pi}^{\pi} \frac{1}{2}[\cos(0) - \cos(m + n)x]\, dx$

$$= \frac{1}{2}\left[x - \frac{1}{m+n}\sin(m + n)x\right]\Big|_{-\pi}^{\pi} = \frac{1}{2}x\Big|_{-\pi}^{\pi} - \frac{1}{2}\left[\frac{1}{m+n}\sin(m + n)x\right]\Big|_{-\pi}^{\pi} = \pi + 0 = \pi$$

**Problem 24.**

$\int_{-\pi}^{\pi} \cos mx \cos nx\, dx = \int_{-\pi}^{\pi} \frac{1}{2}[\cos(m - n)x + \cos(m + n)x]\, dx$

$= \frac{1}{2}\left[\frac{1}{m-n}\sin(m - n)x + \frac{1}{m+n}\sin(m + n)x\right]\Big|_{-\pi}^{\pi} = 0$ if $m \neq n$ as the sin of any multiple of $\pi$ is 0

If $m = n$ then $\int_{-\pi}^{\pi} \frac{1}{2}[\cos(m - n)x + \cos(m + n)x]\,dx = \int_{-\pi}^{\pi} \frac{1}{2}[\cos(0) + \cos(m + n)x]\,dx$

$$= \frac{1}{2}\left[x + \frac{1}{m+n}\sin(m + n)x\right]\Big|_{-\pi}^{\pi} = \frac{1}{2}x\Big|_{-\pi}^{\pi} + \frac{1}{2}\left[\frac{1}{m+n}\sin(m + n)x\right]\Big|_{-\pi}^{\pi} = \pi + 0 = \pi$$

## Problem 26.

Rotating an $x$-slice about line $y = 2$ will form a washer.

$$\int_{\pi/4}^{5\pi/4} \pi[(2 - \cos x)^2 - (2 - \sin x)^2]\,dx = \pi \int_{\pi/4}^{5\pi/4} [4\sin x - 4\cos x + \cos^2 x - \sin^2 x]\,dx$$

$$= \pi \int_{\pi/4}^{5\pi/4} [4\sin x - 4\cos x + \cos 2x]\,dx = \pi \left(-4\cos x - 4\sin x + \frac{1}{2}\sin 2x\right)\Big|_{\pi/4}^{5\pi/4}$$

$$= \pi[(-4(-\tfrac{\sqrt{2}}{2}) - 4(-\tfrac{\sqrt{2}}{2}) + \tfrac{1}{2}) - (-4(\tfrac{\sqrt{2}}{2}) - 4(\tfrac{\sqrt{2}}{2}) + \tfrac{1}{2})] = 8\sqrt{2}\pi$$

## Problem 28.

$$\int \sec^3 x\,dx = \sec x \tan x - \int \sec x \tan x \tan x\,dx = \sec x \tan x - \int \sec x(\sec^2 x - 1)\,dx$$

$$= \sec x \tan x - \int \sec^3 x\,dx + \int \sec x\,dx = \sec x \tan x + \ln|\sec x + \tan x| - \int \sec^3 x\,dx$$

$$\Rightarrow \int \sec^3 x\,dx = \frac{1}{2}\sec x \tan x + \frac{1}{2}\ln|\sec x + \tan x| + C$$

## Problem 30.

Use Parts $u = x^2$   $dv = (9 - x^2)^{-1/2}x\,dx$

$$\int \frac{x^3}{\sqrt{9-x^2}}\,dx = x^2(-(9 - x^2)^{1/2}) - \int -(9 - x^2)^{1/2}2x\,dx = -x^2(9 - x^2)^{1/2} - \int (9 - x^2)^{1/2}(-2x)\,dx$$

$$= -x^2(9 - x^2)^{1/2} - \frac{2}{3}(9 - x^2)^{3/2} + C$$

## Problem 32.

Use $w$-substitution $w = 4 + x^2$ OR Parts $u = x^2$,   $dv = (4 + x^2)^{-1/2}x\,dx$

$$dw = 2x\,dx \quad du = 2x\,dx, \quad v = (4 + x^2)^{1/2}$$

$$\int \frac{x^3}{\sqrt{4+x^2}}\,dx = \int x^2(4 + x^2)^{-1/2}x\,dx = x^2(4 + x^2)^{1/2} - \int (4 + x^2)^{1/2}2x\,dx = x^2(4 + x^2)^{1/2} - \frac{2}{3}(4 + x^2)^{3/2} + C$$

## Problem 34.

Ugly. Use $x = 3\sec\theta$,   $dx = 3\sec\theta\tan\theta$

$$\int x^2\sqrt{x^2 - 9}\,dx = \frac{1}{4}x(x^2 - 9)^{3/2} + \frac{9}{8}x(x^2 - 9)^{1/2} - \frac{81}{8}\ln\left|x + (x^2 - 9)^{1/2}\right| + C$$

**Problem 35.**

Use $x = \frac{3}{2}\sin\theta, \quad dx = \frac{3}{2}\cos\theta\,d\theta$

$$\int_0^{3/4} \sqrt{9 - 4x^2}\,dx = \left(\tfrac{1}{2}x\sqrt{9 - 4x^2} + \tfrac{9}{4}\arcsin(\tfrac{2}{3}x)\right)\Big|_0^{3/4} = \tfrac{9}{16}\sqrt{3} + \tfrac{3}{8}\pi$$

**Problem 37.**

Think $w = 4 - 9x^2, \quad dw = -18x\,dx$

$$\int x\sqrt{4 - 9x^2}\,dx = -\tfrac{1}{27}(2 - 9x^2)^{3/2} + C$$

**Problem 39.**

Use $x = 2\sec\theta, \quad dx = 2\sec\theta\tan\theta\,d\theta$

$$\int \frac{\sqrt{x^2 - 4}}{x}\,dx = \sqrt{x^2 - 4} + 2\arctan\left(2\frac{1}{\sqrt{x^2 - 4}}\right) + C$$

**Problem 41.**

Use $x = \tan\theta, \quad dx = \sec^2\theta\,d\theta$

$$\int_{2/\sqrt{3}}^{2} \frac{\sqrt{x^2 - 1}}{x}\,dx = \sqrt{x^2 - 1} + \arctan\left(\frac{1}{\sqrt{x^2 - 1}}\right)\Big|_{2/\sqrt{3}}^{2} = \tfrac{2}{3}\sqrt{3} - \tfrac{1}{6}\pi$$

**Problem 43.**

Think $w = 4 - x^2, \quad dw = -2x\,dx$

$$\int \frac{x}{4 - x^2}\,dx = -\tfrac{1}{2}\int \frac{1}{4 - x^2}(-2x)\,dx = -\tfrac{1}{2}\ln\left|4 - x^2\right| + C$$

**Problem 44.**

Use $x = k\sin\theta, \quad dx = k\cos\theta\,d\theta$

$$\int \sqrt{k^2 - x^2}\,dx = \tfrac{1}{2}x\sqrt{k^2 - x^2} + \tfrac{1}{2}k^2\arcsin\left(\tfrac{x}{k}\right) + C$$

**Problem 46.**

$\int \sec^3 x\,dx = \sec x\tan x - \int \sec x\tan x\tan x\,dx = \sec x\tan x - \int \sec x(\sec^2 x - 1)\,dx$

$$= \sec x\tan x - \int \sec^3 x\,dx + \int \sec x\,dx = \sec x\tan x + \ln|\sec x + \tan x| - \int \sec^3 x\,dx$$

$$\Rightarrow \int \sec^3 x\,dx = \tfrac{1}{2}\sec x\tan x + \tfrac{1}{2}\ln|\sec x + \tan x| + C$$

**Problem 48.**
Expand the right side and simplify.

## Section 29.3    Integration Using Partial Fractions

**Problem 1.**
(a) $\frac{A}{x} + \frac{B}{x-1} + \frac{C}{x+5}$    (b) $\frac{A}{x} + \frac{Bx+C}{x^2+1}$

**Problem 2.**
(a) $\frac{A}{x} + \frac{B}{x-2} + \frac{C}{x+2}$    (b) $\frac{A}{x} + \frac{Bx+C}{x^2+2}$

**Problem 2.**
(a) $\frac{A}{x+4} + \frac{B}{x-1}$    (b) $\frac{A}{x-2} + \frac{B}{x+2}$

**Problem 5.**
$\frac{A}{x} + \frac{Bx+C}{x^2+4}$

**Problem 7.**
$\int \frac{3x+9}{x^2-6x+5}dx = \int \left( \frac{-3}{x-1} + \frac{6}{x-5} \right)dx = -3\ln|x-1| + 6\ln|x-5| + C$

**Problem 9.**
$\int_0^1 \frac{x^2}{2x+3}dx = \int_0^1 \left( \frac{1}{2}x - \frac{3}{4} + \frac{\frac{9}{4}}{2x+3} \right)dx = \left( \frac{1}{4}x^2 - \frac{3}{4}x + \frac{9}{8}\ln|2x+3| \right)\big|_0^1 = -\frac{1}{2} + \frac{9}{8}\ln 5 - \frac{9}{8}\ln 3$

**Problem 11.**
$\int \frac{3x^2+3}{(x^2-1)(x-2)}dx = \int \left( \frac{-3}{x-1} + \frac{1}{x+1} + \frac{5}{x-2} \right)dx = -3\ln|x-1| + \ln|x+1| + 5\ln|x-2| + C$

**Problem 13.**
Use $w$-substitution, $w = e^x$, $dw = e^x dx$ followed by partial fraction expansion.

$$\int \frac{e^{2x}}{(e^x+2)(e^x-1)^2}dx = \int \frac{w}{(w+2)(w-1)^2}dw = \int \left( \frac{-\frac{2}{9}}{w+2} + \frac{\frac{1}{3}}{(w-1)^2} + \frac{\frac{2}{9}}{(w-1)} \right)dw$$

$$= -\frac{2}{9}\ln|w+2| - \frac{1}{3}\frac{1}{w-1} + \frac{2}{9}\ln|w-1| + C = -\frac{2}{9}\ln|e^x+2| - \frac{1}{3}\frac{1}{e^x-1} + \frac{2}{9}\ln|e^x-1| + C$$

**Problem 15.**
Use parts $u = \arctan x$, $dv = x^2 dx$

$$du = \frac{1}{x^2+1}dx, \quad v = \frac{1}{3}x^3$$

$$\int x^2 \arctan x\, dx = \frac{1}{3}x^3 \arctan x - \int \frac{1}{3}x^3 \frac{1}{x^2+1}dx = \frac{1}{3}x^3 \arctan x - \frac{1}{3}\int \left( x + \frac{-x}{x^2+1} \right)dx$$

$$= \frac{1}{3}x^3 \arctan x - \frac{1}{6}x^2 + \frac{1}{6}\ln|x^2+1| + C$$

**Problem 17.**
Use parts $u = \ln(x^2-1)$, $dv = 1dx$

$$du = \frac{2x}{x^2-1}dx, \quad v = x$$

$$\int \ln(x^2-1)dx = x\ln(x^2-1) - \int x\frac{2x}{x^2-1}dx = x\ln(x^2-1) - \int 2 + \frac{2}{x^2-1}dx = x\ln(x^2-1) + \int -2 + \frac{-2}{x^2-1}dx$$

$$= x\ln(x^2-1) + \int (-2 + \frac{1}{x+1} + \frac{-1}{x-1})dx = x\ln(x^2-1) - 2x + \ln|x+1| - \ln|x-1| + C$$

**Problem 18.**

Ugh! $\int \frac{2x^3-2x^2+4x+8}{(x-2)^2(x^2+3)}dx = \int \left( \frac{\frac{24}{7}}{(x-2)^2} + \frac{\frac{44}{49}}{x-2} + \frac{\frac{54}{49}x}{x^2+3} + \frac{\frac{38}{49}}{x^2+3} \right)dx$

$$= -\frac{24}{7}\frac{1}{x-2} + \frac{44}{49}\ln|x-2| + \frac{27}{49}\ln|x^2+3| + \frac{38\sqrt{3}}{147}\arctan(\frac{\sqrt{3}}{3}x) + C$$

## Section 29.4   Improper Integrals

### Problem 1.

(a) infinite interval

(b) infinite interval, discontinuous integrand at $x = 2$, $\int_0^\infty \frac{1}{x^2-4}dx = \int_0^2 \frac{1}{x^2-4}dx + \int_2^4 \frac{1}{x^2-4}dx + \int_4^\infty \frac{1}{x^2-4}dx$

### Problem 3.

(a) infinite interval, $\int_{-\infty}^\infty \frac{1}{x^2+4}dx = \int_{-\infty}^0 \frac{1}{x^2+4}dx + \int_0^\infty \frac{1}{x^2+4}dx$

(b) infinite interval, discontinuous integrand at $x = -2$, $2$

$$\int_{-\infty}^\infty \frac{1}{x^2-4}dx = \int_{-\infty}^{-3} \frac{1}{x^2-4}dx + \int_{-3}^{-2} \frac{1}{x^2-4}dx + \int_{-2}^0 \frac{1}{x^2-4}dx + \int_0^2 \frac{1}{x^2-4}dx + \int_2^3 \frac{1}{x^2-4}dx + \int_3^\infty \frac{1}{x^2-4}dx$$

### Problem 3.

(a) unbounded integrand at $x = -\pi/2$, $\pi/2$ $\int_{-\pi/2}^{\pi/2} \tan x\,dx = \int_{-\pi/2}^0 \tan x\,dx + \int_0^{\pi/2} \tan x\,dx$

(b) discontinuous integrand at $x = \pi/2$, $\int_0^\pi \tan x\,dx = \int_0^{\pi/2} \tan x\,dx + \int_{\pi/2}^\pi \tan x\,dx$

### Problem 3.

$\int_1^\infty \frac{1}{x^p}dx = \lim_{b\to\infty} \int_1^b x^{-p}dx = \lim_{b\to\infty} \left( \frac{1}{-p+1}x^{-p+1} \right)\Big|_1^b = \lim_{b\to\infty} \left( \frac{1}{-p+1}b^{-p+1} - (\frac{1}{-p+1}) \right)$ if $p \neq 1$
This limit exits if $-p+1 < 0 \Leftrightarrow 1 < p$, it doesn't exist if $-p+1 > 0 \Leftrightarrow 1 > p$
If $p = 1$ then $\int_1^\infty \frac{1}{x^p}dx = \lim_{b\to\infty} \int_1^b \frac{1}{x}dx = \lim_{b\to\infty} (\ln|x|)|_1^b = \lim_{b\to\infty} (\ln|b|)$ which doesn't exit.
Hence this improper integral converges if $p > 1$ and doesn't converge if $p \leq 1$.

### Problem 8.

$\int_{-1}^\infty \frac{1}{x^4}dx = \int_{-1}^0 \frac{1}{x^4}dx + \int_0^1 \frac{1}{x^4}dx + \int_1^\infty \frac{1}{x^4}dx$. By problem #7 we know the middle integral diverges

**Problem 9.**

(a) $\int_0^\infty xe^{-x^2}dx = \lim_{b\to\infty} \int_0^b xe^{-x^2}dx = \lim_{b\to\infty} -\frac{1}{2}e^{-x^2}\Big|_0^b = \lim_{b\to\infty}\left(-\frac{1}{2}\frac{1}{e^{b^2}} - (-\frac{1}{2})\right) = \frac{1}{2}$

(b) $\int_{-\infty}^\infty xe^{-x^2}dx = \int_{-\infty}^0 xe^{-x^2}dx + \int_0^\infty xe^{-x^2}dx$

$$\int_0^\infty xe^{-x^2}dx = \lim_{b\to\infty}\int_0^b xe^{-x^2}dx = \lim_{b\to\infty} -\frac{1}{2}e^{-x^2}\Big|_0^b = \lim_{b\to\infty}\left(-\frac{1}{2}\frac{1}{e^{b^2}} - (-\frac{1}{2})\right) = \frac{1}{2}$$

$$\int_a^0 xe^{-x^2}dx = \lim_{a\to-\infty}\int_a^0 xe^{-x^2}dx = \lim_{a\to-\infty} -\frac{1}{2}e^{-x^2}\Big|_a^0 = \lim_{a\to-\infty}\left((-\frac{1}{2}) - (-\frac{1}{2}\frac{1}{e^{a^2}})\right) = -\frac{1}{2}$$

Therefore $\int_{-\infty}^\infty xe^{-x^2}dx = \int_{-\infty}^0 xe^{-x^2}dx + \int_0^\infty xe^{-x^2}dx = -\frac{1}{2} + \frac{1}{2} = 0$.

**Problem 11.**
$\int_0^\infty e^{-2x}dx = \lim_{b\to\infty}\int_0^b e^{-2x}dx = \lim_{b\to\infty} -\frac{1}{2}e^{-2x}\Big|_0^b = \lim_{b\to\infty}\left(-\frac{1}{2}\frac{1}{e^{2b}} - (-\frac{1}{2})\right) = \frac{1}{2}$ converges.

**Problem 13.**
$\lim_{x\to\infty}\cos x + 1 \neq 0$ hence integral diverges.

**Problem 15.**
$\int_0^\infty xe^{-x}dx = \lim_{b\to\infty}\int_0^b xe^{-x}dx$ (using parts)

$= \lim_{b\to\infty} -xe^{-x} - e^{-x}\Big|_0^b = \lim_{b\to\infty}\left(-\frac{b}{e^b} - \frac{1}{e^b} - (0 - \frac{1}{1})\right) = 1$ converges.

**Problem 17.**
$\int_{-\infty}^\infty \frac{1}{x^3}dx = \int_{-\infty}^{-1}\frac{1}{x^3}dx + \int_{-1}^0 \frac{1}{x^3}dx + \int_0^1 \frac{1}{x^3}dx + \int_1^\infty \frac{1}{x^3}dx$ diverges as $3^{nd}$ integral diverges because $p > 1$.

**Problem 19.**
$\int_1^\infty \ln x\,dx = \lim_{b\to\infty}\int_1^\infty \ln x\,dx$ (using parts)

$= \lim_{b\to\infty}(x\ln x - x)\Big|_1^b = \lim_{b\to\infty}((b\ln b - b) - (-1)) = \infty$, diverges.

**Problem 20.**
$\int_1^\infty \frac{1}{x(x+1)}dx = \lim_{b\to\infty}\int_1^b (\frac{1}{x} - \frac{1}{x+1})dx = \lim_{b\to\infty}(\ln x - \ln(x+1))\Big|_1^b = \lim_{b\to\infty}(\ln\frac{x}{x+1})\Big|_1^b = \lim_{b\to\infty}(\ln\frac{b}{b+1} - \ln\frac{1}{2}) = 0 + \ln 2$
hence converges.

**Problem 21.**
$\int_0^\infty \frac{1}{x(x+1)}dx = \int_0^1 \frac{1}{x(x+1)}dx + \int_1^\infty \frac{1}{x(x+1)}dx$ diverges because
$\int_0^1 \frac{1}{x(x+1)}dx = \int_0^1 \left(\frac{1}{x} - \frac{1}{x+1}\right)dx = \int_0^1 \left(\frac{-1}{x+1}\right)dx + \int_0^1 \left(\frac{1}{x}\right)dx$. The $1^{st}$ is proper but the $2^{nd}$ diverges $(p=1)$

**Problem 23.**
$\int_{e^2}^\infty \frac{1}{x(\ln x)^2}dx = \lim_{b\to\infty}\int_{e^2}^b \frac{1}{x(\ln x)^2}dx = \lim_{b\to\infty} -\frac{1}{\ln x}\Big|_{e^2}^b = \lim_{b\to\infty}\left(-\frac{1}{\ln b} - (-\frac{1}{\ln e^2})\right) = \frac{1}{2}$, converges.

**Problem 25.**
$\int_0^\infty \frac{1}{\sqrt{x+1}}dx = \lim_{b\to\infty}\int_0^b \frac{1}{\sqrt{x+1}}dx = \lim_{b\to\infty}(2\sqrt{x+1})\Big|_0^b = \lim_{b\to\infty}(2\sqrt{b+1} - (2\sqrt{0+1})) = \infty$, diverges.

**Problem 26.**

$\int_{-1}^{1} \frac{1}{\sqrt{x+1}} dx = \lim_{a \to -1} \int_{a}^{1} \frac{1}{\sqrt{x+1}} dx = \lim_{a \to -1} \left(2\sqrt{x+1}\right)\big|_{a}^{1} = \lim_{a \to 0} \left(2\sqrt{1+1} - \left(2\sqrt{a+1}\right)\right) = 2\sqrt{2}$, converges.

**Problem 28.**

$\lim_{x \to \infty} \ln x = \infty \neq 0$ hence integral diverges.

**Problem 30.**

$\int_{1}^{\infty} \frac{\arctan x}{1+x^2} dx = \lim_{b \to \infty} \int_{1}^{b} \frac{\arctan x}{1+x^2} dx = \lim_{b \to \infty} \frac{1}{2} (\arctan x)^2 \big|_{1}^{b}$

$= \lim_{b \to \infty} \left(\frac{1}{2}(\arctan b)^2 - \frac{1}{2}(\arctan 1)^2\right) = \frac{1}{2}\left(\frac{\pi}{2}\right)^2 - \frac{1}{2}\left(\frac{\pi}{4}\right)^2 = \frac{3}{32}\pi^2$, converges

**Problem 32.**

$\lim_{x \to \infty} \arctan x = \frac{\pi}{2} \neq 0$ hence integral diverges.

**Problem 33.**

$\int_{0}^{\pi} \tan x \, dx = \int_{0}^{\pi/2} \tan x \, dx + \int_{\pi/2}^{\pi} \tan x \, dx$, diverges because

$$\int_{0}^{\pi/2} \tan x \, dx = \lim_{b \to \pi/2} \int_{0}^{b} \frac{\sin x}{\cos x} dx = \lim_{b \to \pi/2} -\ln(\cos x)\big|_{0}^{b} = \lim_{b \to \pi/2} \left(-\ln(\cos b) - (-\ln(\cos 0))\right) = \infty$$

**Problem 34.**

$\lim_{x \to \infty} \frac{x^2+3}{x+1} = \infty \neq 0$ hence integral diverges.

**Problem 36.**

$\int_{1}^{e^2} \frac{1}{x\sqrt{\ln x}} dx = \lim_{a \to 1} \int_{a}^{e^2} \frac{1}{x\sqrt{\ln x}} dx = \lim_{a \to 1} 2\sqrt{\ln x} \Big|_{a}^{e^2} = \lim_{a \to 1} \left(2\sqrt{\ln e^2} - 2\sqrt{\ln a}\right) = 2\sqrt{2}$, converges.

**Problem 38.**

$\int_{2}^{\infty} \frac{1}{x(x+1)} dx < \int_{2}^{\infty} \frac{1}{x^2} dx$ which converges as $p = 2 > 1$

**Problem 40.**

$\int_{1}^{\infty} \frac{1}{\sqrt{x^7+1}} dx < \int_{1}^{\infty} \frac{1}{\sqrt{x^7}} dx = \int_{1}^{\infty} \frac{1}{x^{7/2}} dx$ which converges as $p = 7/2 > 1$

**Problem 41.**

$\int_{0}^{\infty} \sin x e^{-x} dx \leq \int_{0}^{\infty} e^{-x} dx = 1$ and $\int_{0}^{\infty} \sin x e^{-x} dx \geq \int_{0}^{\infty} (-1)e^{-x} dx = -1$.

$\int_{0}^{\infty} \sin x e^{-x} dx$ can't oscillate because the tail of $\int_{0}^{\infty} e^{-x} dx$, $\int_{a}^{\infty} e^{-x} dx$ goes to zero as $a \to \infty$. Hence $\int_{0}^{\infty} \sin x e^{-x} dx$ converges.

**Problem 43.**

(a) $\int_{1}^{\infty} \frac{1}{1+x^4} dx < \int_{1}^{\infty} \frac{1}{x^4} dx$ which converges as $p = 4 > 1$

(b) Split your allowable error into two portions, $0.005 + 0.005 = 0.01$. Use the first $0.005$ to make sure the tail of the improper integral is less than $0.005$. Then we know the front part of the integral is within $0.005$ of the whole integral. Next approximate the front part to within $0.005$.

Then that approximation number will be within $0.01$ of the improper integral value.

$\int_a^\infty \frac{1}{1+x^4}dx < \int_a^\infty \frac{1}{x^4}dx = \lim_{b\to\infty} \int_a^b \frac{1}{x^4}dx = \lim_{b\to\infty} \left(-\frac{1}{3}\frac{1}{x^3}\right)\Big|_a^b = \lim_{b\to\infty}\left(-\frac{1}{3}\frac{1}{b^3} - \left(-\frac{1}{3}\frac{1}{a^3}\right)\right) = \frac{1}{3}\frac{1}{a^3}$. To make this less than 0.005 we need $\frac{1}{3}\frac{1}{a^3} < 0.005 \Leftrightarrow \frac{1}{0.015} < a^3 \Leftarrow a > 4.06$. Hence if we take $a = 5$ that is certainly good enough. Now we will approximate $\int_1^5 \frac{1}{1+x^4}dx$ with error no more than 0.005.

For a midpoint approximation we need $\frac{M_2(b-a)^3}{24n^2} < 0.005$ or $\frac{2(4)^3}{24n^2} < 0.005 \Leftrightarrow \frac{2(4)^3}{24(0.005)} < n^2$ for our problem. This happens if $33 \le n$. The midpoint sum approximation is $M_{33} = 0.240$.

Hence 0.240 approximates $\int_1^\infty \frac{1}{1+x^4}dx$ to within 0.01.

**Problem 45.**

$\int_{-\infty}^\infty p(x)\,dx = \int_{-\infty}^0 p(x)\,dx + \int_0^\infty p(x)\,dx = \int_{-\infty}^0 0\,dx + \int_0^\infty \lambda e^{-\lambda x}\,dx = \int_0^\infty \lambda e^{-\lambda x}\,dx$

$$= \lim_{b\to\infty} \int_0^b \lambda e^{-\lambda x}dx = \lim_{b\to\infty} -e^{-\lambda x}\Big|_0^b = \lim_{b\to\infty}\left(-\frac{1}{e^{\lambda b}} - \left(-\frac{1}{1}\right)\right) = 1$$

**Problem 47.**

$\mu = \int_{-\infty}^\infty xp(x)\,dx = \int_{-\infty}^0 xp(x)\,dx + \int_0^\infty xp(x)\,dx = \int_{-\infty}^0 0\,dx + \int_0^\infty x\lambda e^{-\lambda x}\,dx = \int_0^\infty x\lambda e^{-\lambda x}\,dx$

$$= \lim_{b\to\infty} \int_0^b x\lambda e^{-\lambda x}dx = \lim_{b\to\infty}\left(-xe^{-\lambda x} - \frac{1}{\lambda}e^{-\lambda x}\right)\Big|_0^b = \lim_{b\to\infty}\left(-\frac{b}{e^{\lambda b}} - \frac{1}{\lambda}\frac{1}{e^{\lambda b}} - \left(-\frac{0}{1} - \frac{1}{\lambda}\frac{1}{e^0}\right)\right) = \frac{1}{\lambda}$$

# CHAPTER 30

# Series

## Section 30.1    Approximating a Function by a Polynomial

**Problem 1.**

(a) $P_4(x) = 1 - x + \frac{1}{2}x^2 - \frac{1}{6}x^3 + \frac{1}{24}x^4$

(b)

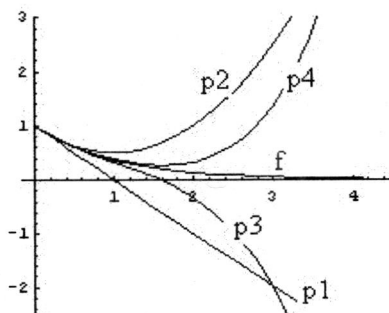

(c)

|   | x=0.1 | x=0.3 |
|---|-------|-------|
| $P_1$ | 0.9 | 0.7 |
| $P_2$ | 0.905 | 0.745 |
| $P_3$ | 0.90483 | 0.74050 |
| $P_4$ | 0.90483750 | 0.74083750 |
| $f$ | 0.90483742 | 0.74081822 |

**Problem 3.**

(a) $P_4(x) = x - \frac{1}{3}x^3$

(b)

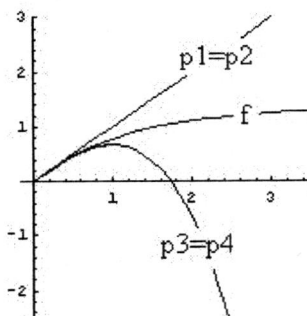

(c)

|       | x=0.1       | x=0.3       |
|-------|-------------|-------------|
| $P_1$ | 0.1         | 0.3         |
| $P_2$ | 0.1         | 0.3         |
| $P_3$ | .09966667   | .29100000   |
| $P_4$ | .09966667   | .29100000   |
| $f$   | .0996686525 | .2914567945 |

**Problem 6.**

(a) $P_4(x) = 1 + \frac{1}{2}x - \frac{1}{8}x^2 + \frac{1}{16}x^3 - \frac{5}{128}x^4$

(b)

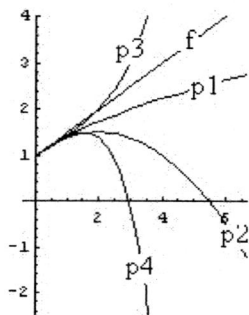

(c)

|       | x=0.1       | x=0.3        |
|-------|-------------|--------------|
| $P_1$ | 1.05        | 1.15         |
| $P_2$ | 1.04875     | 1.13875      |
| $P_3$ | 1.0488125   | 1.1404375    |
| $P_4$ | 1.048808594 | 1.140121094  |
| $f$   | 1.048808848 | 1.140175425  |

**Problem 6.**

(a) $P_4(x) = -1 + x - 3x^2 + 2x^4 = f(x)$

(b)

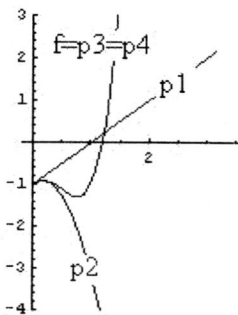

|       | x=0.1 | x=0.3 |
|-------|-------|-------|
| $P_1$ | -.9   | -.7   |
| $P_2$ | -.93  | -.97  |
| $P_3$ | -.93  | -.97  |
| $P_4$ | -.9298 | -.9538 |
| $f$   | -.9298 | -.9538 |

(c)

**Problem 8.**

(a) $Q(0) = 2$ (height is positive), while $f(0) < 0$ ( height is negative)

(b) (b) $Q'(0) = -5$ (slope is negative) while $f'(0) > 0$ ( slope is positive)

(c) $Q''(0) = -\frac{2}{3}$ (concave down) while $f''(0) > 0$ (concave up)

**Problem 11.**

(a) $\tan(0) = 0$
$\tan'(0) = \sec^2(0) = 1$
$\tan''(0) = 2\sec^2(0)\tan(0) = 0$
$\tan'''(0) = 4\sec^2(0)\tan^2(0) + 2\sec^2(0) = 2$
$P_3(x) = x + \frac{2}{3!}x^3 = x + \frac{1}{3}x^3$

(b) $\tan x$ is an "odd" function so it should not have any "even" terms.

**Problem 13.**
$f(x) = \ln(1+x) \Rightarrow f'(x) = \frac{1}{1+x} = (1+x)^{-1} \Rightarrow f''(x) = -1(1+x)^{-2} \Rightarrow f'''(x) = 2(1+x)^{-3}$ *etc.,* so
$f^{(n)}(x) = (-1)^{n+1}(n-1)!(1+x)^{-n} \Rightarrow f^{(n)}(0) = (-1)^{n+1}(n-1)!$
$P_n = x - \frac{1}{2}x^2 + \frac{2!}{3!}x^3 - \frac{3!}{4!}x^4 + \cdots + (-1)^{n+1}\frac{(n-1)!}{n!}x^n$ which simplified is $P_n = x - \frac{1}{2}x^2 + \frac{1}{3}x^3 - \frac{1}{4}x^4 +$
$\cdots + (-1)^{n+1}\frac{1}{n}x^n$

**Problem 15.**
$f(x) = x^{\frac{1}{3}} \Rightarrow f'(x) = \frac{1}{3}x^{-\frac{2}{3}} \Rightarrow f''(x) = -\frac{2}{9}x^{-\frac{5}{3}}$, hence
$f(8) = 2$, $f'(8) = \frac{1}{3}\frac{1}{4}$, $f''(x) = -\frac{2}{9}\frac{1}{32}$.
$P_2(x) = 2 + \frac{1}{12}(x-8) + -\frac{1}{288}(x-8)^2$
$P_2(8.3) = 2 + \frac{1}{12}(0.3) + -\frac{1}{288}(0.3)^2 \approx 2.024688$

**Problem 16.**
$f(x) = x^{\frac{1}{2}} \Rightarrow f'(x) = \frac{1}{2}x^{-\frac{1}{2}} \Rightarrow f''(x) = -\frac{1}{4}x^{-\frac{3}{2}}$,
hence $f(100) = 10$, $f'(100) = \frac{1}{2}\frac{1}{10}$, $f''(x) = -\frac{1}{4}\frac{1}{1000}$.
$P_2(x) = 10 + \frac{1}{20}(x-100) - \frac{1}{8000}(x-100)^2$
$P_2(103) = 10 + \frac{1}{20}(3) - \frac{1}{8000}(3)^2 \approx 10.148875$

**Problem 18.**
$f(x) = x^{\frac{1}{3}} \Rightarrow f'(x) = \frac{1}{3}x^{-\frac{2}{3}} \Rightarrow f''(x) = -\frac{2}{9}x^{-\frac{5}{3}}$
hence $f(27) = 3$, $f'(27) = \frac{1}{3}\frac{1}{9}$, $f''(27) = -\frac{2}{9}\frac{1}{243}$.
$P_2(x) = 3 + \frac{1}{27}(x-27) - \frac{1}{2187}(x-27)^2$
$P_2(29) = 3 + \frac{1}{27}(2) - \frac{1}{2187}(2)^2 \approx 3.072245$

**Problem 20.**

$$f(x) = x^{\frac{1}{2}} \;\Rightarrow\; f'(x) = \tfrac{1}{2}x^{-\frac{1}{2}} \;\Rightarrow\; f''(x) = -\tfrac{1}{4}x^{-\frac{3}{2}} \;\Rightarrow\; f'''(x) = \tfrac{3}{8}x^{-\frac{5}{2}} \;\Rightarrow\; f^{(4)}(x) = -\tfrac{15}{16}x^{-\frac{7}{2}}$$

$\Rightarrow\; f^{(5)}(x) = -\tfrac{105}{32}x^{-\frac{9}{2}}$, hence $f(9) = 3$, $f'(9) = \tfrac{1}{2}\tfrac{1}{3}$, $f''(9) = -\tfrac{1}{4}\tfrac{1}{27}$, $f'''(9) = \tfrac{3}{8}\tfrac{1}{243}$, $f^{(4)}(9) = -\tfrac{15}{16}\tfrac{1}{2187}$, $f^{(5)}(9) = -\tfrac{105}{32}\tfrac{1}{19683}$

$P_5(x) = 3 + \tfrac{1}{6}(x-9) - \tfrac{1}{216}(x-9)^2 + \tfrac{1}{3888}(x-9)^3 - \tfrac{5}{279936}(x-9)^4 + \tfrac{7}{5038848}(x-9)^5$

**Problem 21.**

$$f(x) = (1+x)^{-p} \;\Rightarrow\; f'(x) = -p(1+x)^{-p-1} \;\Rightarrow\; f''(x) = -p(-p-1)(1+x)^{-p-2}$$

$\Rightarrow\; f'''(x) = -p(-p-1)(-p-3)(1+x)^{-p-3}$. So $f(0) = 1$, $f'(0) = -p$, $f''(0) = -p(-p-1)$, $f'''(0) = -p(-p-1)(-p-3)$.

$P_3(x) = 1 - px + \tfrac{-p(-p-1)}{2}x^2 + \tfrac{-p(-p-1)(-p-2)}{6}x^3$

**Problem 23.**

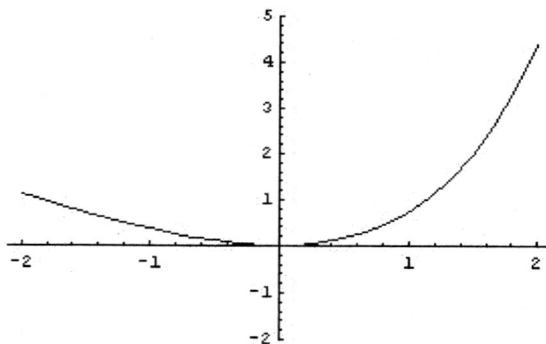

$$R_1(x) = f(x) - P_1(x)$$

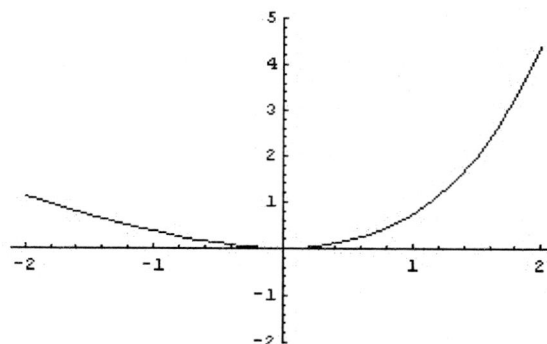

$$R_2(x) = f(x) - P_2(x)$$

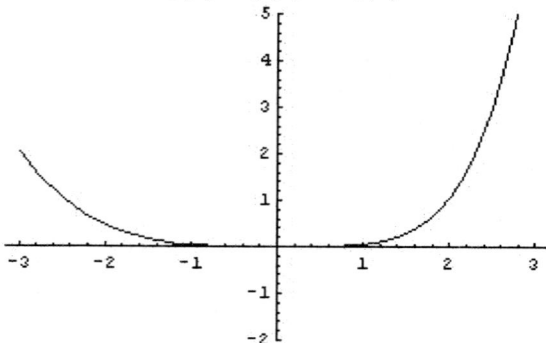

$$R_3(x) = f(x) - P_3(x)$$

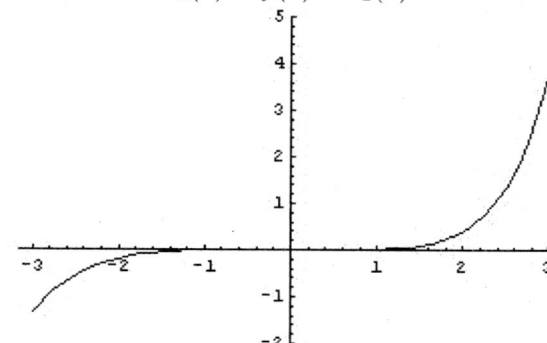

$$R_4(x) = f(x) - P_4(x)$$

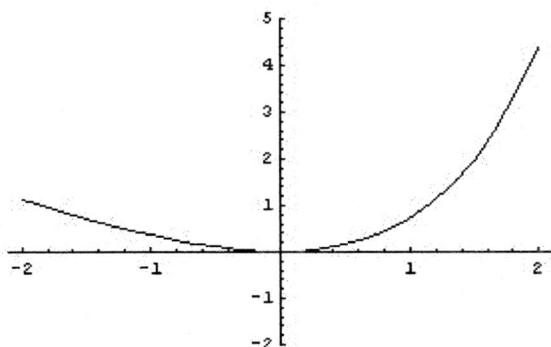

$$R_5(x) = f(x) - P_5(x)$$

**Problem 25.**

(a) $f'(1) = 3$   (b) $f''(1) = 10$   (c) $f'''(1) = 42$   (d) $f(1) = 12$

**Problem 27.**

$f(x) = \sin(x) \Rightarrow f'(x) = \cos(x) \Rightarrow f''(x) = -\sin(x) \Rightarrow f'''(x) = -\cos(x)$
$f^{(4)}(x) = \sin(x) \Rightarrow f^{(5)}(x) = \cos(x) \Rightarrow f^{(6)}(x) = -\sin(x)$, hence $f(\pi) = 0$, $f'(\pi) = -1$, $f''(\pi) = 0$, $f'''(\pi) = 1$, $f^{(4)}(\pi) = 0$, $f^{(5)}(\pi) = -1$, $f^{(6)}(\pi) = 0$
$P_6(x) = -(x - \pi) + \frac{1}{6}(x - \pi)^3 - \frac{1}{120}(x - \pi)^5$

**Problem 28.**

$f(x) = \cos(x) \Rightarrow f'(x) = -\sin(x) \Rightarrow f''(x) = -\cos(x) \Rightarrow f'''(x) = \sin(x) \; f^{(4)}(x) = \cos(x) \Rightarrow f^{(5)}(x) = -\sin(x) \Rightarrow f^{(6)}(x) = -\cos(x)$.
$f(-\frac{\pi}{2}) = 0$, $f'(-\frac{\pi}{2}) = 1$, $f''(-\frac{\pi}{2}) = 0$, $f'''(-\frac{\pi}{2}) = -1$, $f^{(4)}(-\frac{\pi}{2}) = 0$, $f^{(5)}(-\frac{\pi}{2}) = 1$, $f^{(6)}(-\frac{\pi}{2}) = 0$
$P_6(x) = (x + \pi/2) - \frac{1}{6}(x + \pi/2)^3 + \frac{1}{120}(x + \pi/2)^5$

**Problem 25.**

(a) $f(x) = (1 + x)^p \Rightarrow f'(x) = p(1 + x)^{p-1} \Rightarrow f''(x) = p(p - 1)(1 + x)^{p-2}$

$\Rightarrow f'''(x) = p(p-1)(p-2)(1+x)^{p-3}$. So $f(0) = 1$, $f'(0) = p$, $f''(0) = p(p-1)$, $f'''(0) = p(p-1)(p-3)$.
$P_3(x) = 1 + px + \frac{1}{2}p(p - 1)x^2 + \frac{1}{6}p(p - 1)(p - 2)x^3$

(b) $f^{(4)}(x) = p(p-1)(p-2)(p-3)(1+x)^{p-4} \Rightarrow f^{(5)}(x) = p(p-1)(p-2)(p-3)(p-4)(1+x)^{p-5}$ so
$f^{(4)}(0) = p(p-1)(p-2)(p-3)$, $f^{(5)}(0) = p(p-1)(p-2)(p-3)(p-4)$, hence $P_5(x) = 1 + px + \frac{1}{2}p(p-1)x^2 + \frac{1}{6}p(p-1)(p-2)x^3$
$+\frac{1}{24}p(p-1)(p-2)(p-3)x^4 + \frac{1}{120}p(p-1)(p-2)(p-3)(p-4)x^5$.

# Section 30.2   Error Analysis and Taylor's Theorem

**Problem 2.**

For $f(x) = e^x$, $P_3(x) = 1 + x + \frac{1}{2}x^2 + \frac{1}{6}x^3$ so $P_3(\frac{1}{2}) = \frac{79}{48} \approx 1.645833$

$f(x) = e^x \Rightarrow f^{(4)}(c) = e^c \Rightarrow |f^{(4)}(c)| \leq 3 = M$, hence $|R_3(x)| \leq \frac{(3)(\frac{1}{2})^4}{4!} = 0.0078125$

**Problem 4.**
   $P_3(1.5) \approx 0.4166667.$  $f(x) = \ln x \Rightarrow f^{(4)}(x) = -3!x^{-4} \Rightarrow |f^{(4)}(c)| \le 3!$ on $[1,\ 1.5]$,  hence $|R_3(x)| \le$
   $\frac{3!\,(1.5-1)^4}{4!} = \frac{(0.5)^4}{4} = 0.015625$

**Problem 6.**
   On $[0,\ 0.2]$  $\cos(x),\ \sin(x) \le 1,$  hence $|f^{(n+1)}(c)| \ \le 1 = M.$ We want $|R_n(x)| \le \frac{(1)(0.2)^{n+1}}{(n+1)!} \le 10^{-8}.$
   By experimentation $n = 6.$

**Problem 8.**
   $f(x) = \ln(1+x) \Rightarrow f'''(x) = \frac{2}{(1+x)^3} \le 2$ on $[0,\ 0.2]$,  but only $\le 3.90625$ on $[-0.2,\ 0]$

(a) $|R_2(x)| \le \frac{2(0.2)^3}{3!} < 0.002667$

(b) $|R_2(x)| \le \frac{3.90625(0.2)^3}{3!} < 0.005208$

**Problem 10.**
   We cannot presume to know $e^{0.5}$ but we can use easy approximations
   to our advantage. For instance $f'''(.5) = e^{0.5} < \sqrt{3} < 1.74.$
   Hence $|R_3(x)| \le \frac{1.74(0.5)^4}{4!} < 0.00454$
   The graph below is that of the actual $R_3(x)$

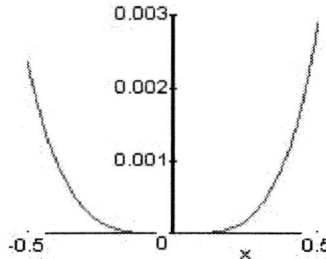

**Problem 12.**
   We cannot presume to know $e^{0.3}$ but we can use easy approximations to our advantage. For instance
   $e^{0.3} < 3^{1/3} < 1.5.$ Hence $|R_n(x)| \le \frac{1.5(0.3)^{n+1}}{(n+1)!} < 10^{-5} \Rightarrow n \ge 5$ by experimentation.

**Problem 13.**

(a) $f(x) = (1-x)^{-1} \Rightarrow f^{(n)}(x) = n!(1-x)^{-(n+1)} \Rightarrow P_n(x) = 1 + x + x^2 + x^3 + \cdots + x^n$

(b) $f^{(n+1)}(c) = (n+1)!(1-c)^{-(n+2)} \Rightarrow |f^{(n+1)}(c)| \le (n+1)!\frac{1}{(0.5)^{n+2}} = M$ on $[0,\ 0.5]$

   $|R_n(x)| \le \frac{(n+1)!(0.5)^{n+1}}{(0.5)^{n+2}(n+1)!} = 2.$ This method is not useful for getting the error less than $10^{-5}.$

   However, $P_n(\frac{1}{2}) = 1 + (\frac{1}{2}) + (\frac{1}{2})^2 + (\frac{1}{2})^3 + \cdots + (\frac{1}{2})^n = \frac{1-(\frac{1}{2})^{n+1}}{1-\frac{1}{2}} = 2 - (\frac{1}{2})^n$ is a partial sum of a geometric

   series that converges to 2. Hence we want the error $(\frac{1}{2})^n < 10^{-5} \Rightarrow n \ge 17$ by experimentation.

## Section 30.3    Taylor Series

**Problem 1.**

$f(x) = \cos(x) \Rightarrow f'(x) = -\sin(x) \Rightarrow f''(x) = -\cos(x) \Rightarrow f'''(x) = \sin(x) \Rightarrow f^{(4)}(x) = \cos(x) \cdot$
$\cdots$etc. $f(0) = 1$, $f'(0) = 0$, $f''(0) = -1$, $f'''(0) = 0$, $f^{(4)}(0) = 1$, $f^{(5)}(0) = 0$, $f^{(6)}(0) = -1$, etc.
Therefore $\cos x = 1 - \frac{x^2}{2!} + \frac{x^4}{4!} - \cdots + (-1)^k \frac{x^{2k}}{(2k)!} + \cdots$
    The analysis to show equality for all $x$ is exactly the same as that in Example 30.14 on page 943.

**Problem 3.**

(a) $\ln(1+u) = u - \frac{1}{2}u^2 + \frac{1}{3}u^3 - \frac{1}{4}u^4 + \cdots + (-1)^{k+1}\frac{1}{k}u^k + \cdots$

(b) $\ln(x) = (x-1) - \frac{1}{2}(x-1)^2 + \frac{1}{3}(x-1)^3 - \cdots + (-1)^{k+1}\frac{1}{k}(x-1)^k + \cdots$

(c) $f(x) = \ln x \Rightarrow f'(x) = x^{-1} \Rightarrow f''(x) = (-1)x^{-2} \Rightarrow f'''(x) = (-1)(-2)x^{-3}$, etc.
    $f(1) = 0$, $f'(1) = 1$, $f''(1) = (-1)$, $f'''(1) = (-1)(-2) = 2$, $f^{(4)}(1) = -3!$, etc.
    $\ln(x) = (x-1) + \frac{-1}{2!}(x-1)^2 + \frac{2}{3!}(x-1)^3 - \cdots + \frac{(-1)^{k+1}(k-1)!}{k!}(x-1)^k + \cdots$
    $= (x-1) - \frac{1}{2}(x-1)^2 + \frac{1}{3}(x-1)^3 - \cdots + (-1)^{k+1}\frac{1}{k}(x-1)^k + \cdots$

(d) $(0, 1]$

**Problem 5.**

$f(x) = \cos(x) \Rightarrow f'(x) = -\sin(x) \Rightarrow f''(x) = -\cos(x) \Rightarrow f'''(x) = \sin(x)$
$f^{(4)}(x) = \cos(x) \Rightarrow f^{(5)}(x) = -\sin(x) \Rightarrow f^{(6)}(x) = -\cos(x)$, etc.
$f(\frac{\pi}{2}) = 0$, $f'(\frac{\pi}{2}) = -1$, $f''(\frac{\pi}{2}) = 0$, $f'''(\frac{\pi}{2}) = 1$, $f^{(4)}(\frac{\pi}{2}) = 0$, $f^{(5)}(\frac{\pi}{2}) = -1$, $f^{(6)}(\frac{\pi}{2}) = 0$, etc.
Hence $2\cos(x) = -2(x - \frac{\pi}{2}) + \frac{2}{3!}(x - \frac{\pi}{2})^3 - \frac{2}{5!}(x - \frac{\pi}{2})^5 + \cdots + (-1)^{k+1}\frac{2}{(2k+1)!}(x - \frac{\pi}{2})^{2k+1} + \cdots$.

**Problem 6.**

$f(x) = 10^x \Rightarrow f'(x) = (\ln 10)10^x \Rightarrow f''(x) = (\ln 10)^2 10^x$, etc. so $f^{(n)}(0) = (\ln 10)^n 10^0 = (\ln 10)^n$.
Hence $10^x = 1 + (\ln 10)x + \frac{(\ln 10)^2}{2!}x^2 + \frac{(\ln 10)^3}{3!}x^3 + \cdots + \frac{(\ln 10)^k}{k!}x^k + \cdots$.

**Problem 8.**

$(3 + 2x)^3 = 27 + 54x + 36x^2 + 8x^3$

**Problem 9.**

$(1 + x)^5 = 1 + 5x + 10x^2 + 10x^3 + 5x^4 + x^5$

**Problem 10.**

Center $b=0$ with radius 5 means series converges for $x \in (-5, 5)$, diverges for $x \notin [-5, 5]$ and we don't
know at the end points without further information.
(a) converges    (b) converges    (c) not enough information    (d) diverges
(e) converges    (f) converges    (g) not enough information    (h) diverges

**Problem 12.**

$(-2, 5]$ has center $b = \frac{3}{2}$ and radius $\frac{7}{2}$.
(a) $R = \frac{7}{2}$

(b) Center is at $b = \frac{3}{2}$

**Problem 14.**

$\sqrt{1 + 3x} = (1 + (3x))^{\frac{1}{2}}$, hence $P_3 = 1 + \frac{3}{2}x - \frac{9}{8}x^2 + \frac{27}{16}x^3$,    $R = \frac{1}{3}$

**Problem 16.**
$$(1-x)^{\frac{2}{3}} = (1+(-x))^{\frac{2}{3}}, \text{ hence} P_3 = 1 - \tfrac{2}{3}x - \tfrac{1}{9}x^2 - \tfrac{4}{81}x^3, \quad R = 1$$

**Problem 17.**
$$\sqrt[3]{1+x^2} = (1+(x^2))^{\frac{1}{3}}, \text{ hence } P_4 = 1 + \tfrac{1}{3}x^2 - \tfrac{1}{9}x^4, \quad R = 1$$

**Problem 19.**
$$\tfrac{1}{(1+x)^2} = (1+x)^{-2}, \text{ hence } P_5 = 1 - 2x + 3x^2 - 4x^3 + 5x^4 - 6x^5, \quad R = 1$$

**Problem 21.**
$$\tfrac{x}{\sqrt{4+x}} = \tfrac{x}{2}(1+(\tfrac{x}{4}))^{-\frac{1}{2}}, \text{ hence } P_3 = \tfrac{1}{2}x - \tfrac{1}{16}x^2 + \tfrac{3}{256}x^3, \quad R = 4$$

**Problem 22.**

(a) $(a+x)^4 = a^4 + 4a^3x + 6a^2x^2 + 4ax^3 + x^4$

(b) $a^4\left(1+\tfrac{x}{a}\right)^4 = a^4\left(1 + 4\left(\tfrac{x}{a}\right) + 6\left(\tfrac{x}{a}\right)^2 + 4\left(\tfrac{x}{a}\right)^3 + \left(\tfrac{x}{a}\right)^4\right) = a^4\left(1 + 4\tfrac{x}{a} + 6\tfrac{x^2}{a^2} + 4\tfrac{x^3}{a^3} + \tfrac{x^4}{a^4}\right)$ =part (a)

**Problem 24.**
$$\tfrac{1}{\sqrt{1-x^2}} = (1+(-x^2))^{-\frac{1}{2}} = 1 - \tfrac{1}{2}(-x^2) + \tfrac{(-\frac{1}{2})(-\frac{3}{2})}{2!}(-x^2)^2 + \tfrac{(-\frac{1}{2})(-\frac{3}{2})(-\frac{5}{2})}{3!}(-x^2)^3 + \cdots$$
$$+ \tfrac{(-\frac{1}{2})(-\frac{3}{2})\cdots(-\frac{2k-1}{2})}{k!}(-x^2)^k + \cdots$$
$$= 1 + \tfrac{1}{2}x^2 + \tfrac{(\frac{1}{2})(\frac{3}{2})}{2!}x^4 + \tfrac{(\frac{1}{2})(\frac{3}{2})(\frac{5}{2})}{3!}x^6 + \cdots + \tfrac{(\frac{1}{2})(\frac{3}{2})\cdots(\frac{2k-1}{2})}{k!}x^{2k} + \cdots, \quad R = 1$$

**Problem 26.**
Use the Maclaurin series for $\sin u$. Substitute $3x$ for $u$. $\sin(3x) = 3x - \tfrac{1}{3!}(3x)^3 + \tfrac{1}{5!}(3x)^5 + \cdots + (-1)^k \tfrac{1}{(2k+1)!}(3x)^{2k+1} + \cdots, \quad R = \infty$

**Problem 28.**
Use the Maclaurin series for $e^u$. Substitute $2x$ for $u$ and multiply all by 3. $3e^{2x} = 3 + 3(2x) + \tfrac{3}{2!}(2x)^2 + \tfrac{3}{3!}(2x)^3 + \cdots + \tfrac{3}{k!}(2x)^k + \cdots$, $\quad R = \infty$

**Problem 30.**
$3^x = (e^{\ln 3})^x = e^{(\ln 3)x}$. Use the Maclaurin series for $e^u$. Substitute $(\ln 3)x$ for $u$. $3^x = 1 + (\ln 3)x + \tfrac{1}{2!}(\ln 3)^2 x^2 + \tfrac{1}{3!}(\ln 3)^3 x^3 + \cdots + \tfrac{1}{k!}(\ln 3)^k x^k + \cdots$, $\quad R = \infty$

**Problem 32.**
$\cos^2(x) = \tfrac{1}{2} + \tfrac{1}{2}\cos(2x)$.
Use the Maclaurin series for $\cos u$. Substitute $2x$ for $u$ and multiply all by $\tfrac{1}{2}$. Finally add $\tfrac{1}{2}$. $\cos^2(x) =$
$$\tfrac{1}{2} + \tfrac{1}{2}\cos(2x) = \tfrac{1}{2} + \tfrac{1}{2}\left(1 - \tfrac{(2x)^2}{2!} + \tfrac{(2x)^4}{4!} - \cdots + (-1)^k \tfrac{(2x)^{2k}}{(2k)!} + \cdots\right)$$
$$= 1 - x^2 + \tfrac{x^4}{3} - \cdots + (-1)^k \tfrac{2^{2k-1}x^{2k}}{(2k)!} + \cdots, \quad R = \infty$$

**Problem 33.**
$(a+x)^p = a^p(1+\tfrac{x}{a})^p$. (Use binomial series.)
$$= a^p\left(1 + p(\tfrac{x}{a}) + \tfrac{p(p-1)}{2!}(\tfrac{x}{a})^2 + \tfrac{p(p-1)(p-2)}{3!}(\tfrac{x}{a})^3 + \cdots + \tfrac{p(p-1)\cdots(p-k+1)}{k!}(\tfrac{x}{a})^k + \cdots\right)$$
$$= a^p + pa^{p-1}x + \tfrac{p(p-1)}{2!}a^{p-2}x^2 + \cdots + \tfrac{p(p-1)\cdots(p-k+1)}{k!}a^{p-k}x^k + \cdots, \quad R = |a|$$

**Problem 36.**
$$\tfrac{1}{\sqrt{e^x}} = e^{-x/2} = 1 - \tfrac{1}{2}x + \tfrac{1}{8}x^2 - \cdots + (-1)^k \tfrac{1}{2^k k!}x^k + \cdots, \quad R = \infty$$

**Problem 38.**

(a) $e^{bi} = 1 + (bi) + \frac{1}{2!}(bi)^2 + \frac{1}{3!}(bi)^3 + \cdots + \frac{1}{k!}(bi)^k + \cdots = 1 + ib - \frac{1}{2!}b^2 - \frac{1}{3!}ib^3 + \frac{1}{4!}b^4 \cdots + (i)^k \frac{1}{k!}b^k + \cdots$

$= 1 - \frac{1}{2!}b^2 + \frac{1}{4!}b^4 + \cdots + (-1)^k \frac{1}{(2k)!}b^{2k} + \cdots \quad + i\left(b - \frac{1}{3!}b^3 + \cdots + (-1)^k \frac{1}{(2k+1)!}b^{2k+1} + \cdots\right)$

$= \cos b + i \sin b$

(b) $e^{\pi i} = \cos \pi + i \sin \pi = -1$

## Section 30.4    Working with Series and Power Series

**Problem 1.**

diverges, individual terms $a_k$ don't converge to zero.

**Problem 2.**

conditionally convergent, alternating harmonic.

**Problem 4.**

diverges, individual terms $a_k$ don't converge to zero.

**Problem 6.**

absolutely convergent, geometric with $|r| = \frac{11}{12}$.

**Problem 7.**

diverges, individual terms $a_k$ don't converge to zero.

**Problem 9.**

diverges, individual terms $a_k$ don't converge to zero.

**Problem 11.**

For $x = \frac{1}{2}$ we need $\frac{(\frac{1}{2})^{k+1}}{k+1} < 10^{-4}$. By experimentation we find $k = 9$ terms are needed.

**Problem 13.**

$\sin x = x - \frac{1}{3!}x^3 + \frac{1}{5!}x^5 + \cdots + (-1)^k \frac{1}{(2k+1)!}x^{2k+1} + \cdots$ Differentiating both sides gives $\cos x =$

$1 - \frac{3x^2}{3!} + \frac{5x^4}{5!} - \cdots + (-1)^k \frac{(2k+1)x^{2k}}{(2k+1)!} + \cdots = 1 - \frac{x^2}{2!} + \frac{x^4}{4!} - \cdots + (-1)^k \frac{x^{2k}}{(2k)!} + \cdots$.

**Problem 14.**

$\sin^{-1} x + C = \int \frac{1}{\sqrt{1-x^2}} dx = \int (1 + (-x^2))^{-1/2} dx$

$= \int \left(1 - \frac{1}{2}(-x^2) + \frac{(-\frac{1}{2})(-\frac{3}{2})}{2!}(-x^2)^2 + \cdots + \frac{(-\frac{1}{2})(-\frac{3}{2}) \cdots (\frac{1}{2} - k)}{k!}(-x^2)^k + \cdots\right) dx$

$= \int \left(1 + \frac{1}{2}x^2 + \frac{3}{8}x^4 + \cdots + \frac{(\frac{1}{2})(\frac{3}{2}) \cdots (k - \frac{1}{2})}{k!}x^{2k} + \cdots\right) dx$

$= x + \frac{1}{6}x^3 + \frac{3}{40}x^5 + \cdots + \frac{(\frac{1}{2})(\frac{3}{2}) \cdots (k - \frac{1}{2})}{k!(2k+1)}x^{2k+1} + \cdots \quad R = 1$

By evaluating both sides at $x = 0$ we find $C$ must be zero.

**Problem 16.**

$\int e^{x^3} dx = \int \left(1 + (x^3) + \frac{1}{2!}(x^3)^2 + \frac{1}{3!}(x^3)^3 + \cdots + \frac{1}{k!}(x^3)^k + \cdots\right) dx = x + \frac{1}{4}x^4 + \frac{1}{2!(7)}x^7 + \frac{1}{3!(10)}x^{10} +$

$\cdots + \frac{1}{k!(3k+1)}x^{3k+1} + \cdots + C$

**Problem 18.**
$$\int_0^{0.5} \sin(x^2)dx = \int_0^{0.5} \left( x^2 - \tfrac{1}{3!}(x^2)^3 + \tfrac{1}{5!}(x^2)^5 - \tfrac{1}{7!}(x^2)^7 + \cdots \right) dx$$
$$= \left( \tfrac{1}{3}x^3 - \tfrac{1}{3!(7)}x^7 + \tfrac{1}{5!(11)}x^{11} - \tfrac{1}{7!(15)}x^{15} + \cdots \right)\Big|_0^{0.5}$$
$$= \tfrac{1}{3}(0.5)^3 - \tfrac{1}{3!(7)}(0.5)^7 + \tfrac{1}{5!(11)}(0.5)^{11} - \tfrac{1}{7!(15)}(0.5)^{15} + \cdots$$
As $\tfrac{1}{7!(15)}(0.5)^{15} < 10^{-8}$ the sum of the first three terms is our <u>over</u>estimate: 0.0414810247

**Problem 20.**
$$\ln(2+x) + C = \int \tfrac{1}{2+x}dx = \tfrac{1}{2} \int \tfrac{1}{1-(-x/2)} \, dx$$
$$= \tfrac{1}{2} \int \left( 1 + (-\tfrac{x}{2}) + (-\tfrac{x}{2})^2 + \cdots + (-\tfrac{x}{2})^k + \cdots \right) dx = \tfrac{1}{2} \int \left( 1 - \tfrac{x}{2} + \tfrac{x^2}{2^2} + \cdots + (-1)^k \tfrac{x^k}{2^k} + \cdots \right) dx$$
$$= \tfrac{1}{2} \left( x - \tfrac{1}{4}x^2 + \tfrac{1}{2^2(3)}x^3 + \cdots + (-1)^k \tfrac{1}{2^k(k+1)}x^{k+1} + \cdots \right) \text{ Substituting } x = 0 \text{ into both sides yields}$$
$$\ln(2+x) = \ln 2 + \tfrac{1}{2}x - \tfrac{1}{8}x^2 + \tfrac{1}{2^3(3)}x^3 + \cdots + (-1)^k \tfrac{1}{2^{k+1}(k+1)}x^{k+1} + \cdots$$

**Problem 21.**

(a) $\ln\left(\tfrac{1+x}{1-x}\right) = \ln(1+x) - \ln(1-x) = \left( x - \tfrac{1}{2}x^2 + \tfrac{1}{3}x^3 - \tfrac{1}{4}x^4 + \tfrac{1}{5}x^5 - \cdots \right) -$
$\left( (-x) - \tfrac{1}{2}(-x)^2 + \tfrac{1}{3}(-x)^3 - \tfrac{1}{4}(-x)^4 + \tfrac{1}{5}(-x)^5 - \cdots \right)$
$= \left( 2x + \tfrac{2}{3}x^3 + \tfrac{2}{5}x^5 + \tfrac{2}{7}x^7 \cdots \right)$

(b) $\dfrac{1+\frac{1}{3}}{1-\frac{1}{3}} = \dfrac{\frac{4}{3}}{\frac{2}{3}} = \tfrac{4}{2} = 2$

(c) $\ln(2) = \ln\left(\dfrac{1+\frac{1}{3}}{1-\frac{1}{3}}\right) \approx 2\left(\tfrac{1}{3}\right) + \tfrac{2}{3}\left(\tfrac{1}{3}\right)^3 + \tfrac{2}{5}\left(\tfrac{1}{3}\right)^5 + \tfrac{2}{7}\left(\tfrac{1}{3}\right)^7 \approx 0.6931347573 \quad \ln(2) = \ln(1+1) \approx 1 - \tfrac{1}{2} + \tfrac{1}{3} - \tfrac{1}{4} \approx$
.583333333
$\ln(2) \approx 0.69314718056$ by calculator.

**Problem 23.**
$$f(x) = \sum_{k=0}^{\infty} a_k x^k \;\Rightarrow\; f'(x) = \sum_{k=0}^{\infty} a_k k\, x^{k-1} = 0 + \sum_{k=1}^{\infty} a_k k\, x^{k-1} = \sum_{k=0}^{\infty} a_{k+1}(k+1)\, x^k$$
It must be that $\displaystyle\sum_{k=0}^{\infty} a_{k+1}(k+1)\, x^k = f'(x) = -f(x) = \sum_{k=0}^{\infty} -a_k\, x^k$
Equating like powers of $x$ yields $a_{k+1}(k+1) = -a_k$.
If $a_0 = 1$, then $a_1 = -1$. Inductively $a_2 = \tfrac{1}{2}$, $a_3 = -\tfrac{1}{3\cdot 2}$, $a_4 = \tfrac{1}{4\cdot 3\cdot 2}$, $\cdots a_k = (-1)^k \tfrac{1}{k!}$.
In this case $f(x) = e^{-x}$. In general $a_k = a_0(-1)^k \tfrac{1}{k!}$ so $f(x) = a_0 e^{-x}$.

**Problem 25.**

(a) Too large. The series is alternating with the $4^{th}$ term ($k = 3$) being negative.

The $4^{th}$ term gives an error bound of $\tfrac{(0.1)^6}{(3!)^2 2^6} \approx 4.34 \times 10^{-10}$.

(b) $\tfrac{(1)^8}{(4!)^2 2^8} \approx 6.78 \times 10^{-6}$ is the first term $< 10^{-4}$. Hence we must use the first four nonzero terms.

# Section 30.5    Convergence Tests

**Problem 1.**
   All $|c| < 1$, as series is geometric.

**Problem 3.**

(a) $\sum a_k$ converges by comparison to $\sum b_k$.

(b) $\sum c_k$ diverges by comparison to $\sum b_k$.

**Problem 5.**
   Converges as it is a multiple of a p-series with $p = \frac{3}{2} > 1$.

**Problem 7.**
   Diverges as it is a p-series with $p = \frac{9}{10} < 1$.

**Problem 9.**
   Converges as it is a multiple of a p-series with $p = \frac{10}{9} > 1$.

**Problem 10.**
   Converges as it is geometric with $(1/e^2) < 1$.

**Problem 12.**
   Converges as it is geometric with $(1/e^{0.1}) < 1$.

**Problem 13.**
   Diverges as the improper integral $\int_2^\infty \frac{1}{\ln x}\frac{1}{x}dx$ diverges.

**Problem 15.**
   Converges by direct comparison. It is $< \sum_{k=1}^\infty \frac{1}{k^2}$ a p-series

**Problem 16.**
   Converges by direct comparison. It is $< 2\sum_{k=1}^\infty \left(\frac{1}{3}\right)^k$ a geometric series.

**Problem 19.**
   Converges by direct comparison. It is $< \sum_{n=1}^\infty \left(\frac{1}{e}\right)^n$ a geometric series.

**Problem 20.**
   $\sum_{k=3}^n a_k < \int_2^n f(x)dx < \sum_{k=2}^{n-1} a_k$. Compare the area of the rectangles and the area under the curve on the interval $[2, n]$.

**Problem 22.**
   $1 = \int_1^\infty \frac{1}{x^2}dx < \sum_{k=1}^\infty \frac{1}{k^2} < 1+\int_1^\infty \frac{1}{x^2}dx = 2$ Draw a picture. For the left inequality think left Riemann sum. For the right inequality think right Riemann sum plus one extra rectangle on the left.

**Problem 24.**

(a) Not enough information. Partial sums could converge to a number $> m$ or keep growing.

(b) Converges. $a_k > 0$ means $S_n$ is an increasing sequence bounded above by $M$.

(c) Converges. Similar to (b) only decreasing bounded below.

(d) Not enough information. $S_n$ sequence could oscillate if some $a_k > 0$ while others are $< 0$.

*There are many ways to show the following. We suggest one way.*

**Problem 25.**

Limit compare to $\sum_{k=1}^{\infty} \left(\frac{1}{e}\right)^k$ which converges. $\lim\limits_{k\to\infty} \frac{\frac{1}{e^k-1}}{\left(\frac{1}{e}\right)^k} = \lim\limits_{k\to\infty} \frac{1}{1-\frac{1}{e^k}} = 1$

**Problem 28.**

Limit compare to $\sum_{k=1}^{\infty} \frac{1}{k^2}$ which converges. $\lim\limits_{k\to\infty} \frac{\frac{2k^2-k}{3k^4+1}}{\frac{1}{k^2}} = \lim\limits_{k\to\infty} \frac{2-\frac{k}{k^3}}{3+\frac{1}{k^4}} = \frac{2}{3}$

**Problem 29.**

Limit compare to $\sum_{k=3}^{\infty} \frac{1}{k^2}$ which converges. $\lim\limits_{k\to\infty} \frac{\frac{k}{2k^3-2}}{\frac{1}{k^2}} = \lim\limits_{k\to\infty} \frac{1}{2-\frac{2}{k^3}} = \frac{1}{2}$

**Problem 31.**

Diverges by n$^{th}$ term test. $\lim\limits_{n\to\infty} \frac{n+1}{\ln n} = \infty$

**Problem 33.**

(a) (i) By direct comparison $\sum_{k=1}^{\infty} a_k^2 < \sum_{k=1}^{\infty} a_k$ which converges.

(ii) By direct comparison $\sum_{k=1}^{\infty} \frac{1}{a_k} > \sum_{k=1}^{\infty} \frac{1}{1}$ which diverges.

(b) As $\sum_{k=1}^{\infty} b_k$ converges, $b_k \to 0$ hence for some $M$, $b_k < 1$ for $k \geq M$ (the tail). Then by the reasoning above $\sum_{k=M}^{\infty} b_k^2 < \sum_{k=M}^{\infty} b_k$ converges. Hence $\sum_{k=1}^{\infty} b_k^2$ converges.

*There are many ways to show the following. We suggest one way.*

**Problem 34.**

Converges using ratio test: $\lim\limits_{k\to\infty} \frac{\frac{3(k+1)}{(k+1)!}}{\frac{3k}{k!}} = \lim\limits_{k\to\infty} \frac{k+1}{k(k+1)} = 0$

**Problem 36.**

Diverges as is geometric with $r = \frac{3}{2}$, OR use root test.

**Problem 38.**

Diverges using ratio test: $\lim\limits_{k\to\infty} \frac{\frac{(k+1)!}{(k+1)^3 3^{k+1}}}{\frac{k!}{k^3 3^k}} = \lim\limits_{k\to\infty} \frac{k+1}{3} \left(\frac{k}{k+1}\right)^3 = \infty \cdot 1 = \infty$

**Problem 39.**

Converges by direct comparison of tails: $\sum_{k=8}^{\infty} \frac{2}{(\ln k)^k} < \sum_{k=8}^{\infty} \frac{2}{2^k} < 2\sum_{k=8}^{\infty} \left(\frac{1}{2}\right)^k$ geometric.

**Problem 41.**

Converges using root test: $\lim\limits_{k\to\infty} \sqrt[k]{\left(\frac{k^2-3k}{5k^2+1}\right)^k} = \lim\limits_{k\to\infty} \frac{1-\frac{3}{k}}{5+\frac{1}{k^2}} = \frac{1}{5}$

**Problem 43.**

For any value $r$. Use ratio test: $\lim\limits_{k\to\infty} \frac{\frac{(r)^{k+1}}{(k+1)!}}{\frac{r^k}{k!}} = \lim\limits_{k\to\infty} \frac{r}{(k+1)} = 0$

**Problem 45.**

Absolutely convergent by direct comparison: $\sum_{k=1}^{\infty} \left| \frac{(-1)^{k+1}}{k\sqrt{2k}} \right| < \frac{1}{2} \sum_{k=1}^{\infty} \frac{1}{k^{3/2}}$ p-series

## Problem 47.

Absolutely convergent by limit comparison: $\sum_{k=1}^{\infty} \left| \frac{\sin(2k)}{2^k} \right| < \sum_{k=1}^{\infty} \left( \frac{1}{2} \right)^k$ geometric series

## Problem 49.

Absolutely convergent using ratio test: $\lim_{k \to \infty} \frac{\frac{5^{k+1}}{(k+1)!}}{\frac{5^k}{k!}} = \lim_{k \to \infty} \frac{5}{(k+1)} = 0$

## Problem 50.

Diverges by $\text{n}^{th}$ term test: $\frac{k^k}{k!} > 1$ for $k > 1$.

## Problem 52.

The series converges by the alternating series test. With absolute values added the series diverges by comparison to the tail of the harmonic series, $\frac{\ln k}{k} > \frac{1}{k}$ for $k \geq 3$. Hence the original series is conditionally convergent.

## Problem 54.

$R = \infty$. Ratio test: $\lim_{k \to \infty} \frac{\left| (-1)^{k+1} \frac{(2x)^{k+1}}{(k+1)!} \right|}{\left| (-1)^k \frac{(2x)^k}{(k)!} \right|} = \lim_{k \to \infty} \left| \frac{2x}{k+1} \right| = 0 < 1$ for all $x$.

## Problem 56.

$R = 2$. Root test: $\lim_{k \to \infty} \sqrt[k]{\left| k(\frac{x}{2})^k \right|} = \lim_{k \to \infty} \sqrt[k]{k} \left| \frac{x}{2} \right| = \left| \frac{x}{2} \right| < 1 \Leftrightarrow |x| < 2$

## Problem 58.

$R = \infty$. Ratio test: $\lim_{k \to \infty} \frac{\left| \frac{(x-1)^{2k+2}}{(k)!} \right|}{\left| \frac{(x-1)^{2k}}{(k-1)!} \right|} = \lim_{k \to \infty} \left| \frac{(x-1)^2}{k} \right| = 0 < 1$ for all $x$.

## Problem 59.

$R = 1$. Ratio test: $\lim_{n \to \infty} \frac{\left| \frac{(x+2)^{n+1}}{(n+1)(2n+5)} \right|}{\left| \frac{(x+2)^n}{(n)(2n+3)} \right|} = \lim_{n \to \infty} \left| \frac{(n)}{(n+1)} \frac{(2n+3)}{(2n+5)} (x+2) \right| = |x+2| < 1 \Leftrightarrow x \in (-3, -1)$.

## Problem 60.

$\left( -\frac{2}{3}, \frac{2}{3} \right)$ By root test $R = \frac{2}{3}$: $\lim_{k \to \infty} \sqrt[k]{\left| \frac{(3x)^k}{2^k} \right|} = \lim_{k \to \infty} \left| \frac{3x}{2} \right| = \left| \frac{3x}{2} \right| < 1 \Leftrightarrow |x| < \frac{2}{3}$.

Endpoints: $x = -\frac{2}{3}$, $\sum (-1)^k$ diverges; $x = \frac{2}{3}$, $\sum 1$ diverges.

## Problem 62.

$[-2, 0)$ By root test $R = 1$: $\lim_{k \to \infty} \sqrt[k]{\left| \frac{(x+1)^k}{3k} \right|} = \lim_{k \to \infty} \left| \frac{x+1}{\sqrt[k]{3} \sqrt[k]{k}} \right| = |x+1| < 1 \Leftrightarrow x \in (-1, 0)$.

Endpoints: $x = -2$, $\sum (-1)^k \frac{1}{3k}$ converges; $x = 0$, $\sum \frac{1}{3k}$ diverges.

## Problem 64.

$(1, 5)$ By root test $R = 2$: $\lim_{k \to \infty} \sqrt[k]{\left| \frac{(x-3)^k}{2^k} \right|} = \lim_{k \to \infty} \left| \frac{x-3}{2} \right| = \left| \frac{x-3}{2} \right| < 1 \Leftrightarrow x \in (1, 5)$.

Endpoints: $x = 1$, $\sum (-1)^k$ diverges; $x = 5$, $\sum 1$ diverges.

## Problem 65.

$(-\infty,\ \infty)$ By ratio test $R = \infty$: $\displaystyle\lim_{k\to\infty} \frac{\left|\frac{(x-3)^{k+1}}{(k+1)!}\right|}{\left|\frac{(x-3)^k}{(k)!}\right|} = \lim_{k\to\infty}\left|\frac{x-3}{k+1}\right| = 0 < 1$ for all $x$.

## Problem 66.

$\{0\}$ By root test $R = 0$: $\displaystyle\lim_{k\to\infty}\sqrt[k]{|(kx)^k|} = \lim_{k\to\infty}|kx| = \pm\infty$ unless $x = 0$

## Problem 68.

$[0,\ 2]$ By root test $R = 1$: $\displaystyle\lim_{k\to\infty}\sqrt[k]{\left|\frac{(x-1)^k}{k^5}\right|} = \lim_{k\to\infty}\left|\frac{x-1}{(\sqrt[k]{k})^5}\right| = |x-1| < 1 \ \Leftrightarrow\ x \in (-1,\ 0)$.

Endpoints: $x = 0$, $\sum(-1)^k\frac{1}{k^5}$ converges; $x = 2$, $\sum\frac{1}{k^5}$ converges.

## Problem 70.

$(-\infty,\ \infty)$ By ratio test $R = \infty$: $\displaystyle\lim_{k\to\infty}\frac{\left|\frac{(2x)^{k+1}}{(k+1)!}\right|}{\left|\frac{(2x)^k}{(k)!}\right|} = \lim_{k\to\infty}\left|\frac{2x}{k+1}\right| = 0 < 1$ for all $x$.

## Problem 71.

$(-1,\ 1]$ By ratio test $R = 1$: $\displaystyle\lim_{k\to\infty}\frac{\left|\frac{x^{k+1}}{2\ln(k+1)}\right|}{\left|\frac{x^k}{2\ln k}\right|} = \lim_{k\to\infty}\left|\frac{\ln k}{\ln(k+1)}x\right| = |x| < 1 \ \Leftrightarrow\ x \in (-1,\ 1)$.

Endpoints: $x = -1$, $\sum\frac{1}{2\ln k}$ diverges; $x = 1$, $\sum(-1)^n\frac{1}{2\ln k}$ converges.

## Problem 74.

For $\sum_{k=0}^{\infty}a_kx^k$ suppose $\displaystyle\lim_{k\to\infty}\frac{|a_{k+1}|}{|a_k|} = Q$. Hence by the ratio test this series converges if

$Q\,|x| < 1 \ \Leftrightarrow\ |x| < \frac{1}{Q}$. Similarly $\sum_{k=0}^{\infty}a_k(x-b)^k$ converges if $Q\,|x-b| < 1 \ \Leftrightarrow\ |x-b| < \frac{1}{Q}$. We have the same radius of convergence $\frac{1}{Q}$ but the center of the interval of convergence is different.

## Problem 76.

Using the Ratio test we get: (See binomial series terms on page 947.) $\displaystyle\lim_{k\to\infty}\frac{\left|\frac{p(p-1)\cdots(p-k+1)(p-k)x^{k+1}}{(k+1)!}\right|}{\left|\frac{p(p-1)\cdots(p-k+1)x^k}{(k)!}\right|} =$

$\displaystyle\lim_{k\to\infty}\left|\frac{p-k}{k+1}x\right| = |x| < 1 \ \Leftrightarrow\ x \in (-1,\ 1)$.

## Problem 77.

(i) For series $\sum_{k=0}^{\infty}a_k$ of not necessarily positive terms we check the convergence of $\sum_{k=0}^{\infty}|a_k|$ to test for absolute convergence of the original. The regular ratio test applied to $\sum_{k=0}^{\infty}|a_k|$ yield that if $L < 1$ then $\sum_{k=0}^{\infty}|a_k|$ converges. Hence it follows that $\sum_{k=0}^{\infty}a_k$ converges absolutely.

(ii) If $L > 1$ for $\sum_{k=0}^{\infty}|a_k|$, the regular ratio test proof (page 973) would say that $\displaystyle\lim_{k\to\infty}|a_k| \neq 0$ which implies that $\displaystyle\lim_{k\to\infty}a_k \neq 0$. Hence $\sum_{k=0}^{\infty}a_k$ diverges.

(iii) The examples given on page 974 also work for the generalized test.

# CHAPTER 31

# Differential Equations

## Section 31.1    Introduction to Modelling with Differential Equations

**Problem 2.**
$$\frac{dP}{dt} = kP(N - P)$$

**Problem 4.**
$$\frac{dG}{dt} = T - kG$$

**Problem 6.**

(a) $\frac{dP}{dt} = 0.02P + (0.1 - 0.3) \Leftrightarrow \frac{dP}{dt} = 0.02P - 0.2$

(b) The population will decrease to zero. At $t = 0$, $P = 9$.  $\frac{dP}{dt} = 0.02P - 0.2$ will be negative and will continue to be more negative as the population decreases.

(c) $0 = 0.02P - 0.2 \Rightarrow P = 10$

**Problem 7.**

(a) $\frac{dN}{dt} = kN(600 - N) = k(600N - N^2)$

(b) $\frac{d^2N}{dt^2} = k(600\frac{dN}{dt} - 2N\frac{dN}{dt}) = 2k\frac{dN}{dt}(300 - N) = 2k^2N(600 - N)(300 - N)$

(c) $\frac{d^2N}{dt^2} = 0$ when $N = 300$ (fastest conversion), (slowest when $N = 0$ or $600$)

## Section 31.2    Solutions to Differential Equations: An Introduction

**Problem 2.**
    d For $y = \sin(5t)$, $LHS = y''(t) = -25\sin(5t)$, while $RHS = -25y = -25\sin(5t)$. DE is satisfied.

**Problem 4.**
    For $y = \frac{\ln x}{x}$, $LHS = \frac{dy}{dx} = \frac{1 - \ln x}{x^2}$ while $RHS = \frac{-y}{x} + \frac{y^2}{(\ln x)^2} = -\frac{\frac{\ln x}{x}}{x} + \frac{(\frac{\ln x}{x})^2}{(\ln x)^2} = -\frac{\ln x}{x^2} + \frac{1}{x^2} = \frac{1 - \ln x}{x^2}$

**Problem 5.**
    ($i$) a, b
        (ii) e
        (iii) d
        (iv) c, f, g, h

**Problem 7.**
    e

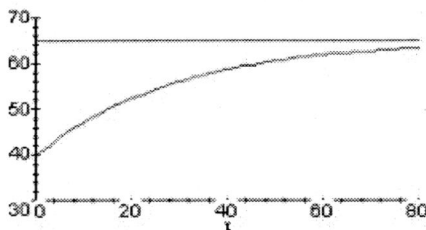

## Problem 9.

For $y = \frac{1}{2}xe^x + \frac{1}{3}\frac{e^x}{x}$, $x\frac{dy}{dx} + (1-x)y = x\left(\frac{1}{2}(e^x + xe^x) + \frac{1}{3}\left(\frac{xe^x - e^x}{x^2}\right) + \frac{1}{2}xe^x + \frac{1}{3}\frac{e^x}{x} - x\left(\frac{1}{2}xe^x + \frac{1}{3}\frac{e^x}{x}\right)\right)$

$$= \frac{1}{2}xe^x + \frac{1}{2}x^2e^x + \frac{1}{3}e^x - \frac{1}{3}\frac{e^x}{x} + \frac{1}{2}xe^x + \frac{1}{3}\frac{e^x}{x} - \frac{1}{2}x^2e^x - \frac{1}{3}e^x = xe^x.$$

Hence DE is satisfied.

## Problem 11.

(a) (i) Use $u = x - 3$. The DE becomes $\frac{du}{dt} = -2u \Rightarrow u = Ce^{-2t} \Rightarrow x(t) = 3 + Ce^{-2t}$.

(ii) Use $v = 6 - 2x$. The DE becomes $-\frac{1}{2}\frac{dv}{dt} = v \Rightarrow \frac{dv}{dt} = -2v \Rightarrow v = Be^{-2t}$ for some constant $B$.

$\Rightarrow 6 - 2x = Be^{-2t} \Rightarrow -2x = -6 + Be^{-2t} \Rightarrow x = 3 - \frac{1}{2}Be^{-2t} = 3 + Ce^{-2t}$ if $C = -\frac{1}{2}B$.

(b) Use $u = x - \frac{7}{3}$. The DE becomes $\frac{du}{dt} = 3u \Rightarrow u = Ce^{3t} \Rightarrow x(t) = \frac{7}{3} + Ce^{3t}$.

(c) Use $u = y + \frac{B}{k}$. The DE becomes $\frac{du}{dt} = ku \Rightarrow u = Ce^{kt} \Rightarrow y(t) = Ce^{kt} - \frac{B}{k}$.

## Problem 13.

(a) $\frac{dP}{dt} = 0.03P - 600 = 0.03(P - 200000)$

(b) Use $u = P - 200000$. The DE becomes $\frac{du}{dt} = 0.03u \Rightarrow u = Ce^{0.03t} \Rightarrow P(t) = 200000 + Ce^{0.03t}$.
At $t = 0$, $3000000 = 200000 + Ce^0 \Rightarrow C = 2800000 \Rightarrow P(t) = 200000 + 2800000e^{0.03t}$

## Problem 15.

(a) $\frac{dT}{dt} = k(T - F)$. This is the same as in example 31.3 with $F$ instead of 65 for surround temp.

(b) As the change is negative and $T - F$ is positive, $k$ must be negative.

(c) (i)

(*ii*) Increasing at a decreasing rate. (*i*)

(*iii*) $\frac{dL}{dt} = k(L - 65)$,     $k < 0$

(*iv*) Use $u = L - 65$. Then the DE becomes

$\frac{du}{dt} = ku \Rightarrow u = Ce^{kt} \Rightarrow L(t) = 65 + Ce^{kt}$. At $t = 0$,

$40 = 65 + C \Rightarrow C = -25 \Rightarrow L(t) = 65 - 25e^{kt}$

At $t = 15$, $50 = 65 - 25e^{k(15)} \Rightarrow \frac{15}{25} = e^{k(15)} \Rightarrow k = \frac{1}{15}\ln(\frac{3}{5}) \approx -0.034055$

$$L(t) = 65 - 25e^{-0.034055t}$$

(*v*) $55 = 65 - 25e^{-0.034055t} \Rightarrow \frac{2}{5} = e^{-0.034055t} \Rightarrow t = \frac{1}{-0.034055}\ln(\frac{2}{5}) \approx 26.9\text{min}.$

## Problem 17.

(a) $\frac{dP}{dt} = kP - E = k(P - \frac{E}{k})$. Use $u = P - \frac{E}{k}$.

The DE becomes $\frac{du}{dt} = ku \Rightarrow u = Ce^{kt} \Rightarrow P(t) = \frac{E}{k} + Ce^{kt}$.

(b) $LHS = \frac{dP}{dt} = Ce^{kt}(k)$ while $RHS = kP - E = k(Ce^{kt} + \frac{E}{k}) - E = kCe^{kt}$, hence the DE is satisfied.

## Problem 19.

$\frac{dM}{dt} = rM - w = r(M - \frac{w}{r})$ hence $\frac{d^2M}{dt^2} = r\frac{dM}{dt} = r^2(M - \frac{w}{r})$

The threshold value is when $\frac{dM}{dt} = 0$ hence $M_t = \frac{w}{r}$. If $M_0 < \frac{w}{r}$ then $(M_0 - \frac{w}{r}) < 0 \Rightarrow \frac{dM}{dt} < 0$ and $\frac{d^2M}{dt^2} < 0$ hence $M(t)$ will be decreasing and concave down. If $M_0 > \frac{w}{r}$ then $(M_0 - \frac{w}{r}) > 0 \Rightarrow \frac{dM}{dt} > 0$ and $\frac{d^2M}{dt^2} > 0$ hence $M(t)$ will be increasing and concave up.

## Problem 21.

(a) $\frac{dN}{dt} = kN$

(b) $N(t) = N_0 e^{kt}$

(c) $\frac{dN}{dt} = kN - S$

(d) For $N(t) = N_0 e^{kt} - St$, $\frac{dN}{dt} = kN_0 e^{kt} - S$ while $kN - S = k(N_0 e^{kt} - St) - S$ which are not equal.

(e) The $-St$ term in fact is right for saying how many fruit flies are siphoned off in $t$ weeks. The problem is that the $N_0 e^{kt}$ term is not correct if some of the population is siphoned off. It would be correct if the original $N_0$ fly population kept growing with no siphoning.

## Problem 22.

These solutions can be found by taking antiderivatives.

(a) $y = -\frac{1}{3}\cos 3t + C$

(b) $y = \frac{5}{\ln 2}2^t + C$

(c) $x = t + \ln t + C$ (Hint: $\frac{t+1}{t} = 1 + \frac{1}{t}$)

(d) $x = \ln t - \frac{1}{t} + C$ (Hint: $\frac{t+1}{t^2} = \frac{1}{t} + t^{-2}$)

## Problem 25.

(a) I   (b) III   (c) II   (d) II   (e) III

## Problem 26.

(a) None   (b) None   (c) $\beta = 4$ or $-4$

## Problem 28.

(a) $y = e^{-3t}$ or $y = e^{-4t}$, $\lambda_1 = -3$, $\lambda_2 = -4$

(b) $y = C_1 e^{-3t} + C_2 e^{-4t} \Rightarrow y' = C_1 e^{-3t}(-3) + C_2 e^{-4t}(-4) \Rightarrow y'' = C_1 e^{-3t}(9) + C_2 e^{-4t}(16)$

$\Rightarrow y'' + 7y' + 12y = C_1 e^{-3t}(9) + C_2 e^{-4t}(16) + 7\left(C_1 e^{-3t}(-3) + C_2 e^{-4t}(-4)\right) + 12\left(C_1 e^{-3t} + C_2 e^{-4t}\right)$

$$= C_1 e^{-3t}(9 - 21 + 12) + C_2 e^{-4t}(16 - 28 + 12) = 0$$

**Problem 29.**

(a) $\lambda = -2$

(b) $y = te^{-2t} \Rightarrow y' = e^{-2t} + te^{-2t}(-2) \Rightarrow y'' = e^{-2t}(-2) + (-2)(e^{-2t} + te^{-2t}(-2)) = -4e^{-2t} + 4te^{-2t}$

$\quad y'' + 4y' + 4y = -4e^{-2t} + 4te^{-2t} + 4\left(e^{-2t} - 2te^{-2t}\right) + 4te^{-2t} = (-4+4)e^{-2t} + (4-8+4)te^{-2t} = 0$

## Section 31.3    Qualitative Analysis of Solutions to Autonomous Differential Equations

**Problem 1.**

(a) $\frac{dy}{dt} = 4y - 8 = 4(y-2)$
    For $y > 2$    $fracdydt > 0 \Rightarrow y(t)$ is increasing.
    For $y < 2$, $\frac{dy}{dt} < 0 \Rightarrow y(t)$ is decreasing.
    $y = 2$ is an unstable equilibrium.

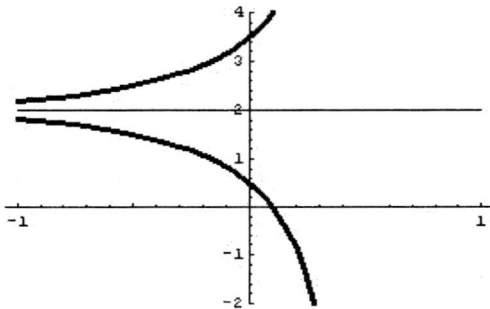

(b) $\frac{dy}{dt} = y^2 - 4 = (y+2)(y-2)$
    For $y > 2$, $\frac{dy}{dt} > 0 \Rightarrow y(t)$ is increasing.
    For $-2 < y < 2$, $\frac{dy}{dt} < 0 \Rightarrow y(t)$ is decreasing.
    For $y < -2$, $\frac{dy}{dt} > 0 \Rightarrow y(t)$ is increasing.
    $y = 2$ is an unstable equilibrium. $y = -2$ is a stable equilibrium.

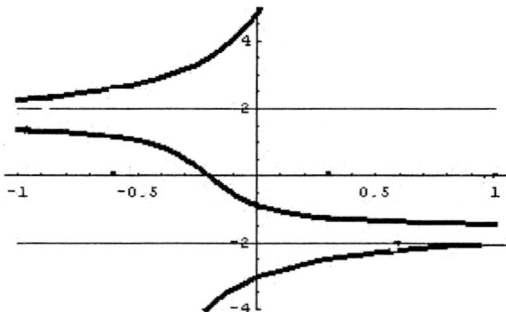

(c) $\frac{dy}{dt} = (y-1)(y-2)(y+1)$
    For $y > 2$, $\frac{dy}{dt} > 0 \Rightarrow y(t)$ is increasing.

For $1 < y < 2$, $\frac{dy}{dt} < 0 \Rightarrow y(t)$ is decreasing.

For $-1 < y < 1$, $\frac{dy}{dt} > 0 \Rightarrow y(t)$ is increasing.

For $y < -1$, $\frac{dy}{dt} < 0 \Rightarrow y(t)$ is decreasing.

$y = -1$ and $2$ are unstable equilibriums. $y = 1$ is a stable equilibrium.

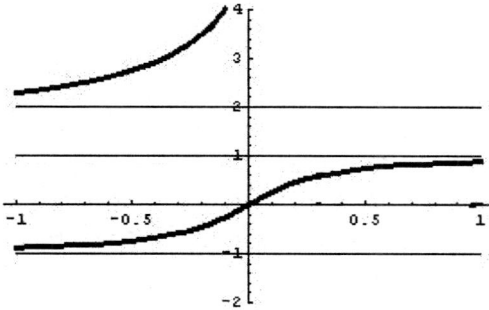

(d) $\frac{dy}{dt} = y^2 + 5y - 6 = (y - 1)(y + 6)$

For $y > 1$, $\frac{dy}{dt} > 0 \Rightarrow y(t)$ is increasing.

For $-6 < y < 1$, $\frac{dy}{dt} < 0 \Rightarrow y(t)$ is decreasing.

For $y < -6$, $\frac{dy}{dt} > 0 \Rightarrow y(t)$ is increasing.

$y = 1$ is an unstable equilibrium. $y = -6$ is a stable equilibrium.

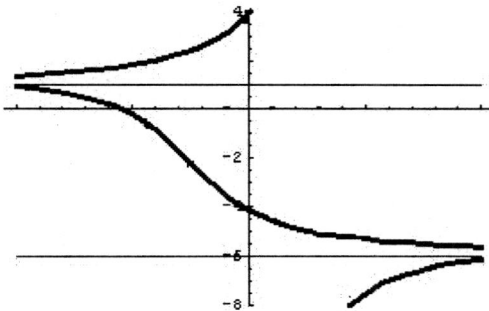

## Problem 3.

(a) $y = 3$ is a stable equilibrium.

(b) $y = 3$ is an unstable equilibrium.

(c) $y = 0$ is an unstable equilibrium. $y = 2$ is a stable equilibrium.

(d) $y = 2$ is an unstable equilibrium. $y = -2$ is a stable equilibrium.

## Problem 5.

(a) $0.02(100) - N = 0 \Leftrightarrow N = 2 \Rightarrow 2000$ people.

(b) $0.02(P) - 1 = 0 \Leftrightarrow P = 50 \Rightarrow 50000$ people.

## Problem 6.

$\frac{dy}{dt} = y^2 - 1 = (y + 1)(y - 1) \Rightarrow y = 1$ is an unstable equilibrium. $y = -1$ is a stable equilibrium.

(a) Graph shows no equilibrium.

(b) Graph shows no equilibrium OR $y = 0$ is wrong equilibrium.

(c) For $0 < y < 1$, $\frac{dy}{dt} < 0$ contrary to the shape of this graph.

(d) CORRECT ANSWER

(e) Graph shows no equilibrium.

**Problem 8.**

(a) For any potential solution function take the appropriate derivatives. Make the appropriate substitutions into the left side of the DE. Separately make the appropriate substitutions into the right side of the DE. If after simplification the left side equals the right side we indeed have a solution.

(b) A general solution is a family of functions each of which satisfies the DE. A particular solution is one member of that family. It is the only member of the family whose graph contains any one of the points on its graph. (No two distinct particular solutions of the same DE have any points in common.)

**Problem 9.**
$$\frac{dyP}{dyt} = 0.01P - 0.0025P^2 = 0.0025P(4 - P) = 0 \Leftrightarrow P = 0 \text{ or } P = 4$$

(a) The carrying capacity is 4.

(b) $\frac{dy^2 P}{dyt^2} = 0.01\frac{dyP}{dyt} - 0.0025(2)P\frac{dyP}{dyt} = 0.005(2 - P)(.0025P(4 - P))$
$= 0.0000125P(2 - P)(4 - P)$

(c) $\frac{dy^2 P}{dyt^2} = 0 \Leftrightarrow P = 0,\ 2,\ 4.$ $P = 2$ is when the population is increasing

(d)

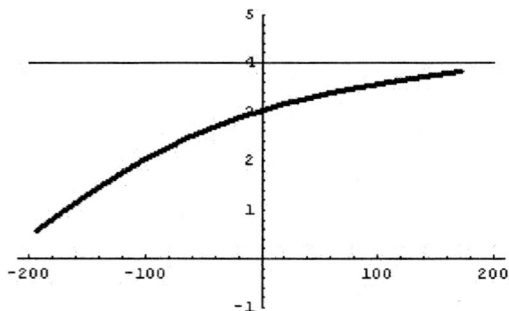

**Problem 11.**

(a)  $\frac{dyy}{dyt} = 2y - 6$

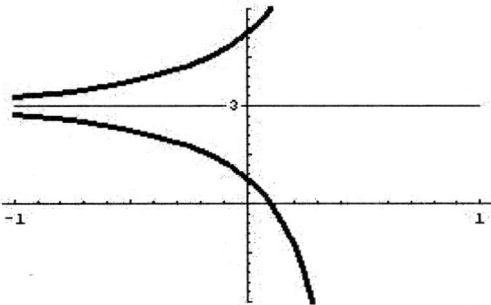

(b)  $\frac{dyy}{dyt} = 6 - 2y$

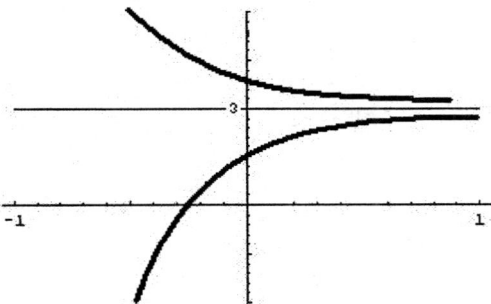

**Problem 13.**

$\frac{dyy}{dyt} = \tan y$ As y approaches an odd multiple of $\frac{\pi}{2}$ the slopes become infinitely steep (but not asymptot-

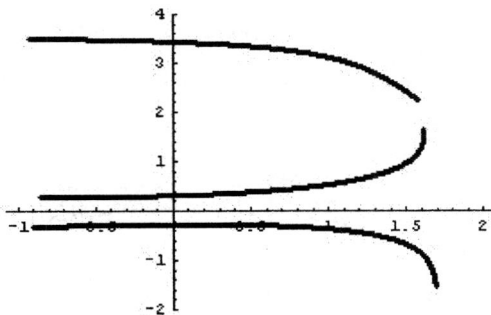

ical).

**Problem 15.**

(a)  $\frac{dy}{dt} = t^2 - 1$

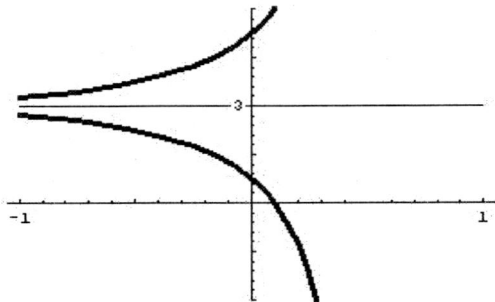

(b)  $\frac{dy}{dt} = y^2 - 1$

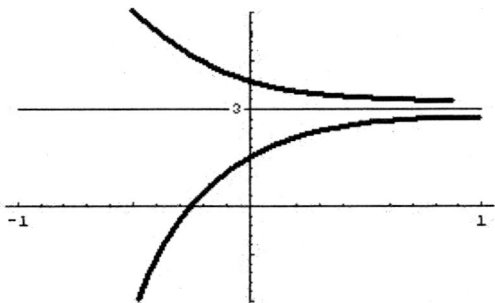

**Problem 17.**

(a)  $\frac{dy}{dt} = (y+1)(y-4)$

(b)  $\frac{dy}{dt} = -(y+1)(y-4)$

(c)  $\frac{dy}{dt} = (y+1)(y-1)(y-4)$ or $\frac{dy}{dt} = (y+1)(y-4)^2$

**Problem 20.**

(a)  $\frac{dM}{dt} = \frac{1}{2}(0.2) - \frac{1}{2}\left(\frac{M}{5}\right) = 0.1 - 0.1M, \quad M(0) = 0.5$

(b)

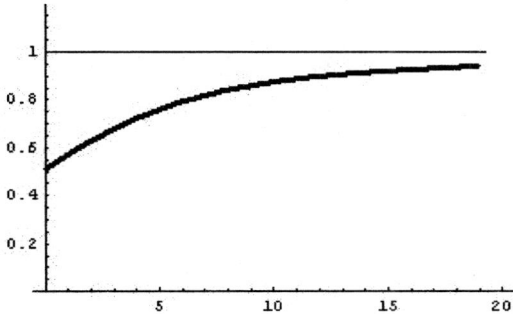

(c) $\frac{dyM}{dyt} = -0.1(M - 1)$,   $M(0) = 0.5$ has solution
$M(t) = 1 + Ce^{-0.1t}$ with $0.5 = 1 + Ce^0 \Rightarrow C = -0.5$
$M(2) = 1 - 0.5e^{-0.1(2)} \approx 0.59$

**Problem 20.**

Logistic Growth model. The population grew 1.25 million from 1961 to 1978 (17 years) and grew 0.5 million from 1978 to 1991 (13 years). If it was exponential growth the growth should be accelerating. Instead it seems to be tapering off, indicating a logistic growth.

**Problem 23.**

(a) (*i*)   $\frac{dyB}{dyt} = 0.05B + 4000 = 0.05(B + 80000)$,   $B(0) = 0$.

(*ii*) Solution for above equation is $B = 80000e^{0.05t} - 80000$.

After 30 years, $B(30) = 80000e^{0.05(30)} - 80000 \approx \$278535.13$.

We could have used $\int_0^{30} \left( \frac{1}{B+8000} \right) \frac{dyB}{dyt} \, dt = \int_0^{30} (0.05)dt$ to find $B(30)$.

(b) (*i*)   $S = 4000 + 4000(1.05) + \cdots + 4000(1.05)^{29}$

(*ii*) $S = \frac{4000((1.05)^{30}-1)}{0.05} \approx \$265755.39$

## Section 31.4    Solving Separable First Order Differential Equations

**Problem 2.**
$\frac{dyy}{dyx} = x^2y \Rightarrow \int \frac{1}{y} \, dy = \int x^2 \, dx \Rightarrow \ln|y| = \frac{1}{3}x^3 + C_1 \Rightarrow |y| = e^{\frac{1}{3}x^3+C_1} \Rightarrow y = \pm e^{\frac{1}{3}x^3+C_1} = Ce^{\frac{1}{3}x^3}$

**Problem 4.**
$\frac{dyy}{dyx} = \frac{y}{x} \Rightarrow \int \frac{1}{y} \, dy = \int \frac{1}{x} \, dx \Rightarrow \ln|y| = \ln|x| + C_1 \Rightarrow |y| = e^{\ln|x|+C_1} = |x| C_2 \Rightarrow y = Cx$. $C$ absorbs $\pm$.

**Problem 5.**
$\frac{dyy}{dyx} = \frac{x-1}{2y+1} \Rightarrow \int (2y + 1) \, dy = \int (x - 1) \, dx \Rightarrow y^2 + y = \frac{1}{2}x^2 - x + C \Rightarrow y^2 + y - \frac{1}{2}x^2 + x - C = 0$

$$\Rightarrow y = \left(-1 \pm \sqrt{1 - 4\left(-\frac{1}{2}x^2 + x - C\right)}\right) /2$$

CHAPTER 31    DIFFERENTIAL EQUATIONS

**Problem 7.**

$$\frac{dy}{dx} - y^2 = 1 \;\Rightarrow\; \frac{dy}{dx} = y^2 + 1 \;\Rightarrow\; \int \frac{1}{y^2+1}\,dy = \int 1\,dx \;\Rightarrow\; \arctan y = x + C \;\Rightarrow\; y = \tan(x+C)$$

**Problem 9.**

$$2\frac{dy}{dx} - 3xy = 0 \;\Rightarrow\; \int \frac{1}{y}\,dy = \int \frac{3}{2}x\,dx \;\Rightarrow\; \ln|y| = \frac{3}{4}x^2 + C_1 \;\Rightarrow\; |y| = e^{\frac{3}{4}x^2 + C_1} \;\Rightarrow\; y = Ce^{\frac{3}{4}x^2}$$

**Problem 10.**

$$\frac{dy}{dx} = \frac{\cos x}{-\sin y} \;\Rightarrow\; \int -\sin y\,dy = \int \cos x\,dx \;\Rightarrow\; \cos y = \sin x + C. \text{ (Be careful with arccos.)}$$

**Problem 12.**

(a) $\frac{dC}{dt} = 3(\frac{2}{5}) - 3(\frac{C}{5}) = -0.6(C-2) \;\Rightarrow\; \int \frac{1}{C-2}\,dC = \int -0.6\,dt \;\Rightarrow\; \ln|C-2| = -0.6t + C_1$

$$\Rightarrow\; C - 2 = \pm e^{-0.6t + C_1}. \quad C(0) = 1 \Rightarrow C(t) = 2 - e^{-0.6t}$$

(b) $\frac{dA}{dt} = 3(\frac{3}{5}) - 3(\frac{A}{5}) = -0.6(A-3) \;\Rightarrow\; \int \frac{1}{A-3}\,dA = \int -0.6\,dt \;\Rightarrow\; \ln|A-3| = -0.6t + C_1$

$$\Rightarrow\; A - 3 = \pm e^{-0.6t + C_1}. \quad A(0) = 4 \Rightarrow A(t) = 3 + e^{-0.6t}$$

## Section 31.5    Systems of Differential Equations

**Problem 1.**

(a) The system is competitive. Both species suffer from interaction. In the absence of the other they each demonstrate logistic growth.

(b) The system is a predator-prey relationship. Species $x$ suffers while $y$ is helped by interaction.

   $x$ grows exponentially in the absence of $y$. $y$ dies out in the absence of $x$.

(c) The system is symbiotic. Both are helped by interaction. In the absence of the other they each demonstrate logistic growth.

**Problem 3.**

   $x$ nullclines: $\frac{dx}{dt} = 0.02x - 0.01xy = 0 \;\Leftrightarrow\; x = 0$ or $y = 2$

   $y$ nullclines: $\frac{dy}{dt} = -0.01y + 0.08xy = 0 \;\Leftrightarrow\; y = 0$ or $x = \frac{1}{8}$

   Equilibrium points are $(0, 0)$ and $(\frac{1}{8}, 2)$.

**Problem 7.**

   $x$ nullclines: $\frac{dx}{dt} = 0.04x - 0.02x^2 - 0.01xy = 0$

   $\Leftrightarrow\; x = 0$ or $y = 4 - 2x$

   $y$ nullclines: $\frac{dy}{dt} = 0.04y - 0.01y^2 - 0.01xy = 0$

   $\Leftrightarrow\; y = 0$ or $x = 4 - y$

   Equilibrium points are $(0, 0)$, $(0, 4)$ and $(2, 0)$.

   If the $y$ population is 0 then the $x$ population will approach
   2 thousand unless it is 0 where it will remain.

   For all other initial population numbers the $x$ population will
   die out and the $y$ population will approach 4 thousand.

**Problem 9.**

$x$ nullclines: $\frac{dyx}{dyt} = x(1 - 0.5x - y) = 0$

$\Leftrightarrow$ $x = 0$ or $y = 1 - 0.5x$

$y$ nullclines: $\frac{dyy}{dyt} = y(-1 - 0.5y + x) = 0$

$\Leftrightarrow$ $y = 0$ or $y = -2 + 2x$

Equilibrium points are $(0, 0)$, $(\frac{6}{5}, \frac{2}{5})$ and $(2, 0)$.

If the $y$ population is 0 then the $x$ population will approach two unless it is 0 where it will remain.

If the $x$ population is 0 then the $y$ population will approach zero.

For all other initial population numbers the solution will spiral with the $x$ population approaching $\frac{6}{5}$ and the $y$ population approaching $\frac{2}{5}$.

## Problem 11.

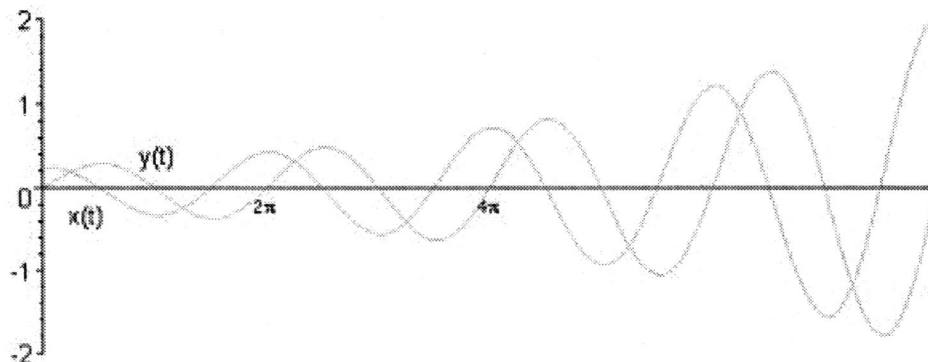

## Problem 13.

(a) None of the graphs are correct.

$\frac{dyx}{dyt} = -2$ $\Rightarrow$ all solution curves move to the left, never to the right.

$\frac{dyy}{dyt} = -4x$ $\Rightarrow$ solution curves move up if $x$ is left of the $y$ axis and down if $x$ is right of the $y$ axis.

item *ix*

$\frac{dyx}{dyt} = 3y$ $\Rightarrow$ solution curves move left if y is negative and right if y is positive.

$\frac{dyy}{dyt} = -3x$ $\Rightarrow$ solution curves move up if $x$ is negative and down if $x$ is positive.

(b) *vii*

$\frac{dyx}{dyt} = 10x$ $\Rightarrow$ solution curves move left if $x$ is negative and right if $x$ is positive.

$\frac{dyy}{dyt} = 10y$ $\Rightarrow$ solution curves move up if $y$ is positive and down if $y$ is negative.

## Section 31.6    Second Order Homogeneous Differential Equations

**Problem 1.**

$y = C_1 e^t + C_2 e^{3t}$ (Since section is for homogeneous equations assume $y'' - 4y' = -3y$.)

**Problem 3.**

$y = C_1 \cos(5t) + C_2 \sin(5t)$

**Problem 5.**

$y = C_1 + C_2 e^{-5t}$

**Problem 6.**

$y = C_1 e^{-t} + C_2 e^{2t}$

**Problem 9.**

$x = -\frac{2}{3} e^t - \frac{1}{3} e^{-2t}$

**Problem 11.**

$x = e^{\frac{1}{2}t} + \frac{3}{2} t e^{\frac{1}{2}t}$

**Problem 13.**

$x(t)$ must be periodic. It's size cannot grow or shrink. Hence $b^2 - 4c < 0$ and $b = 0$.

We also need the period $\frac{4\pi}{\sqrt{4c}} = 1 \Rightarrow 4\pi = \sqrt{4c} \Rightarrow 16\pi^2 = 4c \Rightarrow c = 4\pi^2$.

Hence $x'' + 4\pi^2 x = 0$ has solution $x(t) = 5\cos(2\pi t)$. $x(t) = 5\cos(4\pi t)$, $c = 16\pi^2$, also works.

**Problem 15.**

Throughout as $b < 0$, $\lim_{t\to\infty} e^{-\frac{b}{2}t} = \infty$

If $b^2 - 4c < 0$ then $x(t) = e^{-\frac{b}{2}t}[C_1 \cos(\frac{\sqrt{4c-b^2}}{2}t) + C_2 \sin(\frac{\sqrt{4c-b^2}}{2}t)]$. As $\lim_{t\to\infty} e^{-\frac{b}{2}t} = \infty$, $\lim_{t\to\infty} x(t) \neq L$.

If $b^2 - 4c = 0$ then $x(t) = C_1 e^{-\frac{b}{2}t} + C_2 t e^{-\frac{b}{2}t}$ so as $\lim_{t\to\infty} e^{-\frac{b}{2}t} = \infty$ and $\lim_{t\to\infty} t e^{-\frac{b}{2}t} = \infty$, $\lim_{t\to\infty} x(t) \neq L$.

If $b^2 - 4c > 0$ then $x(t) = C_1 e^{\frac{-b+\sqrt{b^2-4c}}{2}t} + C_2 e^{\frac{-b-\sqrt{b^2-4c}}{2}t}$. As $\frac{-b+\sqrt{b^2-4c}}{2} > 0$ and $\frac{-b-\sqrt{b^2-4c}}{2} > 0$ again $\lim_{t\to\infty} x(t) \neq L$.

**Problem 17.**

$e^t = e^{-\frac{b}{2}t} \Rightarrow -\frac{b}{2} = 1 \Rightarrow b = -2$. $\sin(t) = \sin(\frac{\sqrt{4c-b^2}}{2}t) \Rightarrow 1 = \frac{\sqrt{4c-(-2)^2}}{2} \Rightarrow 4 = 4c - 4 \Rightarrow c = 2$.
Hence equation is $y'' - 2y' + 2 = 0$.

**Problem 18.**

(a) $e^{2\pi i} = \cos(2\pi) + i\sin(2\pi) = 1$

(b) $e^{-\pi i} = \cos(-\pi) + i\sin(-\pi) = -1$

# APPENDIX A
# Algebra

## A .1    Introduction to Algebra: Expressions and Equations

**Problem 1.**

(a) undefined for $x = 0$

(b) undefined for $x < 1$

(c) undefined for $x = -3$

(d) undefined for $x = 1$

**Problem 2.**

(a) No, $\frac{x}{3x} = \begin{cases} \frac{1}{3} & \text{if } x \neq 0 \\ \text{undefined} & \text{if } x = 0 \end{cases}$

(b) No, $\frac{x+1}{x(x+1)} = \begin{cases} \frac{1}{x} & \text{if } x \neq -1 \\ \text{undefined} & \text{if } x = -1 \end{cases}$

(c) Yes

(d) No, $-3^2 = -9$ but $(-3)^2 = 9$

(e) No, see examples like $\frac{2}{3} \neq \frac{4}{9}$

(f) No, take any negative x, like $x = -2$: $\sqrt{(-2)^2} = 2 \neq -2$

(g) Yes (multiplication is commutative)

(h) No, $(xy)^2 = x^2 y^2$

**Problem 4.**

(a) $3x^2$

(b) $(3x)^2 = 9x^2$

(c) $2\sqrt{y - 2x}$

(d) Either $\sqrt[3]{3} \cdot \frac{1}{x}$ or $\sqrt[3]{3 \cdot \frac{1}{x}}$.

**Problem 5.**

(a) $= -4 - 3 \cdot (2 - 8 \cdot 2 + 9) = -4 - 3 \cdot (-5) = -4 + 15 = 11$

(b) $= -x^2 + x^2 - x^2 + 3x^2 = 2x^2$

(c) $= -\frac{x}{y} - \frac{x}{y} - \frac{2x}{y} = -\frac{4x}{y}$

(d) $= \left(3 - \frac{2}{3} - 1\right)\sqrt{y} = \frac{4}{3}\sqrt{y}$

**Problem 7.**

(a) $f(1) = -(1)^3 - 2(1)^2 + (-(1))^2 + 1 = -1 - 2 + 1 + 1 = -1,$

$$f(-1) = -(-1)^3 - 2(-1)^2 + (-(-1))^2 + (-1) = 1 - 2 + 1 - 1 = -1$$

(b) $f(2) = -\frac{1}{2} - \frac{2}{(2)^2} + \frac{-3}{-(2)^3} = -\frac{1}{2} - \frac{2}{4} + \frac{3}{8} = -\frac{5}{8},\ f(-2) = -\frac{1}{-2} - \frac{2}{(-2)^2} + \frac{-3}{-(-2)^3} = \frac{1}{2} - \frac{2}{4} + \frac{-3}{8} = -\frac{3}{8}$

# A .2    Working with Expressions

**Problem 1.**

(a) $\frac{1}{x^2} + \frac{3-x}{x} + \frac{x}{3+x} = \frac{1\cdot(3+x)+(3-x)\cdot x(3+x)+x\cdot x^2}{x^2(3+x)} = \frac{3+x+9x-x^3+x^3}{x^2(3+x)} = \frac{3+10x}{x^2(3+x)}$

(b) $\frac{\frac{1}{x+w}-\frac{1}{w}}{w} = \frac{1}{w}\cdot\frac{w-(x+w)}{(x+w)\cdot w} = \frac{1}{w}\cdot\frac{w-x-w}{(x+w)\cdot w} = \frac{-x}{(x+w)\cdot w^2}$

(c) $\left[\frac{2}{(y+z)^2} - \frac{2}{y^2}\right]\cdot\frac{1}{y} = \frac{y^2-(y+z)^2}{(y+z)^2\cdot y^2}\cdot\frac{2}{y} = \frac{y^2-(y^2+2yz+z^2)}{(y+z)^2\cdot y^2}\cdot\frac{2}{y} = \frac{-2z(2y+z)}{(y+z)^2\cdot y^3}$

**Problem 3.**

(a) $x^2 - x - 6 = (x-3)(x+2)$

(b) $2x^2 - x - 3 = (2x-3)(x+1)$

(c) $6x^3 + 6x^2 - x =$

(d) $16x^2y^4 - 1 = \left(4xy^2\right)^2 - 1^2 = (4xy^2+1)(4xy^2-1)$

(e) $(x-1)xy + 3(1-x) = (x-1)xy - 3(x-1) = (x-1)(xy-3)$

(f) $x^4 - 3x^2 - 10 = \left(x^2\right)^2 - 3\left(x^2\right) - 10 = \left(x^2-5\right)\left(x^2+2\right)$

(g) $x^4 + 3x^2 - 4x^3 = x^2(x^2-4x+3) = x^2(x-3)(x-1)$

**Problem 4.**

(a) $b^{w+2} - b^w = b^w b^2 - b^w = b^w(b^2-1) = b^w(b+1)(b-1)$

(b) $b^{2w} - b^w = (b^w)^2 - b^w = b^w(b^w-1)$

(c) $x^3 b^{x+2} - x^3 b^x =$

(d) $b^{2x} - 4 = (b^x)^2 - 2^2 = (b^x+2)(b^x-2)$

(e) $b^{2w} - b^w - 6 = (b^w)^2 - b^w - 6 = (b^w-3)(b^w+2)$

# A .3    Solving Equations

**Problem 2.**

(a) $\frac{\lambda}{1+x} + \frac{2}{\lambda} = \frac{1}{\lambda\beta} \Rightarrow \frac{\lambda^2+2(1+x)}{(1+x)\lambda} = \frac{1}{\lambda\beta} \Rightarrow \lambda\beta = \frac{(1+x)\lambda}{\lambda^2+2(1+x)} \Rightarrow \beta = \frac{1+x}{\lambda^2+2x+2}$

(b) $\frac{\lambda}{1+x} = \frac{1}{\lambda\beta} - \frac{2}{\lambda} \Rightarrow \frac{\lambda}{1+x} = \frac{1-2\beta}{\lambda\beta} \Rightarrow \frac{1+x}{\lambda} = \frac{\lambda\beta}{1-2\beta} \Rightarrow 1+x = \frac{\lambda^2\beta}{1-2\beta} \Rightarrow x = \frac{\lambda^2\beta}{1-2\beta} - 1$

(c) $\frac{\lambda}{1+x} + \frac{2}{\lambda} = \frac{1}{\lambda\beta} \Rightarrow \frac{\lambda}{1+x} = \frac{1}{\lambda\beta} - \frac{2}{\lambda} \Rightarrow \frac{\lambda}{1+x} = \frac{1-2\beta}{\lambda\beta} \Rightarrow \lambda^2 = \frac{(1-2\beta)(1+x)}{\beta}$

$$\Rightarrow \lambda = \pm\sqrt{\frac{(1-2\beta)(1+x)}{\beta}}$$

**Problem 3.**
$$f(g(w)) - 2g(w) = 0 \Rightarrow (1+w)^2 - 2(1+w) = 0 \Rightarrow w^2 - 1 = 0 \Rightarrow w = \pm 1$$

**Problem 4.**

(a) $y(2y-3) = 5 \Rightarrow 2y^2 - 3y - 5 = 0 \Rightarrow (2y-5)(y+1) = 0$ so $y = \frac{5}{2}$ or $y = -1$

(b) $y(2y-3) = -5 \Rightarrow 2y^2 - 3y + 5 = 0 \Rightarrow y = \frac{3\pm\sqrt{9-4\cdot2\cdot5}}{2\cdot2} = \frac{3\pm\sqrt{-31}}{4}$ so no real solution.

(c) $y(y-6) = 9 \Rightarrow y^2 - 6y - 9 = 0 \Rightarrow y = \frac{6\pm\sqrt{6^2-4\cdot(-9)}}{2} = \frac{6\pm\sqrt{2\cdot36}}{2} = \frac{6\pm6\sqrt{2}}{2} = 3\pm3\sqrt{2}$ so $y = 3\pm3\sqrt{2}$

(d) $y(y-6) = -9 \Rightarrow y^2 - 6y + 9 = 0 \Rightarrow (y-3)^2 = 0$ so $y = 3$

(e) $\frac{y}{y-6} = -9 \Rightarrow y = -9(y-6) \Rightarrow y = -9y + 54 \Rightarrow 10y = 54$ so $y = \frac{54}{10} = \frac{27}{5}$

(f) $\frac{1}{y-1} = \frac{3}{y} \Rightarrow 1\cdot y = 3\cdot(y-1) \Rightarrow y = 3y - 3 \Rightarrow 2y = 3$ so $y = \frac{3}{2}$

**Problem 6.**

(a) $w(z+w) = z \Rightarrow wz + w^2 = z \Rightarrow w^2 = z(1-w) \Rightarrow z = \frac{w^2}{1-w}$

(b) $w(z+w) = z \Rightarrow wz + w^2 = z \Rightarrow w^2 + wz - z = 0 \Rightarrow w = \frac{-z\pm\sqrt{z^2+4z}}{2}$

(c) $\frac{z+w}{w} = z \Rightarrow z + w = zw \Rightarrow w = z(w-1) \Rightarrow z = \frac{w}{w-1}$

(d) $\frac{z+w}{w} = z \Rightarrow z + w = zw \Rightarrow z = w(z-1) \Rightarrow w = \frac{z}{z-1}$

(e) $z^3 + 3z^2 + 2z = 0 \Rightarrow z(z^2 + 3z + 2) = 0 \Rightarrow z(z+2)(z+1) = 0 \Rightarrow z = 0, -2, -1$

(f) $z^4 + 3z^2 + 2 = 0 \Rightarrow (z^2)^2 + 3(z^2) + 2 = 0 \Rightarrow (z^2+2)(z^2+1) = 0 \Rightarrow z^2 = -2, -1$ so no real sol for $z$

(g) $z^5 = 16z \Rightarrow z(z^4 - 16) = 0 \Rightarrow z(z^2+4)(z^2-4) = 0 \Rightarrow z(z^2+4)(z+2)(z-2) = 0$ so $z = 0, -2, 2$

# APPENDIX D

# Proof by Induction

**Problem 1.**

Let $P(n)$ be the statement: $1 + 2 + 3 + \cdots + n = \frac{n(n+1)}{2}$

$P(1)$ is true because $1 = \frac{1(1+1)}{2}$.

Assume that $P(k)$ is true. Then $1 + 2 + 3 + \cdots + k = \frac{k(k+1)}{2} \Rightarrow$

$1 + 2 + 3 + \cdots + k + (k+1) = \frac{k(k+1)}{2} + (k+1) = (k+1)\left(\frac{k}{2} + 1\right) = (k+1)\left(\frac{k+2}{2}\right) = \frac{(k+1)((k+1)+1)}{2}$

$\Rightarrow \quad P(k+1)$ is true. Hence by induction, $P(n)$ is true for all positive integers $n$.

**Problem 3.**

Error in the text, it should read: Prove that the sum of the first $n$ nonzero even integers is $n(n+1)$

Let $P(n)$ be the statement: $1 \cdot 2 + 2 \cdot 2 + 3 \cdot 2 + 4 \cdot 2 + \ldots + k \cdot 2 = k(k+1)$

$P(1)$ is true because $1 \cdot 2 = 1(1+1)$.

Assume that $P(k)$ is true. Then $1 \cdot 2 + 2 \cdot 2 + 3 \cdot 2 + 4 \cdot 2 + \ldots + k \cdot 2 = k(k+1) \Rightarrow$

$1 \cdot 2 + 2 \cdot 2 + 3 \cdot 2 + 4 \cdot 2 + \ldots + k \cdot 2 + (k+1) \cdot 2 = k(k+1) + (k+1) \cdot 2 = (k+1)(k+2) = (k+1)\left((k+1)+1\right)$

$\Rightarrow \quad P(k+1)$ is true. Hence by induction, $P(n)$ is true for all positive integers $n$.

**Problem 4.**

Let $P(n)$ be the statement that $1 + 2 + 2^2 + 2^3 + \ldots + 2^n = 2^{n+1} - 1$

$P(1)$ is true because $1 + 2^1 = 2^{1+1} - 1$.

Assume that $P(k)$ is true. Then $1 + 2 + 2^2 + 2^3 + \ldots + 2^k = 2^{k+1} - 1 \Rightarrow$

$1 + 2 + 2^2 + 2^3 + \ldots + 2^k + 2^{k+1} = (2^{k+1} - 1) + 2^{k+1} = 2 \cdot 2^{k+1} - 1 = 2^{(k+1)+1} - 1 \quad \Rightarrow \quad P(k+1)$

is true.

Hence by induction, $P(n)$ is true for all positive integers $n$.

**Problem 6.**

Let $P(n)$ be the statement that $\left(1 + \frac{1}{1}\right) \cdot \left(1 + \frac{1}{2}\right) \cdot \left(1 + \frac{1}{3}\right) \cdot \ldots \cdot \left(1 + \frac{1}{n}\right) = n + 1$

$P(1)$ is true because $1 + \frac{1}{1} = 1 + 1$.

Assume that $P(k)$ is true. Then $\left(1 + \frac{1}{1}\right) \cdot \left(1 + \frac{1}{2}\right) \cdot \left(1 + \frac{1}{3}\right) \cdot \ldots \cdot \left(1 + \frac{1}{k}\right) = k + 1 \Rightarrow$

$\left(1 + \frac{1}{1}\right) \cdot \left(1 + \frac{1}{2}\right) \cdot \left(1 + \frac{1}{3}\right) \cdot \ldots \cdot \left(1 + \frac{1}{k}\right) \cdot \left(1 + \frac{1}{k+1}\right) = (k+1) \cdot \left(1 + \frac{1}{k+1}\right) = (k+1) \cdot \left(\frac{(k+1)+1}{k+1}\right) = (k+1) + 1$

$\Rightarrow \quad P(k+1)$ is true. Hence by induction, $P(n)$ is true for all positive integers $n$.

# APPENDIX E

## Conic Sections

**Problem 1.**

(a) Every line perpendicular to the directrix contains one point on the parabola. Hence the line through $(0, c)$ perpendicular to the directrix contains a point $(0, x)$ on the parabola. $(0, -c)$ is the point on the directrix and this line. Hence the distance from $(0, x)$ to $(0, c)$ must equal the distance from $(0, x)$ to $(0, -c)$. Thus $(0, 0)$ is that point.

(b) The distance between $(x, y)$ and the focus$(0, c)$ is $\sqrt{(x-0)^2 + (y-c)^2}$. The distance from $(x, y)$ to the directrix is perpendicular distance. Hence it is the distance from $(x, y)$ to $(x, -c)$ or $\sqrt{(y-(-c))^2} = \sqrt{(y+c)^2}$. Equating these gives $\sqrt{x^2 + (y-c)^2} = \sqrt{(y+c)^2}$.

(c) $x^2 + (y-c)^2 = (y+c)^2 \Rightarrow x^2 + y^2 - 2yc + c^2 = y^2 + 2yc + c^2 \Rightarrow x^2 = 4yc \Rightarrow y = \frac{1}{4c}x^2$.

**Problem 3.**

$\frac{1}{4c} = 2 \Rightarrow c = \frac{1}{8}$. Hence the focus is at $(0, \frac{1}{8})$ and the directrix is $y = -\frac{1}{8}$.

**Problem 5.**

The focus at $(0, 6) \Rightarrow c = 6 \Rightarrow 4c = 24 \Rightarrow y = \frac{1}{24}x^2$.

**Problem 6.**

The directrix $y = 5 \Rightarrow c = -5 \Rightarrow 4c = -20 \Rightarrow y = -\frac{1}{20}x^2$.

**Problem 8.**

(a) The focus at $(0, 1.5) \Rightarrow c = 1.5 = \frac{3}{2} \Rightarrow 4c = 6 \Rightarrow y = \frac{1}{6}x^2$.

(b) Denote depth by $d$. Point $(5, d)$ on $y = \frac{1}{6}x^2 \Rightarrow d = \frac{1}{6} \cdot 5^2 = \frac{25}{6}$. The reflector is $\frac{25}{6}$" deep.

**Problem 9.**

Note: *For the work in this problem, the foci should be $(c, 0)$ and $(-c, 0)$ not $(0, c)$ and $(0, -c)$.*

(a) The distance between $(x, y)$ and $(-c, 0)$ is $\sqrt{(x+c)^2 + (y-0)^2}$. The distance between $(x, y)$ and $(c, 0)$ is $\sqrt{(x-c)^2 + (y-0)^2}$. Adding these distances gives $\sqrt{(x+c)^2 + y^2} + \sqrt{(x-c)^2 + y^2} = 2a$.

Isolate the term $\sqrt{(x-c)^2 + y^2}$ (instead of $\sqrt{(x+c)^2 + y^2}$ as the text suggests) and square both sides.

$\left(\sqrt{(x-c)^2 + y^2}\right)^2 = \left(2a - \sqrt{(x+c)^2 + y^2}\right)^2$

$\Rightarrow \quad (x-c)^2 + y^2 = 4a^2 - 4a\sqrt{(x+c)^2 + y^2} + (x+c)^2 + y^2$

$x^2 - 2xc + c^2 + y^2 = 4a^2 - 4a\sqrt{(x+c)^2 + y^2} + x^2 + 2xc + c^2 + y^2$

$\Rightarrow \quad 4a\sqrt{(x+c)^2 + y^2} = 4a^2 + 4xc \quad \Rightarrow \quad a\sqrt{(x+c)^2 + y^2} = a^2 + xc$

(b) $\left(a\sqrt{(x+c)^2 + y^2}\right)^2 = (a^2 + xc)^2 \quad \Rightarrow \quad a^2\left((x+c)^2 + y^2\right) = a^4 + 2a^2xc + x^2c^2$

$\Rightarrow \quad a^2x^2 + 2a^2xc + a^2c^2 + a^2y^2 = a^4 + 2a^2xc + x^2c^2 \quad \Rightarrow \quad a^2x^2 + a^2c^2 + a^2y^2 = a^4 + x^2c^2$

$\Rightarrow \quad a^2x^2 - x^2c^2 + a^2y^2 = a^4 - a^2c^2 \quad \Rightarrow \quad x^2(a^2 - c^2) + a^2y^2 = a^2(a^2 - c^2)$

(c) Divide both sides of the equation in part (b) by $a^2(a^2 - c^2)$.

$$\frac{x^2(a^2 - c^2)}{a^2(a^2 - c^2)} + \frac{a^2 y^2}{a^2(a^2 - c^2)} = \frac{a^2(a^2 - c^2)}{a^2(a^2 - c^2)} \quad \Rightarrow \quad \frac{x^2}{a^2} + \frac{y^2}{a^2 - c^2} = 1.$$

Using $b^2 = a^2 - c^2$, we get $\frac{x^2}{a^2} + \frac{y^2}{b^2} = 1$

**Problem 11.**

Note: *For the work in this problem, the foci should be* $(c, 0)$ *and* $(-c, 0)$ *not* $(0, c)$ *and* $(0, -c)$

(a) The distance between $(x, y)$ and $(-c, 0)$ is $\sqrt{(x + c)^2 + (y - 0)^2}$. The distance between $(x, y)$ and $(c, 0)$ is $\sqrt{(x - c)^2 + (y - 0)^2}$. Subtracting these distances gives

$$\sqrt{(x + c)^2 + y^2} - \sqrt{(x - c)^2 + y^2} = \pm 2a.$$

(b) $\sqrt{(x + c)^2 + y^2} = \pm 2a + \sqrt{(x - c)^2 + y^2}$

$$\Rightarrow \quad (x + c)^2 + y^2 = 4a^2 \pm 4a\sqrt{(x - c)^2 + y^2} + (x - c)^2 + y^2$$

$$x^2 + 2xc + c^2 + y^2 = 4a^2 \pm 4a\sqrt{(x - c)^2 + y^2} + x^2 - 2xc + c^2 + y^2$$

$$\Rightarrow \quad \pm 4a\sqrt{(x - c)^2 + y^2} = 4a^2 - 4xc \quad \Rightarrow \quad \pm a\sqrt{(x - c)^2 + y^2} = a^2 - xc$$

$$\Rightarrow \quad \left(a\sqrt{(x - c)^2 + y^2}\right)^2 = \left(a^2 - xc\right)^2 \quad \Rightarrow \quad a^2\left((x - c)^2 + y^2\right) = a^4 - 2a^2 xc + x^2 c^2$$

$$\Rightarrow \quad a^2 x^2 - 2a^2 xc + a^2 c^2 + a^2 y^2 = a^4 - 2a^2 xc + x^2 c^2 \quad \Rightarrow \quad a^2 x^2 + a^2 c^2 + a^2 y^2 = a^4 + x^2 c^2$$

$$\Rightarrow \quad a^2 c^2 - a^4 = x^2 c^2 - a^2 x^2 - a^2 y^2 \Rightarrow \quad a^2(c^2 - a^2) = x^2(c^2 - a^2) - a^2 y^2$$

(c) The distance between the two foci is $2c$. The difference between the lengths of the other two sides $(2a)$ of the triangle cannot be greater than $2c$ if the triangle is to be connected. Hence

$2c > 2a \Rightarrow c > a \Rightarrow c^2 > a^2 \Rightarrow c^2 - a^2 > 0$. Since this is positive we can let $b^2 = c^2 - a^2$.

Now divide both sides of the equation in part (b) by $a^2(c^2 - a^2)$.

$$\frac{a^2(c^2 - a^2)}{a^2(c^2 - a^2)} = \frac{x^2(c^2 - a^2)}{a^2(c^2 - a^2)} - \frac{a^2 y^2}{a^2(c^2 - a^2)} \quad \Rightarrow \quad 1 = \frac{x^2}{a^2} - \frac{y^2}{c^2 - a^2}.$$

Using $b^2 = c^2 - a^2$ we obtain $\frac{x^2}{a^2} - \frac{y^2}{b^2} = 1$.

# APPENDIX F

# L'Hopital's Rule

**Problem 2.**

$\lim\limits_{x \to 0^+} \frac{\ln x}{x} = -\infty$ since it is of the form $\frac{-\infty}{0^+}$.

**Problem 4.**

$\lim\limits_{x \to 0} \frac{x}{e^{-x}} = \frac{0}{1} = 0$.

**Problem 5.**

Use L'Hopital's rule three times: $\lim\limits_{x \to \infty} \frac{100x^3}{e^x} = \lim\limits_{x \to \infty} \frac{300x^2}{e^x} = \lim\limits_{x \to \infty} \frac{600x}{e^x} = \lim\limits_{x \to \infty} \frac{600}{e^x} = 0$.

**Problem 7.**

(a) $\lim\limits_{t \to 0} \frac{t^2+3}{2t^3+100t+1} = \frac{0+3}{0+0+1} = 3$.

(b) Use L'Hopital's rule twice: $\lim\limits_{t \to \infty} \frac{t^2+3}{2t^3+100t+1} = \lim\limits_{t \to \infty} \frac{2t}{6t^2+100} = \lim\limits_{t \to \infty} \frac{2}{12t} = 0$.

**Problem 9.**

Rewrite and then use L'Hopital's rule: $\lim\limits_{x \to \infty} 2x \cdot e^{-x} = \lim\limits_{x \to \infty} \frac{2x}{e^x} = \lim\limits_{x \to \infty} \frac{2}{e^x} = 0$.

**Problem 11.**

$\lim\limits_{x \to \infty} \frac{r^x}{x} = 0$ since it is of the form $\frac{0}{\infty}$ (note: $\lim\limits_{x \to \infty} r^x = 0$ because $0 < r < 1$).

**Problem 12.**

$\lim\limits_{x \to 0.5} \frac{\ln(1-2x)}{2x-1}$ is undefined. Note: $\lim\limits_{x \to 0.5^+} \frac{\ln(1-2x)}{2x-1} = -\infty$ since it is of type $\frac{-\infty}{0^+}$

**Problem 14.**

$\lim\limits_{t \to 3} \frac{\ln(t-3)}{t^2-t-6}$ is undefined. $\lim\limits_{t \to 3^+} \frac{\ln(t-3)}{t^2-t-6} = -\infty$ since it is of the form $\frac{-\infty}{0^+}$

**Problem 15.**

(a) Use L'Hopital's rule and simplify: $\lim\limits_{x \to \infty} \frac{\ln x}{x^n} = \lim\limits_{x \to \infty} \frac{\frac{1}{x}}{n \cdot x^{n-1}} = \lim\limits_{x \to \infty} \frac{1}{n \cdot x^n} = 0$.

(b) Use L'Hopital's rule n times:

$$\lim\limits_{x \to \infty} \frac{e^{nx}}{x^n} = \lim\limits_{x \to \infty} \frac{n \cdot e^{nx}}{n \cdot x^{n-1}} = \lim\limits_{x \to \infty} \frac{n \cdot n \cdot e^{nx}}{n(n-1) \cdot x^{n-2}} = \lim\limits_{x \to \infty} \frac{n \cdot n \cdot n \cdot e^{nx}}{n(n-1)(n-2) \cdot x^{n-3}} = \ldots = \lim\limits_{x \to \infty} \frac{n^n \cdot e^{nx}}{n!} = \infty.$$

**Problem 17.**

(a) Use L'Hopital's rule: $\lim\limits_{x \to \infty} \frac{x}{\ln x} = \lim\limits_{x \to \infty} \frac{1}{\frac{1}{x}} = \lim\limits_{x \to \infty} x = \infty$.

(b) Since $\lim\limits_{k \to \infty} \frac{k}{\ln k} = \infty \neq 0$ the series must diverge.

**Problem 19.**

$\lim\limits_{x \to 0^+} e^x \ln x = -\infty$ since it is of type $1 \cdot (-\infty)$.

**Problem 20.**

Rewrite $\left(1 + \frac{5}{x}\right)^{3x} = \left(\left(1 + \frac{5}{x}\right)^{\frac{x}{5}}\right)^{15}$ and substitute $k = \frac{x}{5}$,

then $\lim\limits_{x \to \infty} \left(1 + \frac{5}{x}\right)^{3x} = \lim\limits_{x \to \infty} \left(\left(1 + \frac{5}{x}\right)^{\frac{x}{5}}\right)^{15} = \lim\limits_{k \to \infty} \left(\left(1 + \frac{1}{k}\right)^{k}\right)^{15} = e^{15}.$

**Problem 21.**

Rewrite and then use L'Hopital's rule: $\lim\limits_{x \to \infty} e^{-x} \ln x = \lim\limits_{x \to \infty} \frac{\ln x}{e^x} = \lim\limits_{x \to \infty} \frac{\frac{1}{x}}{e^x} = \lim\limits_{x \to \infty} \frac{1}{x \cdot e^x} = 0$

# APPENDIX G

## Newton's Method

**Problem 2.**

$f(x) = e^x - x - 3$, $f'(x) = e^x - 1$, if $x_0 = 2$ (other first guesses are possible) then

$x_1 = x_0 - \frac{f(x_0)}{f'(x_0)} \approx 2 - \frac{2.3890560989}{6.3890560989} \approx 1.626070571$

$x_2 = x_1 - \frac{f(x_1)}{f'(x_1)} \approx 1.626070571 - \frac{.457788186}{4.083858757} \approx 1.513973603$

$x_3 = x_2 - \frac{f(x_2)}{f'(x_2)} \approx 1.513973603 - \frac{.030780407}{3.544754010} \approx 1.505290234$

$x_4 = x_3 - \frac{f(x_3)}{f'(x_3)} \approx 1.505290234 - \frac{.000170845}{3.505461079} \approx 1.505241497$ The last two agree to three decimal places, so we can stop.

**Problem 4.**

$f(x) = x^3 - 20$, $f'(x) = 3x^2$, if $x_0 = 2.5$ (other first guesses are possible) then

$x_1 = x_0 - \frac{f(x_0)}{f'(x_0)} = 2.5 - \frac{-4.375}{18.75} \approx 2.733333333$

$x_2 = x_1 - \frac{f(x_1)}{f'(x_1)} \approx 2.733333333 - \frac{.42103703}{22.4133333} \approx 2.714548219$

$x_3 = x_2 - \frac{f(x_2)}{f'(x_2)} \approx 2.714548219 - \frac{.00288700}{22.10631610} \approx 2.714417623$ The last two agree to three decimal places, so we can stop.

**Problem 6.**

$f(x) = x^4 - 3x^2 + 2$, $f'(x) = 4x^3 - 6x$, if $x_0 = -1.5$ (other first guesses are possible) then

$x_1 = x_0 - \frac{f(x_0)}{f'(x_0)} = -1.5 - \frac{.3125}{-.45} \approx -1.430555556$

$x_2 = x_1 - \frac{f(x_1)}{f'(x_1)} \approx -1.430555556 - \frac{.048650444}{-3.127132644} \approx -1.414998031$

$x_3 = x_2 - \frac{f(x_2)}{f'(x_2)} \approx -1.414998031 - \frac{.002224353}{-2.842558004} \approx -1.414215513$

$x_4 = x_3 - \frac{f(x_3)}{f'(x_3)} \approx -1.414215513 - \frac{.000005518}{-2.828462242} \approx -1.414213562$ The last two agree to five decimal places, so we can stop.

**Problem 8.**

$f(x) = \cos x - x$, $f'(x) = -\sin x - 1$, if $x_0 = \frac{1}{2}$ (other first guesses are possible) then

$x_1 = x_0 - \frac{f(x_0)}{f'(x_0)} \approx .5 - \frac{.3775825619}{-1.479425539} \approx .7552224170$

$x_2 = x_1 - \frac{f(x_1)}{f'(x_1)} \approx .7552224170 - \frac{-.0271033117}{-1.685450632} \approx .7391416661$

$x_3 = x_2 - \frac{f(x_2)}{f'(x_2)} \approx .7391416661 - \frac{-.0000946153}{-1.673653811} \approx .7390851339$

$x_4 = x_3 - \frac{f(x_3)}{f'(x_3)} \approx .7390851339 - \frac{.0000000011}{-1.673612030} \approx .7390851332$ The last two agree to five decimal places, so we can stop.

**Problem 9.**

$f(x) = \sin x - x^2$, $f'(x) = \cos x - 2x$, if $x_0 = 1$ (other first guesses are possible) then

$x_1 = x_0 - \frac{f(x_0)}{f'(x_0)} \approx 1 - \frac{-.15852901519}{-1.4596976941} \approx .8913959953$

$x_2 = x_1 - \frac{f(x_1)}{f'(x_1)} \approx .8913959953 - \frac{-.0166371741}{-1.154465366} \approx .8769848448$

$x_3 = x_2 - \frac{f(x_2)}{f'(x_2)} \approx .8769848448 - \frac{-.0002881492}{-1.114497548} \approx .8767262985$

$x_4 = x_3 - \frac{f(x_3)}{f'(x_3)} \approx .8767262985 - \frac{-.0000000926}{-1.113781703} \approx .8767262154$ The last two agree to four decimal places, so we can stop.